WITHDRAWN FROM STOCK

SOC (System-on-a-Chip) Testing for Plug and Play Test Automation

Edited by

Krishnendu Chakrabarty
Duke University

Reprinted from a Special Issue of the
Journal of Electronic Testing (JETTA)
Volume 18, Nos. 4 & 5

Kluwer Academic Publishers
Boston/Dordrecht/London

FRONTIERS IN ELECTRONIC TESTING

Consulting Editor

Vishwani D. Agrawal

Books in the series:

Test Resource Partitioning for System-on-a-Chip
K. Chakrabarty, Iyengar & Chandra
ISBN: 1-4020-7119-1

A Designers' Guide to Built-in Self-Test
C. Stroud
ISBN: 1-4020-7050-0

Boundary-Scan Interconnect Diagnosis
J. de Sousa, P. Cheung
ISBN: 0-7923-7314-6

Essentials of Electronic Testing for Digital, Memory, and Mixed Signal VLSI Circuits
M.L. Bushnell, V.D. Agrawal
ISBN: 0-7923-7991-8

Analog and Mixed-Signal Boundary-Scan: A Guide to the IEEE 1149.4 Test Standard
A. Osseiran
ISBN: 0-7923-8686-8

Design for At-Speed Test, Diagnosis and Measurement
B. Nadeau-Dosti
ISBN: 0-79-8669-8

Delay Fault Testing for VLSI Circuits
A. Krstic, K.-T. Cheng
ISBN: 0-7923-8295-1

Research Perspectives and Case Studies in System Test and Diagnosis
J.W. Sheppard, W.R. Simpson
ISBN: 0-7923-8263-3

Formal Equivalence Checking and Design Debugging
S.-Y. Huang, K.-T. Cheng
ISBN: 0-7923-8184-X

On-Line Testing for VLSI
M. Nicolaidis, Y. Zorian
ISBN: 0-7923-8132-7

Defect Oriented Testing for CMOS Analog and Digital Circuits
M. Sachdev
ISBN: 0-7923-8083-5

Reasoning in Boolean Networks: Logic Synthesis and Verification Using Testing Techniques
W. Kunz, D. Stoffel
ISBN: 0-7923-9921-8

Introduction to I_{DDQ} Testing
S. Chakravarty, P.J. Thadikaran
ISBN: 0-7923-9945-5

Multi-Chip Module Test Strategies
Y. Zorian
ISBN: 0-7923-9920-X

Testing and Testable Design of High-Density Random-Access Memories
P. Mazumder, K. Chakraborty
ISBN: 0-7923-9782-7

From Contamination to Defects, Faults and Yield Loss
J.B. Khare, W. Maly
ISBN: 0-7923-9714-2

SOC (SYSTEM-on-a-Chip) Testing for Plug and Play Test Automation

A Special Issue of Journal of Electronic Testing

Vol. 18, Nos. 4/5, August/October 2002

Distributors for North, Central and South America:
Kluwer Academic Publishers
101 Philip Drive
Assinippi Park
Norwell, Massachusetts 02061 USA
Telephone (781) 871-6600
Fax (781) 681-9045
E-Mail: kluwer@wkap.com

Distributors for all other countries:
Kluwer Academic Publishers Group
Post Office Box 322
3300 AH Dordrecht, THE NETHERLANDS
Telephone 31 786 576 000
Fax 31 786 576 474
E-Mail: services@wkap.nl

Electronic Services < http://www.wkap.nl>

Library of Congress Cataloging-in-Publication Data
SOC (System-on-a-Chip) Testing for Plug and Play Test Automation edited by Krishnendu Chakrabarty
ISBN: 1-4020-7205-8

Printed on acid-free paper.

Printed in the United States of America.

JOURNAL OF ELECTRONIC TESTING: Theory and Applications 18, v, 2002

Foreword

"As custom-configured chips carve out a larger portion of the microprocessor market, major changes are in store for the way chip fabrication facilities operate. To understand why, consider the underlying trends. One is the ever-increasing abundance of transistors on a chip, which portends a day when the number of chips per product will approach one. . ." This quotation from a recent article by Bass and Christensen (*IEEE Spectrum*, vol. 39, no. 4, p. 38, April 2002) describes just one of the changes in technology that lead us to the system-on-a-chip (SOC). They call the core-based SOCs as the "disruptive technology" and line up arguments showing that these are likely to dislodge today's high-performance microprocessor chips from their industry-favorite position.

This is the beginning of exciting times. Looking back over a period of fifty years we see printed-circuit boards going through phases of development from discrete components to discrete-chips and multi-chip modules. With each phase came new materials, design algorithms, performance criteria, heat dissipation problems, and manufacturing and test technologies. We can see the history repeating. To cite an example, we learned a great deal about signal propagation on the board interconnects, but must do still more to solve the problems of signal propagation on SOC interconnects. Of course, we can build upon the previous knowledge.

Interconnects may be just one problem. To test the SOCs we have to deal with new situations. Access to embedded cores, test data volume, at-speed test, fault diagnosis, power dissipation, test time, and test scheduling are some of the other problems for which solutions have already started emerging.

We have timed this special issue of *JETTA* for those readers who want to catch the new wave. The guest editor, Krishnendu Chakrabarty, issued an open call for papers about a year ago. We had an enthusiastic response from authors. He then selected the papers appearing here through the journal's peer review process. Unfortunately, not all the submitted papers could be accepted. A large variety of topics are covered in the articles appearing here and we believe this collection will have a lasting value. Therefore, this special issue is also being published as a book in the *Frontiers in Electronic Testing* series.

I am grateful to Krishnendu for meticulously handling all editorial responsibilities. My thanks to the authors and reviewers for their excellent contributions.

Vishwani D. Agrawal
Editor-in-Chief
va@agere.com

JOURNAL OF ELECTRONIC TESTING: Theory and Applications 18, vii, 2002

Preface

System-on-a-chip (SOC) integrated circuits composed of embedded cores are now commonplace. Nevertheless, there remain several roadblocks to rapid and efficient system integration. Test development is seen as a major bottleneck in SOC design, and test challenges are a major contributor to the widening gap between the design and manufacturing capabilities. Testing SOCs is especially challenging in the absence of standardized test structures, test automation tools, and test protocols. In addition, long interconnects, high density, and high-speed designs lead to new types of faults involving crosstalk and signal integrity.

This special issue of JETTA contains thirteen papers that address various aspects of SOC testing. These papers were selected after a thorough peer review process in which every paper received at least three reviews. The first paper by Marinissen et al. describes the proposed IEEE P1500 standard on embedded core testing. It provides an overview as well as technical details underlying the scalable wrapper architecture and the core test language envisaged in P1500. Papers 2–8 are on various aspects of test scheduling and test access mechanisms for core-based SOCs. Larsson and Peng describe an optimization framework for test scheduling, test access mechanism design, test set selection, and test resource placement. Huang et al. describe a power-constrained test scheduling approach based on two-dimensional bin-packing. In the third paper in this group, Koranne presents a reconfigurable wrapper design and a test scheduling algorithm for an SOC that employs reconfigurable wrappers. Marinissen shows how test protocols can be used for translating core-level tests to chip-level tests, and for test scheduling. In the next paper, Benabdenbi et al. describe a reconfigurable test access mechanism based on the use of a core access switch and test pattern compression/decompression. Basu et al. show how a test access mechanism switch can be used to test embedded cores and interconnect in an SOC. In the final paper of this group, Yoneda and Fujiwara show how transparent test data propagation in core-based SOCs can be achieved using the concept of consecutive testability.

Papers 9 and 10 propose solutions to the problem of high test data volume in SOCs. Jas and Touba describe how an embedded processor can be used for deterministic test vector compression and decompression. Next, Li et al. describe two schemes for compressing diagnostic test data for memory BIST. The last three papers are devoted to crosstalk, signal integrity, and faults affecting clock signals. Chen et al. describe a software-based self-test methodology that uses an embedded processor to detect interconnect crosstalk faults. Nourani and Attarha present fault models for signal integrity loss on interconnects in high-speed SOCs, and describe a BIST technique to detect noise and skew on the interconnects of such circuits. In the last paper, Metra et al. describe the design of an on-chip detector for on-line testing of faults affecting clock signals.

Finally, I take this opportunity to thank Editor-in-Chief Vishwani Agrawal for his strong support and encouragement for this special issue. I also thank all the authors for their timely submissions and the reviewers for their help.

Krishnendu Chakrabarty
Guest Editor
Department of Electrical and Computer Engineering
Duke University
Durham, NC
krish@ee.duke.edu

JOURNAL OF ELECTRONIC TESTING: Theory and Applications 18, 365–383, 2002

On IEEE P1500's Standard for Embedded Core Test

ERIK JAN MARINISSEN
Philips Research Laboratories, Prof. Holstlaan 4, 5656 AA Eindhoven, The Netherlands
Erik.Jan.Marinissen@philips.com

ROHIT KAPUR
Synopsys, Inc., 455 N. Mary Avenue, Sunnyvale, CA 94087, USA
RKapur@synopsys.com

MAURICE LOUSBERG
Philips Research Laboratories, Prof. Holstlaan 4, 5656 AA Eindhoven, The Netherlands
Maurice.Lousberg@philips.com

TERESA McLAURIN
ARM, Inc., 1250 S. Capital of Texas Highway, Austin, TX 78746, USA
Teresa.McLaurin@arm.com

MIKE RICCHETTI
Intellitech Corp., 70 Main Street, Durham, NH 03824, USA
MikeR@intellitech.com

YERVANT ZORIAN
LogicVision, Inc., 101 Metro Drive, 3rd Floor, San Jose, CA 951100, USA
Zorian@logicvision.com

Received September 21, 2001; Revised January 16, 2002

Editor: Krishnendu Chakrabarty

Abstract. The increased usage of embedded pre-designed reusable cores necessitates a core-based test strategy, in which cores are tested as separate entities. IEEE P1500 Standard for Embedded Core Test (SECT) is a standard-under-development that aims at improving ease of reuse and facilitating interoperability with respect to the test of core-based system chips, especially if they contain cores from different sources. This paper briefly describes IEEE P1500, and illustrates through a simplified example its scalable wrapper architecture, its test information transfer model described in a standardized Core Test Language, and its two compliance levels. *The standard is still under development, and this paper only reflects the view of six active participants of the standardization committee on its current status.*

Keywords: embedded cores, standardization, core test wrapper, core test language, compliance levels

1. Introduction

Until recently, most electronic systems consisted of one or multiple printed circuit boards, containing multiple integrated circuits (ICs) each. Advances in IC design methods and manufacturing technologies allow to integrate these complete systems onto a single IC. These so-called *system chips* offer advantages such as higher performance, lower power consumption, and smaller volume and weight, when compared to their traditional multi-chip equivalents. System chips are typically very heterogeneous, in the sense that they contain a mix of various types of circuitry, such as digital logic, memories in various flavors, and analog [5]. Many system chips are designed by embedding large reusable building blocks, commonly called *cores*. Design reuse [12] speeds up the design and allows import of external design expertise. The functions provided by cores include CPUs and DSPs, serial interfaces, modules for interconnect standards such as PCI, USB, and IEEE 1394, graphics computation functions such as MPEG and JPEG, and memories [2].

Usage of embedded cores divides the IC design community into two groups: *core providers* and *core users*. In traditional System-on-Board (SOB) design, the components that go from provider to user are ICs, which are designed, manufactured, and tested. The user of these components is only concerned with the design, manufacturing, and testing of his/her system, while using the components as fault-free building blocks. Testing SOBs is limited to detecting manufacturing defects in the interconnect between the components. In System-on-Chip (SOC) design, the components are cores. Independent of whether these components are delivered as *soft* (RTL code), *firm* (netlist), or *hard* (layout) cores, they are design descriptions only, not yet manufactured nor tested for manufacturing defects [18]. This makes the core user responsible for manufacturing and testing the entire system chip, i.e., not only the interconnect in between the cores, but also the cores themselves. However, in many cases assistance of the core provider to the core user in the test development trajectory is indispensable, as often only the core provider is familiar with the implementation details of a core. The most preferred form of assistance is that the core provider delivers pre-defined tests with the core.

This paper describes IEEE P1500 Standard for Embedded Core Test (SECT). IEEE P1500 SECT is an IEEE standard-under-development that intends to facilitate core-based testing, i.e., testing embedded cores as separate entities [26]. The motivation behind this industry-wide standard is to enable the reuse of tests when a core gets embedded in multiple different SOCs, as well as to enable interoperable core-based testing of SOCs that contain multiple cores from distinct core providers. IEEE P1500 SECT standardizes a test information transfer model as well as (part of) the on-chip test access hardware that enables core-based testing.

- The total SOC test development now is a (either in time or place) distributed effort, which requires transfer of information regarding the core tests from core provider to core user. IEEE P1500's standardized information model, based on IEEE 1450.6 CTL [11], allows to express this information.
- A core is typically deeply embedded in the system-chip design. In order to execute its tests, which are defined at the core terminals, we need a test access infrastructure to link the test pattern *source* (either an off-chip ATE or on-chip BIST) to the core inputs, and vice versa to link the core outputs to the test pattern *sink* (again, either ATE or BIST) [27]. IEEE P1500 standardizes the core-specific part of this infrastructure, i.e., a test access wrapper around (or embedded into) the core.

IEEE P1500 does *not* cover the core's internal test methods or DfT, nor SOC test integration and optimization. These are completely in the hands of core provider or core user respectively, and not suited for standardization due to the fact that their requirements differ for the different technologies and design styles of different cores and SOCs.

The prime targets of IEEE P1500 SECT are black-boxed third-party cores, of which the implementation details are hidden from the core user by the core provider. The majority of today's core market is (still) formed by black-boxed hard and firm cores. In such cases, the core user cannot create tests for the core, simply because he/she does not know what needs to be tested. Therefore, in the case of black-boxed cores, it is mandatory that the core provider delivers tests with the core. The core user assembles an SOC-level test out of the pre-defined tests for the various cores and additional tests for non-core chip circuitry. The test approach that is induced by IEEE P1500 lends itself also for modular testing of large system chips that consist entirely of modules of which all implementation details are known. These modules can be part of the growing legion of soft cores, in-house cores, or SOC-specific modules. Also in these cases, a modular test approach is often an attractive proposition, as it allows the core user to abstract from the module's implementation details. Overall system chip test development becomes a process of *divide-and-conquer*, allowing concurrent engineering, and hence a drastically reduced test development time [22]. Furthermore, *test reuse*, i.e., reusing a test for the same module integrated in different SOCs, becomes a realistic option. In this paper, we use the term *core-based testing* loosely for all forms of modular testing, and hence cover both black-boxed, as well as non-black-boxed modules.

IEEE P1500 SECT supports two compliance levels, commonly referred to as IEEE 1500 *Unwrapped* and IEEE 1500 *Wrapped*. In both cases, the core comes with a CTL program that describes the core tests. In the case of a 1500 Wrapped core, the core incorporates a complete P1500 wrapper function, while for a 1500 Unwrapped core, the wrapper is not present yet, but the CTL program contains the information on the basis of which a compliant wrapper can be added. Although the benefits of modular testing, test interoperability, and test reuse only become apparent when indeed the P1500 wrapper is used, the two compliance levels provide flexibility in the usage of the standard.

The first version of IEEE P1500 SECT focuses on non-merged digital logic and memory cores. This standard is currently in its development phase; the latest internal draft standard document (IEEE P1500/D0.5) has been released in October 2001. After completion of this standard, P1500 also intends to cover analog and mixed-signal cores, as well as DfT guidelines for mergeable cores. Note that the letter "P" for "Project" in the name P1500 indicates that the standard is not yet completed; once the standard has been accepted, the "P" will drop out of the name.

This paper provides a general view of the current direction of the standardization activity. Note that the technical solution presented is owned by the IEEE P1500 Working Group and that its presentation in this paper is an unapproved preliminary view on the standard. It is unapproved because it only reflects the interpretation of the authors of this paper and not the full IEEE P1500 Working Group. It is preliminary, because the standardization work is not finalized at this point of time.

The sequel of the paper is organized as follows. Section 2 describes the relation of IEEE P1500 SECT to other standards. Section 3 gives an overview of the core test wrapper and how it fits into an SOC-level test access architecture and Section 4 outlines the core test language CTL. Section 5 explains the dual compliance concept that P1500 uses to obtain choice and flexibility. In Section 6, IEEE P1500 is shown in action, by means of a simplified example core; the originally 'bare' core is taken through two subsequent compliance levels of IEEE P1500, and the corresponding CTL descriptions and architecture are shown. Section 7 concludes this paper.

2. P1500's Relation to Other Standards

IEEE P1500 SECT relates to several other IEEE standards and industry-wide efforts. These relations are listed below.

The *Virtual Socket Interface Alliance* (VSIA) [24] is a business alliance of over 200 companies. It is meant to promote core-based SOC design by specifying interface standards for design reuse of *virtual components*, the VSIA term for embedded cores. Typically, VSIA endorses existing standards and evaluates emerging ones; if nothing else exists VSIA also develops its own standards or specifications. VSIA covers various areas of core-based SOC design. Testing is covered by the Manufacturing-Related Test Development Working Group. This group has in the past specified common test data formats and design-for- testability guidelines for core providers [15]. The VSIA worked in parallel with the IEEE P1500 Working Group to create a gateway for test interoperability meant to be compatible with IEEE 1500. In 2001, it published a specification for a test access infrastructure, which is a prelude to the IEEE 1500 standard and is meant for temporary use until the complete IEEE 1500 standard is eventually finalized and approved.

The information model of P1500 uses CTL. CTL is built on IEEE Std. 1450.0, also known as *Standard Test Interface Language* (STIL) [6, 7, 21]. STIL is a language that can describe test vectors and waveforms; its original purpose was to be a generic interface language between EDA tools and ATE machines. In P1500, CTL has been developed as an extension of STIL, by adding core-test specific language constructs [11]. In May 2001, it has been decided to bring the standardization of CTL's syntax and semantic under IEEE P1450.6, whereas the P1500 information model using CTL remains under auspices of P1500.

IEEE P1500 also relates to IEEE Std. 1149.1, the Standard for Boundary Scan Test (also referred to as 'JTAG') [8]. IEEE 1149.1 specifies a chip-level 'wrapper' to facilitate board-level interconnect testing. The IEEE 1149.1 wrapper has served in some ways as a role model for the P1500 wrapper and hence the two have certain similarities. On the other hand, the two have important differences too, most of which stem from either the increased flexibility required for cores and SOCs or the relatively low costs of additional wrapper terminals [27]. Embedded cores have a wide range in test volume, and hence P1500 wrappers support a test access infrastructure of user-scalable width. Embedded cores are being utilized in both high-performance, as well as low-cost SOCs, and hence the wrapper itself can be scaled in various directions. Also, unlike IEEE 1149.1, the P1500 wrapper does not have a hard-wired finite state machine for control purposes. P1500 does recognize that P1500-wrapped cores might need to interface with (legacy) cores and/or SOC top-levels equipped with IEEE 1149.1 circuitry and is designed to do so [25].

3. Scalable Core Test Wrapper

In [27], a generic access architecture for testing embedded cores was introduced. This architecture is depicted schematically in Fig. 1. It consists of the following elements: an (off- or on-chip) source that generates test stimuli, a similar sink that evaluates the test responses, test access mechanisms (TAMs) that transport the stimuli from source to core and from core to sink, and a wrapper that, amongst other things, connects the TAMs to the core. From these elements, IEEE P1500 SECT only standardizes the wrapper [1], and leaves the

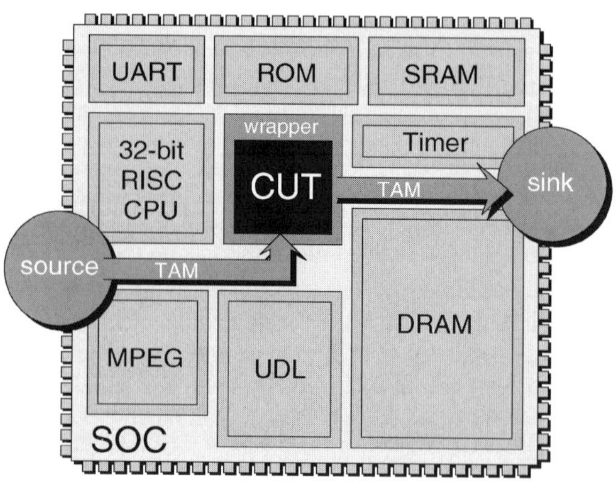

Fig. 1. Generic test access architecture for embedded cores, consisting of source and sink, TAMs, and wrapper.

source of source, sink, and TAMs up to the system chip designer. The wrapper has TAM ports, which, if properly connected to TAMs which in turn are properly connected to sources and sinks, provide controllability and observability for test purposes. The wrapper is standardized but scalable, as it has to be able to wrap all kinds of cores with potentially different test access requirements, as well as connect to different numbers and types of TAMs.

3.1. *Wrapper Architecture Overview*

The P1500 wrapper is a shell around a core, that allows that core to be tested as a stand-alone entity by shielding it off from its environment. Likewise, the wrapper allows the environment to be tested independent from the state of the core. The wrapper has three main types of modes: (1) functional operation, in which the wrapper is transparent and operates as if not existing, (2) *inward-facing* test modes, in which test access is provided to the core itself, and (3) *outward-facing* test modes, in which test access is provided to the circuitry outside the core.

Fig. 2 gives an overview of the main elements of the P1500 wrapper architecture. The P1500 wrapper is shown as a shell around the core. It has functional input and output ports, matching those of the unwrapped core. Furthermore it has a mandatory one-bit input/output port pair, `WSI` ('Wrapper Serial Input') and `WSO` ('Wrapper Serial Output'), and optionally one or more multi-bit input/output port pairs; in Fig. 2, one is drawn, named `WPI` ('Wrapper Parallel Input') and `WPO` ('Wrapper Parallel Output').

For control purposes, the wrapper has a *Wrapper Interface Port* (WIP) and an internal *Wrapper Instruction Register* (WIR). The WIP consists of six control signals (cf. Section 3.3). The WIP controls the WIR. The WIP allows the WIR to be loaded with an instruction via `WSI`. The operation of the remainder of the wrapper is controlled by both the WIP signals, as well as the instruction loaded into the WIR.

Once an instruction is loaded into the WIR, the corresponding (test) mode becomes active. In case of an inward-facing test mode, controllability needs to be provided at the core input terminals and observability needs to be provided at the core output terminals, such that the *core-internal tests* (e.g., the ones described in the CTL program that comes with the core) can be executed. The *Wrapper Boundary Register* (WBR) provides this type of controllability and observability.

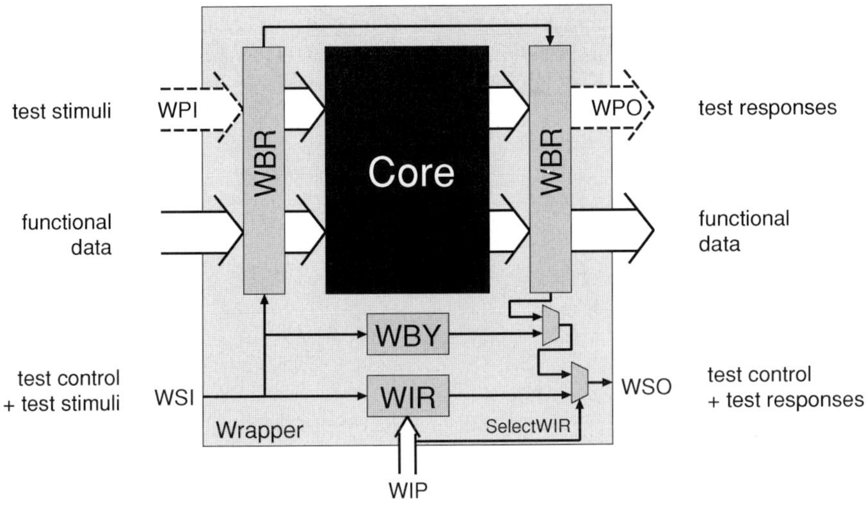

Fig. 2. Overview of IEEE P1500's wrapper architecture.

`WSI` is used to load test stimuli and `WSO` to unload test responses. Obviously, a one-bit access mechanism provides limited bandwidth, and that is why the wrapper can be extended with one or more multi-bit test data ports of scalable, user-defined width (such as `WPI` and `WPO` in Fig. 2).

An outward-facing test mode requires controllability of core outputs and observability of core inputs to facilitate the testing of circuitry external of the core. For that reason the *Wrapper Boundary Cells* from which the WBR is built all provide both controllability and observability.

The P1500 wrapper also contains a *Wrapper Bypass Register* (WBY), which serves as a bypass for the serial test data mechanism through `WSI` and `WSO`. In case multiple P1500-wrapped cores are daisychained into one serial TAM, WBY enables a shortened test access path to other cores up- and downstream. Similarly, in case parallel test access is provided, also a parallel bypass register can be implemented (not drawn in Fig. 2).

3.2. Wrapper Instruction Register

The WIR provides control over the wrapper operation. It determines whether the wrapper is in an inward-facing or outward-facing test mode, whether a serial or parallel access mode is utilized, and which wrapper data register is selected for test access. The WIR may also provide test signals to the core for certain instructions, such as those that enable inward-facing test modes used for internal testing of the core.

When loading an instruction, the WIR is exclusively connected in between `WSI` and `WSO`. When `WSI` and `WSO` are used for transporting instructions, they cannot be used for test data. This mutual exclusion is obtained by the `SelectWIR` signal of the WIP (cf. Section 3.3).

The WIR has a dual register implementation, consisting of a Shift Register and an equally long Update Register. Instructions are scanned in into the Shift Register, and become active only when clocked into the Update Register. Parallel inputs to the WIR may optionally be provided in order to capture data into the Shift Register during WIR capture operations. This captured data may be used for test control, testing of the WIR, or testing of other P1500 circuitry.

3.3. Wrapper Interface Port

The WIP consists of six signals. These signals control the operation of the WIR. For example, they determine whether the WIR is uniquely connected between `WSI` and `WSO` for loading an instruction, or whether these terminals are free for test data. The overall operation of the wrapper is controlled from the WIP signals and the instruction loaded into the WIR.

- `WRCK` ('Wrapper Clock') is the dedicated P1500 clock signal for WIR, WBY, and WBR.
- `WRSTN` ('Wrapper Reset') is the dedicated P1500 asynchronous, active-low reset signal. When asserted, it resets the WIR, which puts the wrapper in normal functional mode.

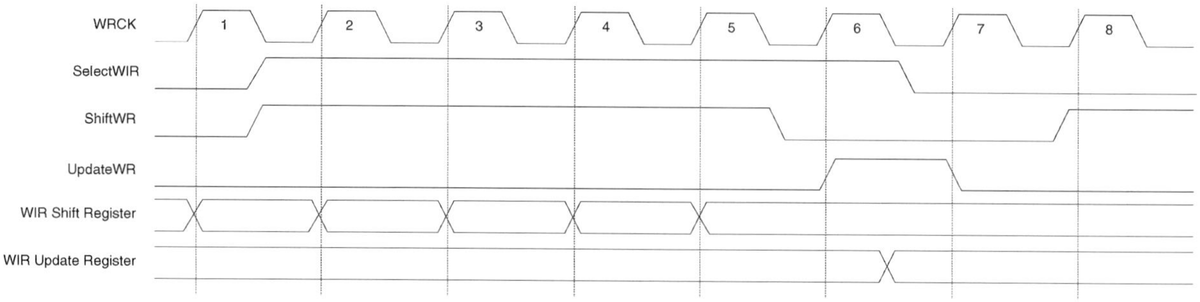

Fig. 3. Example WIP timing diagram.

- SelectWIR selects whether WSI and WSO are used for instructions or test data. When asserted, the WIR is exclusively connected between WSI and WSO. When SelectWIR is de-asserted, WBR, WBY, or any core-internal data register can be selected by means of an instruction in the WIR.
- CaptureWR, ShiftWR, UpdateWR. When the corresponding WIP signal is asserted, a capture, shift, or update operation will be enabled for the selected wrapper register, i.e., either the WIR (in case SelectWIR is asserted) or a wrapper data register identified by the instruction in the WIR (only if SelectWIR is de-asserted). These operations are synchronous to WRCK.

The timing, with respect to the WIP, for a WIR shift operation followed by a WIR update operation is illustrated in Fig. 3. The WIP timing is such that WIP input signals SelectWIR, CaptureWR, and ShiftWR are set up prior to the rising edge of WRCK, and the signal UpdateWR is set up prior to the falling edge of WRCK. In addition, new WSI data will be captured on the rising edge of WRCK, while WSO data changes on the falling edge of WRCK.

For the WIR, multiple operations are not permitted to occur simultaneously. Accordingly, when the SelectWIR signal is asserted, only one of CaptureWR, ShiftWR or UpdateWR may also be asserted. When the CaptureWR, ShiftWR, and UpdateWR signals are de-asserted, the WIR Shift Register and WIR Update Register will hold their current states.

In Fig. 3, the SelectWIR and ShiftWR WIP signals are asserted prior to the first rising edge of WRCK and held for example for the next four WRCK cycles. This initiates a WIR shift operation and subsequently shifts four bits of WIR data, shifting data in on WSI on the rising edge of WRCK and shifting data out on WSO with the falling edge of WRCK. In the sixth WRCK cycle, the ShiftWR signal has been de-asserted and the UpdateWR signal asserted. Consequently, on the next falling edge of WRCK the data from the WIR Shift Register will be clocked into the WIR Update Register. This sequence has now updated a new wrapper instruction in the WIR. Following the WIR update, the SelectWIR signal is de-asserted. Starting with the next WRCK cycle, a shift of the data register selected by the wrapper instruction will occur.

3.4. *Wrapper Boundary Register*

The WBR provides test access at the core terminals. The WBR is built up from *Wrapper Boundary Cells*. There is one cell per functional digital core terminal. The basic functionality of a wrapper boundary cell is that it needs to provide (1) functional pass-through, (2) controllability from the test data ports WSI and WPI, and (3) observability to the test data ports WSO and WPO. Fig. 4 shows simple implementation examples of wrapper input and output cells that provide such functionality [19]. The dark-gray circles in the multiplexers show which path is enabled if the multiplexer control signal is set to '1'. Note that IEEE P1500 does only define the behavior of such cells, and *not* the implementation. P1500 allows for extension of the functionality of the wrapper cells, for example with an additional gate or flip-flop for 'ripple-while-shift' protection, additional flip-flops for local storage of additional stimuli, etc.

In a serial test mode, test data is fed through WSI and WSO. For core-external testing, the WBR forms one serial shift register in between WSI and WSO. For core-internal testing, the WBR may be combined with core-internal scan registers between WSI and WSO. Obviously, the bandwidth provided in the serial test modes

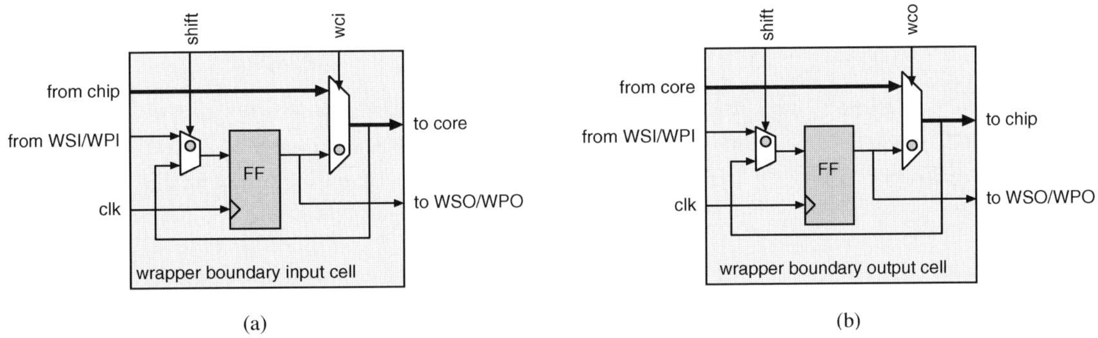

Fig. 4. Wrapper boundary cells for (a) core input and (b) core output terminal.

is limited, as the width of the test access mechanism is limited to one bit only.

In an (optional) parallel test mode, WBR and possibly the core-internal scan chains are fed from the multi-bit `WPI` and `WPO` and hence multiple 'wrapper chains' can be constructed in order to reduce the shift lengths involved. A P1500 wrapper can have zero or more `WPI`/`WPO` port pairs, all of user-defined width. The idea is that the width can be adapted to the bandwidth need of the core and/or the bandwidth availability at SOC level. IEEE P1500 does not mandate optimized wrapper design. Partitioning and ordering of WBR segments and core-internal scan chains, if any, in order to minimize the test application time, is beyond the scope of the standard. Optimization of the wrapper design for test application time is described in [9, 17].

3.5. *Wrapper Instruction Set*

The P1500 wrapper has various modes of operation. There are modes for functional (non-test) operation, inward-facing test operation, and outward-facing test operation. Different test modes also determine whether the serial test data mechanism (`WSI`–`WSO`) or the parallel test data mechanism (`WPI`–`WPO`), if present, is being utilized. In this paper, we distinguish the following modes: Normal, Serial InTest, Serial ExTest, Serial Bypass, Parallel InTest, and Parallel ExTest.

Instructions loaded into the WIR determine, together with the WIP signals, the mode of operation of the wrapper and possibly the core itself. There is a minimum set of instructions and corresponding operations that shall be supplied. Optional instructions and their corresponding behavior are also defined, together with the requirements for extension of the instruction set. All instructions that establish test modes that utilize the parallel ports `WPI` and `WPO` are optional, as the very presence of these ports is optional.

This section discusses the instructions that establish 'serial' test modes, in order to provide an understanding of the structure and reasons behind the instructions; a similar reasoning applies to the optional instructions that establish 'parallel' modes. Table 1 provides an overview of the various instructions.

The mandatory WBYPASS instruction establishes two non-conflicting modes, viz. Normal mode and Serial Bypass mode. The instruction configures the WBR such that the wrapper cells allow functional operation of the core. In addition, it connects WBY in between `WSI` and `WSO`, such that the serial test access mechanism can be utilized to provide test access to other cores.

The instruction WEXTESTS accomplishes controllability to and observability of circuitry external to the core. The serial version of this instruction is mandatory to allow for minimal pin testing of the external logic. It is unknown how many wrappers from separate cores will be needed to test the logic external to the cores on an SOC. The serial version of this instruction minimizes the risk of not having enough pins to activate all of the wrappers. In fact, all of the wrappers can be concatenated into one chain if needed.

P1500 does not define the core tests; the P1500 wrapper only focuses on allowing accessibility to execute the core's test(s). The main core test instruction, WCORETEST, is meant to be flexible enough to allow any core test to execute. The WCORETEST instruction was derived so that it can be defined differently for each core test within the parameters of a template. There are two other core test instructions that are encompassed within the WCORETEST instruction, but are defined

Table 1. IEEE P1500 wrapper instructions.

Instruction	Mode	Type	Description
wBypass	Normal + Serial Bypass	Mandatory	Wrapper allows functional mode, WSI-WSO connected through WBY
wExTestS	Serial ExTest	Mandatory	Test of core-external circuitry through WSI-WSO
wCoreTest	Serial/Parallel InTest	Optional[a]	User-specified core test, either through WSI-WSO or WPI-WPO
wCoreTestS	Serial InTest	Optional[a]	WSI-WSO connected through WBR and core-internal scan chain, internal testing
wCoreTestWS	Serial InTest	Optional[a]	WSI-WSO connected through WBR, internal testing
wPreloadS	Other	Cond. Mand.	Loads data into dedicated shift path of WBR (if existent)
wClamp	Other	Optional	WSI-WSO connected through WBR, outputs static state from all outputs
wSafeState	Other	Optional	WSI-WSO connected through WBY, recommends core in quiet mode, outputs static 'safe' state from all outputs
wExTestP	Parallel ExTest	Optional	Test of core-external circuitry through WPI-WPO
wPreloadP	Parallel Preload	Optional	Loads data into the dedicated shift paths of the WIR using ports in addition to or other than the WIP

[a]At least one of these instructions needs to be implemented.

additionally to help with standardization of these two types of testing. wCoreTestWS is implemented for a core test that uses a single wrapper scan chain using WSI/WSO for test access. wCoreTestS is a core test instruction that concatenates all wrapper cells and core-internal scan chains into one chain in between WSI and WSO. In case of a full-scan core, this instruction can be used for minimal port access testing of a core. Other possible utilizations of this instruction may be debug or burn-in. The user of the standard is allowed to use one or more of these three core test instructions, but the standard mandates that at least one of them must be implemented for the wrapper to be compliant.

In the case that the WBR contains a dedicated, non-shared shift path, the wPreload instruction is mandatory. This instruction preloads the WBR with specified values to be used before the wClamp, wExTestS, or wExTestP instruction.

The wClamp instruction allows the state of the signals driven from wrapper output terminals to be determined from WBR, while WBY is selected as the serial path between WSI and WSO. The signals driven from the wrapper terminals will not change while the wClamp instruction is selected.

The wSafeState instruction allows the state of the signals driven from wrapper output terminals to be determined from WBR, while WBY is selected as the serial path between WSI and WSO. The signals driven from the wrapper terminals will not change while the wSafeState instruction is selected. This instruction is a subset of the wClamp instruction. The wClamp instruction actually can be used to put the wrapper into a safe state. However, wSafeState was created as a separate instruction to allow the core integrator to more easily put various wrappers into safe states while other portions of an SOC are being tested. It is recommended that the core be in reset mode during this instruction. This instruction may also be used to bypass a broken WBR or core to put the core and wrapper into a state that will allow the rest of the SOC to be tested, provided the WIR and WBY are operational.

3.6. *Minimal and Typical Usage*

A minimal implementation of an IEEE P1500 compliant wrapper has no parallel test data ports. This wrapper provides limited test access bandwidth through WSI and WSO only. However, there are cases where more bandwidth is simply not needed. Examples of such cases are cores with built-in stimulus generators and response evaluators.

Perhaps a more typical situation we envision is that a core has a wrapper with both the mandatory one-bit access mechanism and one multi-bit TAM port. This is depicted in Fig. 5. The principal function of the parallel TAM is to test the cores internally. Its actual width is chosen such that it meets the test requirements of the cores, but also is not too expensive to the core integrator, e.g., with respect to silicon area and/or number of SOC pins. Core-external tests might also be applied via the parallel TAM. The serial TAM is meant to support

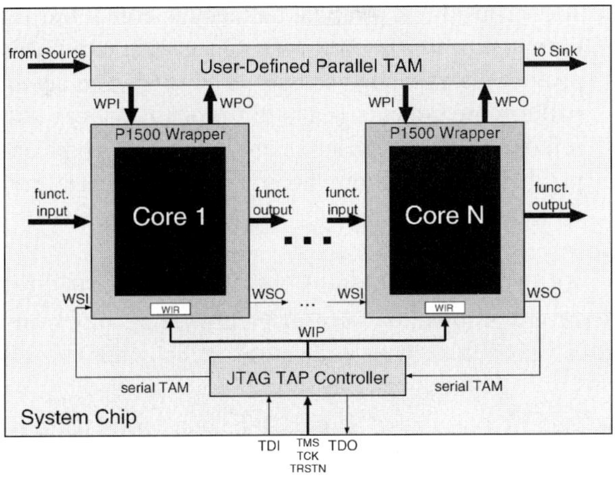

Fig. 5. SOC with typical P1500 usage; cores have both a serial and a parallel TAM connected.

core-level test and debug. In Fig. 5, the IEEE P1500 infrastructure is at SOC top-level connected to an IEEE Std. 1149.1 ('JTAG') Test Access Port (TAP) and Controller [8]. Once the system chip is soldered onto a printed circuit board, the normal SOC pins are not directly accessible, but test access can be provided via the serial data pins `TDI` (Test Data Input) and `TDO` (Test Data Output) of the TAP. Furthermore the serial TAM can be utilized to perform core-external ('interconnect') tests or even core-internal tests. However, depending on the actual tests, it might be that the serial TAM does not provide sufficient bandwidth to perform these tests in a timely and at-speed manner.

4. Core Test Language

CTL is a language for capturing and expressing test-related information for reusable cores [10, 11] which is meant to co-exist with and complement information expressed as a netlist. Test information about a core can be captured in CTL such that the SOC integrator or automation tools could successfully create a complete test for the SOC. Using the CTL description of a core, a wrapper can be constructed, and the appropriate TAM can be determined based on the test constraints in the CTL description. Once the test access infrastructure is in place, the test patterns that are also a part of the CTL description can be translated from the core boundary to the SOC boundary.

CTL is the language to support all the information that the core provider needs to give the system integrator such that the latter can successfully test the embedded core and any user-defined logic around the core. This language is broad enough to describe cores on which P1500 wrappers are to be implemented, 1500-Wrapped cores, their different test methodologies, and the different ways in which they are integrated in the SOC.

An important requirement for CTL is that the patterns, which contain the bulk of the data, are re-usable without any modification whatsoever. This is accomplished by creating the patterns by using macro statements (`M` statements) as opposed to vector statements (`V` statements) as used in traditional STIL [7]. This allows the vector application protocol to be modified by the core integrator in an expeditious manner. Each pattern is identified as to its intended purpose and required test mode, so that a test synthesis tool can select and re-order as desired, again, without actually having to adjust the bulk pattern data.

CTL is both human- and computer-readable, as is STIL. Hence, it can be utilized for documentation purposes, as well as for driving chip test integration tools. Among the tools which might be built using CTL are wrapper generation tools, 1500-Compliance checkers, chip-level test access planning and synthesis tools [9, 23], test expansion tools that automatically translate a core-level test into an SOC-level test [16], test scheduling tools [3, 4, 13, 14, 16, 20], etc. The hope and expectation is that once the IEEE 1500 standard, and hence CTL, is in place, this will accelerate the development of such tools.

Fig. 6 shows the global CTL structure. The circled numbers in the figure correspond to the usage of the various language constructs in the CTL examples in Figs. 8, 9 and 12 in Section 6. Fig. 6 partitions CTL into STIL-dependent information on the right-hand side and

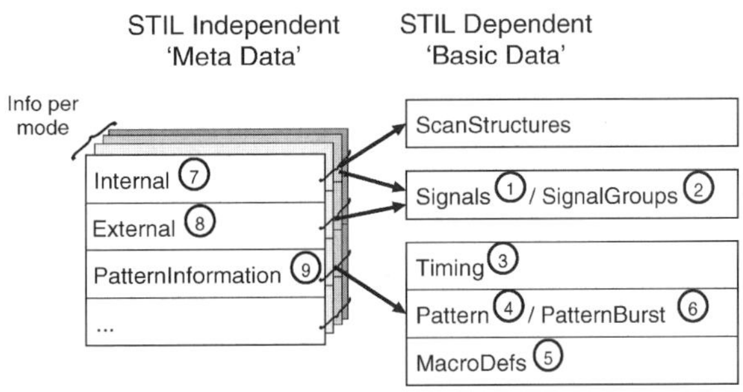

Fig. 6. CTL contains STIL-independent and STIL-dependent information, organized per test mode of the core.

STIL-independent information on the left-hand side. The STIL-independent information is the added value of CTL above STIL. It contains meta data on the more basic STIL information. Fig. 6 also shows that information in CTL is organized around (test) modes of the represented core. The signals of the core are global across all modes. In each mode the signals are used in different ways and are attributed to describe the different needs of the mode. Every mode of the core can have a mechanism to initialize the mode. The initialization sequences are described in CTL on a per-mode basis. Some modes of the core contain test patterns with their associated timing information, constraints, and statistics. Some modes contain structural information such that the structure can be used to create patterns at another level of integration of the design. CTL is designed to describe all these needs of different modes of the core. For any given test mode of a core, typically only a subset of the full CTL syntax will be required to adequately describe the attributes of the mode.

At the basic level, `Signals` and `SignalGroups` are defined with their attributes.

- `Signals`. This block defines each of the signal names of the core. Attributes can be attached to signals for additional information, e.g., `ScanIn` and `ScanOut` length, indicating the length of the scan chain connected to the signal.
- `SignalGroups`. Signals can be clustered into signal groups. Also here attributes can be attached. A signal can occur in multiple signal groups, each time having different attributes. In our example in Section 6, the attributes are used to represent the chain length in various test modes.

Timing, patterns, and protocols are also defined using STIL syntax.

- `Timing`. Each parallel or scan vector in a pattern macro has an associated timing block to define the waveform and corresponding timing on each signal.
- `Pattern`. The pattern blocks themselves contain the parallel and scan data for testing the core. Since pattern data is typically the bulk of the CTL file, it is often split off into a separate file. CTL defines the patterns in a restricted form, using `M` (macro) statements instead of STIL's `V` (vector) statements, such that they can be applied to the core via the wrapper and TAM.
- `MacroDefs`. This block contains the protocol for applying the test pattern data to the core or chip. A core-level test protocol defines the core's test requirements to the core user. Once integrated, this protocol is expanded to chip level [16]. The actual patterns are not changed in this process, so they still reference the same macro name, but in fact a new, expanded macro definition has replaced the core-level definition.

At a higher abstraction level, but still in STIL, patterns are grouped by the `PatternBurst` block and DfT information is given in `ScanStructures`.

- `PatternBurst`. This is a collection of patterns that are to be executed on the core. The patterns contain information as to the purpose of each pattern (normal functional, scan, I_{DDQ}, BIST, other). Information is also contained to indicate any required sequencing of the patterns and which patterns may be run independently.
- `ScanStructures`. This block can be used to describe the details of the internal scan chains of a core. Next to the scan-in and scan-out ports, also a list of scan cells and for example clocking information is specified using the `ScanStructures` block.

At CTL level, the `Internal`, `External`, and `PatternInformation` blocks are defined.

- `Internal`. The `Internal` block is used to describe the internal characteristics of the core signals. This information is provided to allow the core integrator to know what is happening with each signal of the core without needing full access to the design information. Examples of this information are: wrapper type, timing accuracy required, electrical characteristics such as analog or digital, test data volume, and more.
- `External`. The `External` block is used to describe the external characteristics that are expected from the perspective of the core boundary. Examples are: connect to chip pins (input, output, bidirectional), connect to another named core, connect to TAM, connect to user-defined logic, and more.
- `PatternInformation`. The `PatternInformation` block defines the purpose of each of the test patterns provided and the test mode necessary for the execution of each pattern. Other information like the fault model used and achieved coverage can also be given.

5. Two Compliance Levels

The market of embedded cores is a very heterogeneous one. Cores comprise very different functions, implemented in digital logic, memory, analog, RF, FPGAs, or combinations of the above. Cores come in many different flavors: 'hard', 'firm', and 'soft'. Cores are being used and/or sold within companies, as well as between companies. Within its current focus on non-merged digital logic and memory cores, IEEE P1500 SECT intends to support all types and forms of cores and associated business models. A means to get the flexibility required for this was to introduce a concept of two compliance levels into IEEE P1500.

- *IEEE 1500 Unwrapped Core.*[1] This compliance level refers to a core which does not have a (complete) IEEE 1500 wrapper, but does have a IEEE 1500 CTL program on the basis of which the core could be made '1500-Wrapped', either manually or by using dedicated tools. The CTL program describes the core test knowledge at the bare core terminals.
- *IEEE 1500 Wrapped Core.*[1] This compliance level refers to a core that incorporates an IEEE 1500 wrapper function, and comes with an IEEE 1500 CTL program. The CTL program describes the core test knowledge, including how to operate the wrapper, at the wrapper's external terminals.

The motivation behind the two different levels of compliance is that they provide the flexibility that is required in testing core-based system chips. Only once the P1500 wrapper is added, a core and its environment can be tested as separate entities, and hence the ultimate goal should be to make a core compliant to '1500-Wrapped'. Nevertheless, the two levels of compliance provide the option to become '1500-Wrapped' directly, or via an intermediate step at '1500-Unwrapped'.

The first alternative, i.e., direct generation of a '1500-Wrapped' core, provides the possibility to integrate the wrapper functionality with the core itself and hence minimize the performance and area impact of the wrapper. The second alternative is to first create a '1500-Unwrapped' core, which is then, in a second step, turned into a '1500-Wrapped' core. This allows to take advantage of the scalability of the standardized wrapper and instantiate the wrapper with particular parameter values, which take into account certain aspects of the system chip environment in which this particular core version is used.

In the sequel of this section, we show three examples of use scenarios and corresponding business models between core provider and core user that are supported by IEEE P1500 SECT, exploiting the two compliance levels. Note that this list of examples is not exhaustive and that other scenarios can be supported as well.

1. *Catalogue of Wrapped Cores.* The core provider offers a catalogue of off-the-shelf '1500-Wrapped' cores with fixed wrapper parameters. The core user then selects from that catalogue the core which best matches the system chip needs. This scenario allows the core provider to integrate the design of the wrapper with that of the core, in order to minimize silicon area and/or performance impact. Hence, it seems likely that this scenario will be popular for cores with strict requirements in those domains. Obviously, a larger catalogue with respect to different wrapper parameters will imply more work for the core provider. On the other hand, it provides freedom of choice to the customers and that might be a distinguishing feature.
2. *Unwrapped Core.* A core provider offers a '1500-Unwrapped' core. The CTL program that comes with it contains all relevant core test knowledge. This also includes core-related data for generation of the IEEE 1500 wrapper, such as the number, names, and types of terminals of the core and which terminals are involved at which data rates in which core tests.
   The customer who buys the '1500-Unwrapped' core turns it into a '1500-Wrapped' by adding a wrapper to it and translating the CTL file from the core terminals to the wrapper terminals. He/she can take into account information specific to the particular system chip environment into which the core is integrated and which have impact on the wrapper, such as the number and width of the TAMs that will connect to the wrapper, the necessity of certain optional wrapper instructions, the 'safe' wrapper output values in case the corresponding wrapper instructions are implemented, etc.
   The benefit of the '1500-Unwrapped' predicate of the core provider's product is that it guarantees the core user that all core-related test data is present and, while the wrapper is still adaptable to the specific chip environment needs, the route to a '1500-Wrapped' core is reliable and fast. We expect that, once the IEEE 1500 standard is in place, various '1500-Unwrapped-to-Wrapped' tools that automate this trajectory will come into existence.

3. *Core Wrapped-to-Order.* The core provider offers a '1500-Wrapped' core, which includes a wrapper built-to-order on the basis of parameters provided by customer specification. It might be that internally the core provider has already '1500-Unwrapped' cores on the shelf. In that case, this scenario is very similar to the previous one; the main difference is that the work of converting a '1500-Unwrapped' core into a '1500-Wrapped' one is now carried out by the core provider instead of the core user. Core provider and core user must anyway cooperatively exchange information relating to core functionality, usage, and test. Hence, often it will not be a problem to take the customer specifications regarding the wrapper parameters into this discussion as well. The core provider might have experts with special tools, for whom transforming a '1500-Unwrapped' core into a customer-specified '1500-Wrapped' core is daily business. This 'Wrapping Service' is similar to the 'Hardening Service' offered today by some providers of soft cores.

6. Example

This section features a running example, which starts as a 'bare' core, and is subsequently made '1500-Unwrapped' and '1500-Wrapped' compliant. The core in question is really a 'toy example', not at all representative of the complexity of today's cores. However, we have selected this simplified example because of its educational value. In all its simplicity, it shows a large number of relevant aspects of the IEEE P1500 standard-under-development.

6.1. Bare Core

Our example Core *A* is a scan-testable core with two core-internal scan chains. Scan chain 0 has a length of six flip flops and runs between terminals `si[0]` to `so[0]`. Scan chain 1 has length eight and runs between terminals `si[1]` to `so[1]`. The scan chains are shifting if terminal `sc` is asserted to logic value '1'. Furthermore, the core has five other input terminals, called `d[0]` ... `d[4]`, and three other output terminals, called `q[0]` ... `q[2]`. Finally, Core *A* has one clock input, called `clk`, which clocks all core-internal circuitry, including the scan chains. Core *A* is depicted in Fig. 7.

Core *A* has a set of predefined scan test patterns, which all adhere to the following protocol. Both scan

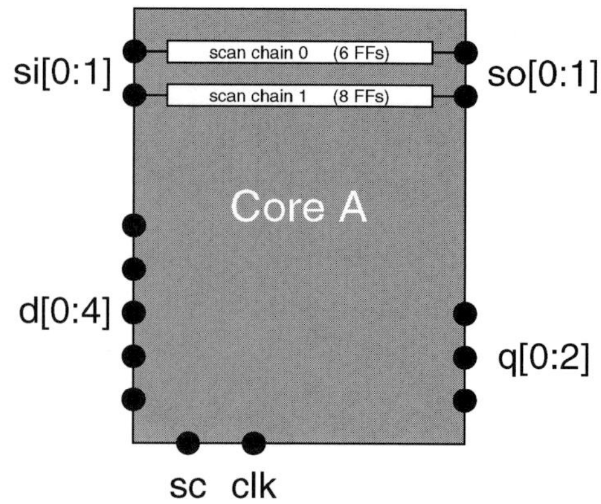

Fig. 7. Example Core *A*.

chains are loaded with stimuli, after which the stimuli in the scan chains, as well as additional stimuli on inputs `d[0..4]` are applied. Subsequently, responses on `q[0..2]` are observed, and additional responses in the two scan chains are shifted out.

6.2. '1500-Unwrapped' Core

A '1500-Unwrapped' core comes with a CTL description containing the information as described in Section 4. The CTL code accompanying our example focuses only on a subset of all that can be described in CTL, viz. the representation of test patterns for core-internal testing, their application protocol, and the instructions that prescribe how to integrate the core in a system chip such that its tests can be applied.

Fig. 9 contains the CTL description for Core *A*, while Fig. 8 shows `pat1.stil`, containing the test pattern data for Core *A*. Together they form the CTL code that makes Core *A* '1500-Unwrapped'. The pattern data is split off from the main program into a separate file, because it will be reused in the subsequent section of this paper by the CTL program for the '1500-Wrapped' core (see Fig. 12). Note that the bracketed numbers in Figs. 8 and 9 correspond to the circled numbers in Fig. 6.

The bare core has only one internal test mode, named `internal_test_mode`. The test patterns in Fig. 8 contain the raw stimuli and responses, i.e., the 0's and 1's to be applied to and observed at the core terminals. This pattern file `pat1.stil` contains five test patterns, which are applied in such a way that scanning out the responses of one pattern is combined with scanning in the stimuli of the next pattern. At which core terminals and time slots these patterns are to be

```
STIL 1.0 {
  Design 2001;
  CTL 2001;
}
Header {
  Title "pat1.stil: Shared CTL Pattern Data";
  Date "Sat Sep 01 2001";
}
Pattern pat1 {                                               (4)
  // initialization
  M setup_intest {}
  // bulk of pattern data follows
  M do_intest { d=00000; si[0]=111000; si[1]=11110000; }
  M do_intest { so[0]=LLHXHH; so[1]=HHHHLLXH; q=LLH;
    d=01101; si[0]=011010; si[1]=01011101; }
  M do_intest { so[0]=HHLLXH; so[1]=HLHHLLLL; q=HHL;
    d=11001; si[0]=110010; si[1]=00011100; }
  M do_intest { so[0]=LHLLLH; so[1]=HXHHLHLL; q=LLX;
    d=01010; si[0]=001101; si[1]=10011101; }
  M do_intest { so[0]=HHXLLL; so[1]=HLHHHLXX; q=LHH;
    d=01111; si[0]=010001; si[1]=10101101; }
  M do_intest { so[0]=LXHHHH; so[1]=LLLLHXHL; q=LLL; }
}
```

Fig. 8. `pat1.stil`, containing the CTL-coded test patterns for Core *A*.

```
STIL 1.0 {
  Design 2001;
  CTL 2001;
}
Header {
  Title "CTL Program for 1500-Unwrapped Core";
  Date "Sat Sep 01 2001";
}
Signals {                                              (1)
  si[0] In { ScanIn 6;}
  si[1] In { ScanIn 8;}
  sc In;
  clk In;
  d[0..4] In;
  q[0..2] Out;
  so[0] Out { ScanOut 6;}
  so[1] Out { ScanOut 8;}
}
SignalGroups {                                         (2)
  d = 'd[0..4]';
  q = 'q[0..2]';
  ins = 'd+si[0..1]+sc+clk';
  outs = 'q+so[0..1]';
  all = 'ins+outs';
}
Timing normal {                                        (3)
  WaveformTable scan_timing { }
  WaveformTable func_timing { }
  // Here the waveforms are defined for WFCs
  // 0 = logic-0, 1 = logic-1, P = Pulse Waveform
}
MacroDefs {                                            (5)
  setup_intest {
    W func_timing;
    V { all=xxxxx xx 00 xxx xx; }
  }
  do_intest {
    W scan_timing;
    C { sc=1;}
    Shift { V { si[0]=#; si[1]=#;
                so[0]=#; so[1]=#; clk=P;}}
    W func_timing;
    C { sc=0;}
    V { d=#; q=#; }
    V { clk=P; }
  }
}
PatternBurst all_pats {                                (6)
  PatList {
    Pat1;
  }
}
PatternExec run_all {
  Timing normal;
  PatternBurst all_pats;
}
Environment unwrapped_core {
  CTL internal_test_mode {
      TestMode InternalTest;
      Internal {                                                (7)
        clk {
          DataType
            MasterClock TestClock ScanMasterClock
            { ActiveState ForceUp; }
          DriveRequirements {
            TimingCritical {
               Period Min 50ns;
               Period Max 100ns;
            }
          }
        }
        sc {
          DataType TestControl ScanEnable
            { ActiveState ForceUp; }
        }
        si[0..1] {
          DataType TestData ScanDataIn
            { ScanDataType Internal; }
        }
        so[0..1] {
          DataType TestData ScanDataOut
            { ScanDataType Internal; }
        }
        'd+q' {
          DataType Functional TestData;
        }
      }
      External {                                                (8)
        si[0..1] { ConnectTo { TAM; In;}}
        so[0..1] { ConnectTo { TAM; Out;}}
        'clk+sc' { ConnectTo { In;}}
        'd+q' {
          ConnectTo {
            WBR; In; Out; Wrapper IEEE1500;
          }
        }
      }
      PatternInformation {                                      (9)
        PatternExec run_all {
          Purpose Production;
          Fault {
            Type StuckAt;
            FaultCoverage 98.7;
          }
        }
        Macro setup_intest {
          Purpose Generic;
          UseByPattern EstablishMode;
        }
        Macro do_intest {
          Purpose DoTestOverlap;
        }
      }
    }
}
Include "pat1.stil";
```

Fig. 9. CTL code for the '1500-Unwrapped' Core *A*.

applied, is defined in a corresponding test protocol in the `MacroDefs` section of the main CTL program in Fig. 9. Two macros are defined, viz. `setup_intest` and `do_intest`. The first macro configures the core in a way that the patterns can be applied to the core terminals, the second defines the actual application of test patterns. The protocol described in macro `do_intest` is a scan test of the bare core that uses its internal scan chains. First it shifts in data in the scan chains, at the same time shifting out data from a previous pattern, and then it applies a functional cycle. The macros refer to signals of the bare core defined in the `Signals` and `SignalGroups` sections of CTL. The patterns and macros fit in a framework defined by `PatternExec` and `PatternBurst`.

The macros contain four different types of statements [7]. All statements are *incremental*, i.e., their data encodes only the differences relative to previous statements.

- *WaveformTable statement* `W`, defining the timing information.
- *Vector statement* `V`, defining broadside stimuli and responses to be applied. For the actual values of stimuli and responses is referred to `pat1.stil`.
- *Scan test statement* `Shift`, defining scan chain stimuli and responses to be applied. For their actual values is again referred to `pat1.stil`.
- *Condition statement* `C`, which defines the default background value of terminals, to be used and possibly overruled in the `V` statements that follow.

Most of the CTL code in this example is actually standard STIL [7]. An example of newly added CTL constructs, i.e., constructs which did previously not exist in standard STIL, are the `ConnectTo` attributes in the `External` section of Fig. 9 (Label (8)). For a '1500-Unwrapped' core, these attributes give directions for wrapping the various core terminals. In our example, these attributes in the `Signals` section of the CTL program specify that `d[0..4]` and `q[0..2]` should be connected to functional circuitry, the scan chain inputs and outputs are to be connected to TAMs, and `sc` (a dedicated test signal) and `clk` (a 'special care' signal) require a direct connection to an SOC pin.

6.3. '1500-Wrapped' Core

In order to make Core *A* '1500-Wrapped' compliant, a wrapper is instantiated and added to the core, and a CTL description is created that describes the core's test(s) at the wrapper boundary. In this section we describe both. '1500-Wrapped' compliance can be reached either directly from the bare core, or via an intermediate step at '1500-Unwrapped', as described in Section 5.

Core plus wrapper are depicted in Fig. 10. Note that, whereas this figure shows some implementation details, IEEE P1500 only defines the behavior of the wrapper and leaves the implementation free. In this particular example, the wrapper can connect Core *A* both to a serial TAM, via `WSI` and `WSO`, as well as to a three-bit wide parallel TAM, via `WPI[0:2]` and `WPO[0:2]`. The example wrapper provides six instructions, supporting normal mode, and test modes with connections to either serial or parallel TAM. Mode setting through the WIR actually means setting the controls of the wrapper's multiplexers. In Fig. 10, the multiplexers that do occur in the wrapper are named $m1, m2, \ldots, m12$. Fig. 4(a) and (b) show additional multiplexers *wci* and *wco* in the wrapper input and output cells, respectively. The dark-gray circles in the multiplexers indicate which path is enabled if the multiplexer control signal is set to '1'. Table 2 lists the six instructions and the corresponding multiplexer settings they cause. Fig. 11 depicts the wrapper in each of its six modes by bold highlighting of the active wires in a particular mode. Note that both the multiplexer settings in Table 2 as well as the highlighted active wires

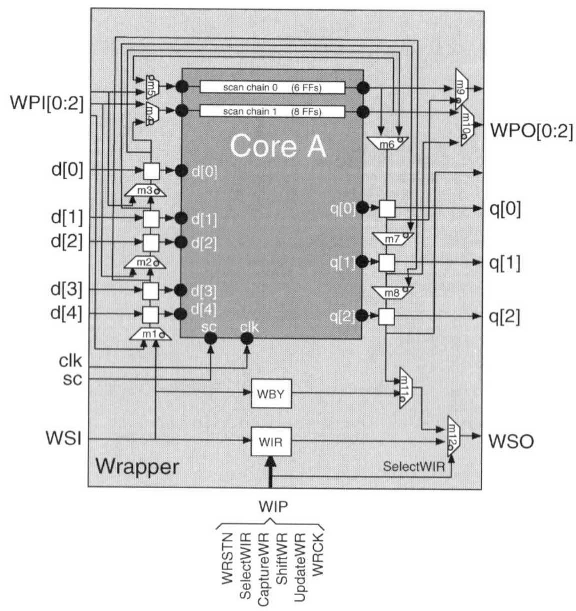

Fig. 10. Example Core *A* and its IEEE 1500 compliant wrapper.

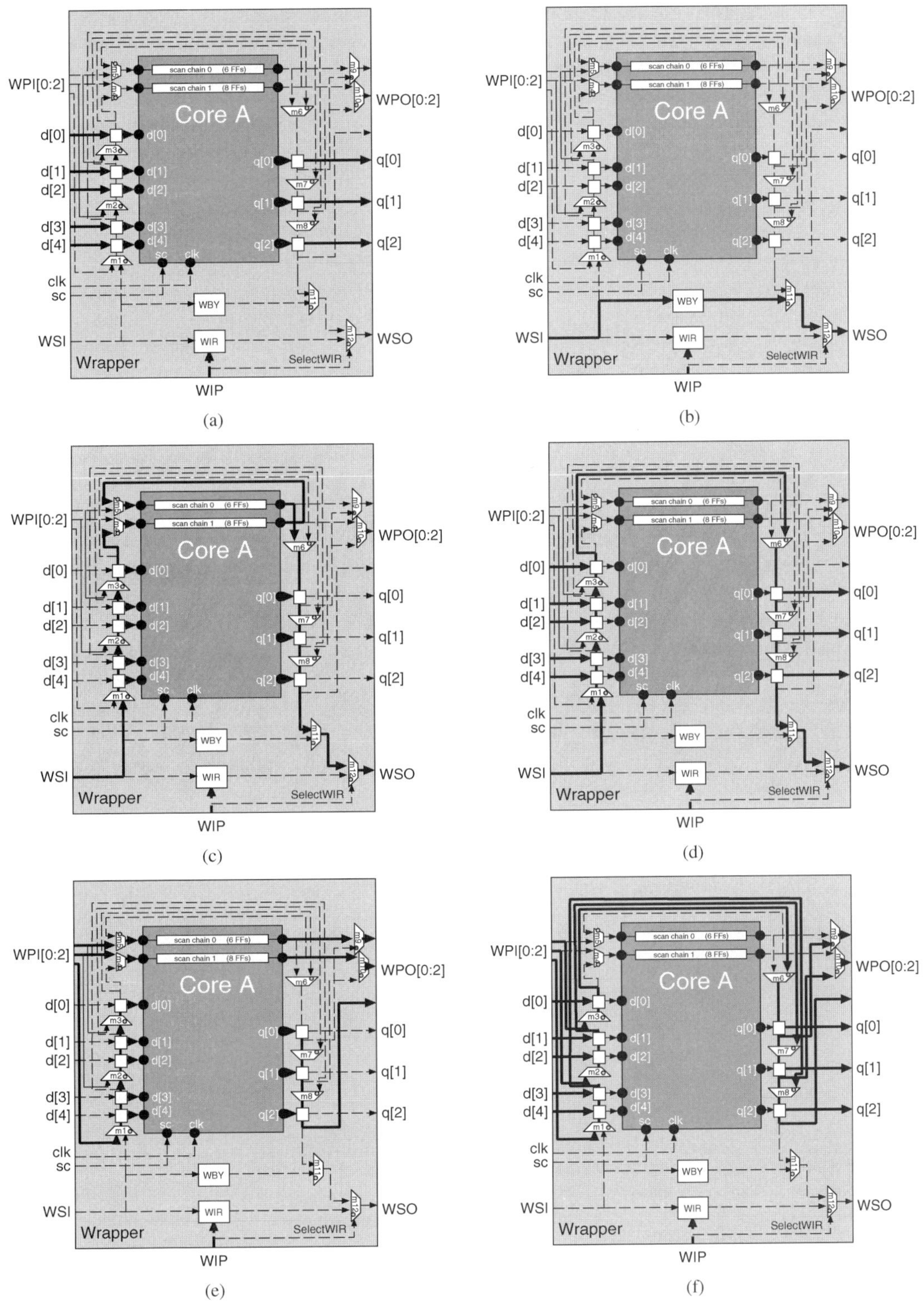

Fig. 11. The wrapper in action in its various modes. (a) Normal Mode with WBYPASS; (b) Serial Bypass Mode with WBYPASS; (c) Serial Internal Test Mode with WCORETESTS; (d) Serial External Test Mode with WEXTESTS; (e) Parallel Internal Test Mode with WCORETEST; (f) Parallel External Test Mode with WEXTESTP.

```
STIL 1.0 {
  Design 2001;
  CTL 2001;
}
Header {
  Title "CTL Program for 1500-Compliant Core";
  Date "Sat Sep 01 2001";
}
Signals {                                          (1)
  WSI In { DefaultState ForceOff; }
  sc In { DefaultState ForceDown; }
  clk In { DefaultState ForceDown; }
  d[0..4] In { DefaultState ForceOff; }
  WPI[0..2] In { DefaultState ForceOff; }
  WPO[0..2] Out { DefaultState ExpectOff; }
  q[0..2] Out { DefaultState ExpectOff; }
  WSO Out { DefaultState ExpectOff; }
  WRSTN In;
  SelectWIR In;
  CaptureWR In;
  ShiftWR In;
  UpdateWR In;
  WRCK In;
}
SignalGroups {                                     (2)
  si_m123 = 'WSI' {ScanIn 22;}
  si_wir = 'WSI' { ScanIn 4;}
  so_m123 = 'WSO' { ScanOut 22; }
  d = 'd[0..4]';
  q = 'q[0..2]';
  wip = 'WRSTN+SelectWIR+CaptureWR+ShiftWR+
         UpdateWR+WRCK';
  si[0..1] SignalVariable;
  so[0..1] SignalVariable;
}
Timing normal {                                    (3)
  WaveformTable setup_timing { }
  WaveformTable do_intest_timing { }
  // waveforms for signals are defined here
  // 0 = logic-0, 1 = logic-1, P = Pulse
}
MacroDefs {                                        (5)
  setup_intest {
    W setup_timing;
    V { WRSTN=0; }
    V { wip=110100; }
    Shift { V { si_wir=1010; WRCK=P;}}
    V { ShiftWR=0; UpdateWR=1;}
    V { WRCK=P;}
    V { SelectWIR=0; UpdateWR=0;}
  }
  do_intest {
    W do_intest_timing;
    C { sc=1; ShiftWR=1;}
    Shift { V { si_m123='d si[1] si[0]';
                so_m123='so[1] so[0] q';
                clk=P; WRCK=P;}}
    V { sc=0; CaptureWR=1; ShiftWR=0;}
    V { clk=P; WRCK=P;}
    V { CaptureWR=0; }
  }
}
PatternBurst all_pats {                            (6)
  PatList {
    Pat1;
  }
}
PatternExec run_all {
  Timing normal {
    PatternBurst all_pats;
  }
}
Environment wrapped_core {
  CTL serial_internal_test_mode {
    TestMode InternalTest;
    Wrapper IEEE1500;
    TestModeForWrapper WSCORETEST 1010;
    Internal {                                     (7)
      clk {
        DataType
          MasterClock TestClock ScanMasterClock;
        DriveRequirements {
          TimingCritical {
             Period Min 50ns;
             Period Max 100ns;
          }
        }
      }
      sc {
        DataType TestControl ScanEnable
           { ActiveState ForceUp; }
      }
      si_m123 {
        DataType TestData ScanDataIn
           { ScanDataType Boundary Internal; }
      }
      so_m123 {
        DataType TestData ScanDataOut
           { ScanDataType Boundary Internal; }
      }
      WRSTN {
        Wrapper IEEE1500 PinId WRSTN;
        DataType Asynchronous Reset {
          ActiveState ForceDown;
        }
        DataType TestMode { ActiveState ForceUp; }
      }
      SelectWIR {
        Wrapper IEEE1500 PinId SelectWIR;
        DataType TestControl { ActiveState ForceUp; }
        DataType TestMode { ActiveState ForceDown; }
      }
      CaptureWR {
        Wrapper IEEE1500 PinId CaptureWR;
        DataType TestControl { ActiveState ForceUp; }
      }
      ShiftWR {
        Wrapper IEEE1500 PinId ShiftWR;
        DataType TestControl ScanEnable {
          ActiveState ForceUp;
        }
      }
      UpdateWR {
        Wrapper IEEE1500 PinId UpdateWR;
        DataType TestControl { ActiveState ForceUp; }
      }
      WRCK {
        Wrapper IEEE1500 PinId WRCK;
        DataType TestClock ScanMasterClock {
          ActiveState ForceUp;
        }
      }
      'd+q+WPI[0..2]+WPO[0..2]' { DataType Unused;}
    }
    External {                                     (8)
      WSI { ConnectTo { In; Wrapper IEEE1500 PinId WSO;}
      WSO { ConnectTo { Out; Wrapper IEEE1500 PinId WSI;
      clk { ConnectTo {In; }}
      sc { ConnectTo { In;}}
      wip { ConnectTo {In; WIP;}}
    }
    PatternInformation {                           (9)
      PatternExec run_all {
        Purpose Production;
        Fault {
          Type StuckAt;
          FaultCoverage 98.7;
        }
      }
    }
    Macro setup_intest {
      Purpose Generic;
      UseByPattern EstablishMode;
    }
    Macro do_intest { Purpose DoTestOverlap;}
  }
}
Include "pat1.stil";
```

Fig. 12. CTL code for the '1500-Wrapped' Core *A*.

Table 2. Instruction decoding per multiplexer for the wrapper of Core *A*.

			Multiplexer settings													
Mode	Instruction	Opcode	wci	wco	m1	m2	m3	m4	m5	m6	m7	m8	m9	m10	m11	m12
Normal	wBypass	0000	0	0	X	X	X	X	X	X	X	X	X	X	X	X
Serial Bypass	wBypass	0000	X	X	X	X	X	X	X	X	X	X	X	X	1	0
Serial InTest	wCoreTestS	1010	1	0	1	1	1	1	1	0	0	0	X	X	0	0
Serial ExTest	wExTestS	1001	0	1	1	1	1	X	X	1	0	0	X	X	0	0
Parallel InTest	wCoreTest	1110	1	0	0	1	1	0	0	1	0	0	0	0	X	X
Parallel ExTest	wExTestP	1101	0	1	0	0	0	X	X	1	1	1	1	1	X	X

in Fig. 11(a) and (b) show that the Normal and Serial Bypass modes are non-conflicting, and hence can be established by one combined instruction, wBypass.

Fig. 12 contains the CTL program for Core *A* defined at its wrapper boundary. This '1500-Wrapped' CTL program contains the same information as the original '1500-Unwrapped' CTL program of Fig. 9. Hence, while transitioning from '1500-Unwrapped' to '1500-Wrapped', no information was lost. On top of that, additional information is present in the '1500-Wrapped' CTL code, containing the wrapper terminal names, the WIR instructions implemented, the modified test protocols at the wrapper boundary, and modified system integration instructions.

The wrapper in our example implements six different modes. In the CTL program in Fig. 12 we focus only on the Serial InTest mode, as depicted in Fig. 11(c). The test patterns specified in the file `pat1.stil` (see Fig. 8) remain intact and valid for core-internal testing of Core *A*, and hence are again referenced from our CTL program. However, their application protocol at the wrapper boundary has changed, which is reflected in the modified `MacroDefs` `setup_intest` and `do_intest`. The setup for Serial InTest requires that the Opcode 1010 of instruction wCoreTestS is scanned into the Shift Register of the WIR and subsequently loaded into its Update Register. The changes in macro `do_intest` reflect the extra steps needed to move the test pattern data from `WSI` through the wrapper to the core, and vice versa back out again via `WSO`. Of course, this protocol takes more time now, because the shift path is longer. The chain `si_m123` has length 22, compared to length 8 of the bare core. If test time is of concern, then it is better to use the parallel internal test mode depicted in Fig. 11(e), which provides a higher bandwidth.

After integrating the core in an SOC, the macros can be rewritten again to represent the top-level protocols (see Fig. 5). Here one should also investigate to what degree it is possible to run tests of the various cores in parallel in order to further reduce test time. For all practical situations, the writing and continuous rewriting of CTL descriptions would be too much work for any designer or test engineer under time pressure to finish his/her design. Some companies have already demonstrated that this task can be automated, although quite some effort is still needed to tune these efforts to the flexibility of CTL.

7. Conclusion

In this paper we described IEEE P1500 SECT, a standard-under-development that focuses on interoperability and re-use of tests for pre-designed embedded cores.

P1500 defines a core test wrapper that supports functional modes as well as inward-facing and outward-facing test modes. Basic test access is guaranteed by the mandatory hardware elements and instructions of the wrapper; flexibility is obtained by the scalability of many aspects of the wrapper. Examples of scalability are the optional ports `WPI` and `WPO` of user-defined width, and the elements of the WBR that can have optional functionality such as additional storage elements, ripple protection, etc.

P1500 also defines a Core Test Language, built on top of IEEE 1450.0 (STIL). This language is meant for transfer of 'core test knowledge' from core provider to core user. P1500 supports two compliance levels. '1500-Unwrapped' cores come with a CTL description which contains both the test patterns, as well as data on the basis of which a wrapper can easily be generated. '1500-Wrapped' cores both have a compliant wrapper and a full CTL description. These two compliance levels are required to support the various use scenarios of P1500.

All elements of the P1500 standard were illustrated by means of a simplified example, which aims at providing an early introduction to the standard-under-development.

Acknowledgments

We thank our fellow members of the IEEE P1500 Standard for Embedded Core Test (SECT) Working Group for co-developing the standard with us and we look forward to our continued cooperation with them in the future. We thank Benoit Nadeau-Dostie, Harald Vranken, Tom Waayers, and Lee Whetsel for providing useful comments on an earlier draft version of this paper.

Note

1. Previously, the two compliance levels were referred to as 'IEEE 1500 Ready' and 'IEEE 1500 Compliant', respectively [19]. In May 2000, the IEEE P1500 Working Group decided on the naming 'IEEE 1500 Unwrapped' and 'IEEE 1500 Wrapped', as used in this paper.

References

1. S. Adham et al., "Preliminary Outline of IEEE P1500 Scalable Architecture for Testing Embedded Cores," in *Proceedings IEEE VLSI Test Symposium (VTS)*, Dana Point, CA, April 1999, pp. 483–488.
2. T. Anderson, "This is Hard Core," *Test—The European Test Industry Journal*, vol. 25, no. 5, pp. S5–S6, June 1999.
3. K. Chakrabarty, "Test Scheduling for Core-Based Systems," in *Proceedings International Conference on Computer-Aided Design (ICCAD)*, San Jose, CA, Nov. 1999, pp. 391–394.
4. K. Chakrabarty, "Test Scheduling for Core-Based Systems Using Mixed-Integer Linear Programming," *IEEE Transactions on Computer-Aided Design*, vol. 19, no. 10, pp. 1163–1174, Oct. 2000.
5. R.K. Gupta and Y. Zorian, "Introducing Core-Based System Design," *IEEE Design & Test of Computers*, vol. 14, no. 4, pp. 15–25, Dec. 1997.
6. IEEE 1450 Web Site. http://grouper.ieee.org/groups/1450/.
7. IEEE Computer Society, *IEEE Standard Test Interface Language (STIL) for Digital Test Vector Data—Language Manual—IEEE Std. 1450.0-1999*. New York: IEEE, Sept. 1999.
8. IEEE Computer Society, *IEEE Standard Test Access Port and Boundary-Scan Architecture—IEEE Std. 1149.1-2001*. New York: IEEE, July 2001.
9. V. Iyengar, K. Chakrabarty, and E.J. Marinissen, "Co-Optimization of Test Wrapper and Test Access Architecture for Embedded Cores," *Journal of Electronic Testing: Theory and Applications*, vol. 18, no. 2, pp. 213–230, April 2002.
10. R. Kapur et al., "P1500-CTL: Towards a Standard Core Test Language," in *Proceedings IEEE VLSI Test Symposium (VTS)*, Dana Point, CA, April 1999, pp. 489–490.
11. R. Kapur et al., "CTL—The Language for Describing Core-Based Test," in *Proceedings IEEE International Test Conference (ITC)*, Baltimore, MD, Oct. 2001, pp. 131–139.
12. M. Keating and P. Bricaud, *Reuse Methodology Manual for System-on-a-Chip Designs*, Norwell, Massachusetts: Kluwer Academic Publishers, June 1999.
13. E. Larsson and Z. Peng, "An Integrated System-on-Chip Test Framework," in *Proceedings Design, Automation, and Test in Europe (DATE)*, Munich, Germany, March 2001, pp. 138–144.
14. E. Larsson and Z. Peng, "Test Scheduling and Scan-Chain Division Under Power Constraint," in *Proceedings IEEE Asian Test Symposium (ATS)*, Kyoto, Japan, Nov. 2001, pp. 259–264.
15. Manufacturing-Test Related DWG. *Test Data Interchange Formats and Guidelines for VC Providers*. VSI Alliance, Los Gatos, CA, Jan. 2001.
16. E.J. Marinissen, "The Role of Test Protocols in Automated Test Generation for Embedded-Core-Based System ICs," *Journal of Electronic Testing: Theory and Applications*, vol. 18, no. 4, pp. 435–454, Aug. 2002.
17. E.J. Marinissen, S.K. Goel, and M. Lousberg, "Wrapper Design for Embedded Core Test," in *Proceedings IEEE International Test Conference (ITC)*, Atlantic City, NJ, Oct. 2000, pp. 911–920.
18. E.J. Marinissen and Y. Zorian, "Challenges in Testing Core-Based System ICs," *IEEE Communications Magazine*, vol. 37, no. 6, pp. 104–109, June 1999.
19. E.J. Marinissen, Y. Zorian, R. Kapur, T. Taylor, and L. Whetsel, "Towards a Standard for Embedded Core Test: An Example," in *Proceedings IEEE International Test Conference (ITC)*, Atlantic City, NJ, Sept. 1999, pp. 616–627.
20. M. Sugihara, H. Date, and H. Yasuura, "A Novel Test Methodology for Core-Based System LSIs and a Testing Time Minimization Problem," in *Proceedings IEEE International Test Conference (ITC)*, Washington, DC, Oct. 1998, pp. 465–472.
21. T. Taylor and G. Maston, "Standard Test Interface Language (STIL): A New Language for Patterns and Waveforms," in *Proceedings IEEE International Test Conference (ITC)*, Washington, DC, Nov. 1996. IEEE Computer Society Press, pp. 565–570.
22. J. van Beers and H. van Herten, "Test Features of a Core-Based Co-Processor Array for Video Applications," in *Proceedings IEEE International Test Conference (ITC)*, Atlantic City, NJ, Sept. 1999, pp. 638–647.
23. P. Varma and S. Bhatia, "A Structured Test Re-Use Methodology for Core-Based System Chips," in *Proceedings IEEE International Test Conference (ITC)*, Washington, DC, Oct. 1998, pp. 294–302.
24. VSI Alliance Web Site. http://www.vsi.org/.
25. L. Whetsel and M. Ricchetti, "Tapping into IEEE P1500 Domains," in *Digest of Papers of IEEE International Workshop on Testing Embedded Core-Based Systems (TECS)*, Marina del Rey, CA, May 2001, pp. 3.2-1–7.
26. Y. Zorian, "Test Requirements for Embedded Core-Based Systems and IEEE P1500," in *Proceedings IEEE International Test Conference (ITC)*, Washington, DC, Nov. 1997, pp. 191–199.
27. Y. Zorian, E.J. Marinissen, and S. Dey, "Testing Embedded-Core Based System Chips," in *Proceedings IEEE International Test Conference (ITC)*, Washington, DC, Oct. 1998, pp. 130–143.

Erik Jan Marinissen is Principal Scientist at Philips Research Laboratories in Eindhoven, The Netherlands. He holds an M.Sc. degree in Computing Science (1990) and an MTD (Master of Technological Design) degree in Software Technology (1992), both from Eindhoven University of Technology. Marinissen's research interests include all topics in the domain of test and debug of digital VLSI circuits. He has published over 50 journal and conference papers, holds two US patents, and has several US and EP patents pending in the domain of core test and other digital test fields. He is the recipient of the Best Paper Award of the Chrysler-Delco-Ford Automotive Electronics Reliability Workshop 1995. Marinissen is a Senior Member of IEEE, and member of VIE, XOOTIC, and Philips CTAG. He serves as Editor-in-Chief of the IEEE P1500 Standard for Embedded Core Test and is a member of the organizing and program committees of DATE, DDECS, ETW, ITSW, LATW, SBCCI, and TECS. He has presented numerous tutorials on core-based testing within Philips and at international conferences. Marinissen serves as Member of the Editorial Board of JETTA.

Rohit Kapur is a Principal Engineer at Synopsys. His research interests are in VLSI test. Kapur has a B.Sc. degree in Engineering from Birla Institute of Technology, Mesra, India, and an M.Sc. and a Ph.D. in Computer Engineering from the University of Texas at Austin, Texas. He is chair of the IEEE P1450.6 standardization activity to create a core test language and the CTL Task Force in IEEE P1500.

Maurice Lousberg is Manager of Philips ED&T/Test, a tool and consultancy provider at Philips Research, specialized in IC and assembly test solutions. His current research interests are in core-based test and testability, especially for memories and mixed-signal blocks, but also in delay fault test and electrical diagnosis and he has authored various papers on these topics. Prior to working at Philips, Lousberg was head of R&D at Sagantec Europe BV. He holds a M.Sc. in Mathematics from Eindhoven University of Technology in Eindhoven, The Netherlands. He is a an active member of the IEEE P1500 CTL task force and of IEEE P1450.6; furthermore he has chaired the Philips core-based test activity CTAG for a number of years and participates in its mixed-signal extension CTAG.AMS.

Teresa L. McLaurin is a Distinguished Member of the Technical Staff at ARM Inc. in Austin, Texas. She is currently leading the Design-for-Test effort at ARM. She is a member of the IEEE as well as a member of the IEEE P1500 Scaleable Architecture Task Force, where she leads the Tiger Team defining the P1500 instruction set. She has authored several papers concerning Design-for-Test on cores and in SoCs. McLaurin received her B.Sc. from the University of Houston in Houston, Texas.

Mike Ricchetti is the Chief Technology Officer at Intellitech Corporation in Durham, New Hampshire. Prior to joining Intellitech, Ricchetti was a Senior Staff Methodology Consultant with Synopsys, Inc. While at Synopsys, he focused on Design-for-Test and System-on-Chip Test methodologies. In addition to DfT consulting, his work included definition and architecture of SOC test solutions and design automation tools. He has also been with several other high-tech companies, including Sunrise Test Systems, Hewlett-Packard, and Apollo Computers. He has nearly 20 years of experience in Design-for-Testability and has authored many papers in the area of Design-for-Test. Ricchetti earned a B.Sc. Degree in Computer Engineering at the Ohio State University in 1982. Ricchetti is a Member of the IEEE and is a working group member of both the IEEE 1149.1 Boundary Scan Standard and the proposed IEEE P1500 Standard for Embedded Core Test. He is also an active member of the P1500 Scaleable Architecture Task Force, and leads the Tiger Team defining the Wrapper Instruction Register (WIR) and Wrapper Interface Port (WIP). He is also a Working Group member of the newly formed AC Extest standardization effort.

Yervant Zorian received the M.Sc. degree in Computer Engineering from the University of Southern California in Los Angeles, California, and a Ph.D. in Electrical Engineering from McGill University in Montreal, Canada. Previously, he was a Distinguished Member of Technical Staff at Bell Labs, Lucent Technologies. His responsibilities include developing self-test strategies for embedded cores, chips and systems. He is currently the Vice President and Chief Scientist of Virage Logic, Inc. and the Chief Technology Advisor of LogicVision, Inc. He serves as the Editor-in-Chief of IEEE Design & Test of Computers and participates in editorial advisory boards of IEEE Spectrum, and JETTA. He chaired a number of symposia and workshops and the Test Technology Technical Council of IEEE Computer Society. He founded and chairs the IEEE Workshops on Testing Embedded Core-based Systems-on-Chips (TECS), and Test Resource Partitioning (TRP), and also the IEEE P1500 SECT Standard Working Group. Zorian has authored over 150 papers and three books, received several best paper awards, and holds over ten U.S. patents. He is a Golden Core Member of IEEE Computer Society, Honorary Doctor of National Academy of Sciences of Armenia, and a Fellow of IEEE.

JOURNAL OF ELECTRONIC TESTING: Theory and Applications 18, 385–400, 2002

An Integrated Framework for the Design and Optimization of SOC Test Solutions*

ERIK LARSSON AND ZEBO PENG
Embedded Systems Laboratory, Department of Computer and Information Science, Linköpings Universitet, Sweden

Received September 6, 2001; Revised January 16, 2002

Editor: Krishnendu Chakrabarty

Abstract. We propose an integrated framework for the design of SOC test solutions, which includes a set of algorithms for early design space exploration as well as extensive optimization for the final solution. The framework deals with test scheduling, test access mechanism design, test sets selection, and test resource placement. Our approach minimizes the test application time and the cost of the test access mechanism while considering constraints on tests and power consumption. The main feature of our approach is that it provides an integrated design environment to treat several different tasks at the same time, which were traditionally dealt with as separate problems. We have made an implementation of the proposed heuristic used for the early design space exploration and an implementation based on Simulated Annealing for the extensive optimization. Experiments on several benchmarks and industrial designs show the usefulness and efficiency of our approach.

Keywords: SOC test, test scheduling, test access mechanism design, test resource partitioning, test resource placement, test conflicts, power consumption

1. Introduction

The testing of System-on-Chip (SOC) is a crucial and time consuming problem due to the increasing design complexity. Therefore it is important to provide the test designer with support to develop an efficient test solution.

The work-flow for a test designer developing a test solution consists typically of two consecutive parts: an early design space exploration and an extensive optimization for the final solution. During the process, conflicts and limitations must be carefully considered. For instance, tests may be in conflict with each other due to the sharing of test resources; and power consumption must be controlled, otherwise the system may be damaged during test. Furthermore, test resources such as external testers support a limited number of scan-chains and have a limited test memory, which also introduce constraints.

Research has been going on in developing techniques for test scheduling, test access mechanism (TAM) design and testability analysis. For example, a framework to support the design of test solutions for SOC with Built-In Self-Test (BIST) has been proposed by Benso et al. [2]. In this paper, we combine and generalize several approaches in order to create an integrated framework for the development of SOC test solutions where:

- tests are scheduled to minimize the total test time,
- a minimal TAM is designed,
- test resources are floor-planned, and
- test sets for each core with test resources are selected.

The above set of tasks is performed in a single algorithm, which considers test conflicts, power limitation

*This work has partially been supported by the Swedish Agency for Innovation System (VINNOVA) and Ericsson.

and test resource constraints [17–19]. The algorithm is suitable for early design space exploration due to its low computational complexity, which is an advantage since it will be used iteratively many times.

Chakrabarty showed that test scheduling is equal to the open-shop scheduling [4], which is known to be NP-complete and the use of heuristics are therefore justified. Several heuristics have been proposed [1, 4, 7, 9, 11, 15, 23, 27]; however, they have been evaluated using rather small benchmarks. For such benchmarks, a technique based on Mixed-Integer Linear-Programming (MILP) can be used [7]. A disadvantage is the complexity of solving the MILP model since the size of it quickly grows with the number of tests making it infeasible for large industrial designs. We have therefore made an implementation based on Simulated Annealing for the extensive optimization of the test schedule and the TAM design for the final solution [16]. We have performed experiments on several benchmarks and on an Ericsson design consisting of 170 tests to show the efficiency and usefulness of our approach.

The rest of the paper is organised as follows. Related work is outlined in Section 2 and the system modelling is introduced in Section 3. Factors affecting the test solution are presented in Section 4. The integrated algorithm is described in Section 5 and how we used Simulated Annealing is outlined in Section 6. The paper is concluded with experimental results in Section 7 and conclusions are in Section 8.

2. Related Work

The basic problem in test scheduling is to assign a start time for all tests. In order to minimize the test application time, tests are scheduled as concurrent as possible, however, various types of constraints must be considered.

A test to be scheduled consists of a set of test vectors produced or stored at a test source (placed on-chip or off-chip). The test response from the test is evaluated at a test sink (placed on-chip or off-chip). When applying a test, a test conflict may occur, which must be considered during the scheduling process. For instance, often a testable unit is tested by several test sets (usually an external test set and a on-chip test set are required to reach high test quality). If several tests are used for a testable unit, only one test can be applied to the testable unit at a time.

Constraints on power consumption must be considered otherwise the system can be damaged. During testing mode, the power consumption is usually higher compared to normal operation [12]. For instance, consider a memory, which often is organized in memory banks. During normal operation, only a single bank is activated. However, during testing mode, in order to test the system in the shortest possible time it is desirable to concurrently activate as many banks as possible [9].

The use of different test resources may entail constraints on test scheduling. For instance, external testers have limitations of bandwidth due to that a scan chain operates usually at a maximum frequency of 50 MHz [13]. External testers can usually only support a maximum of 8 scan chains [13], resulting in long test application time for large designs. Furthermore, an external tester's memory is limited by its size [13].

Zorian proposed a test scheduling technique for fully BISTed systems where test time is minimized while power constraint is considered [27]. The tests are scheduled in sessions where tests at cores placed physically close to each other are grouped in the same test session. In a fully BISTed system, each core has its dedicated test source and test sink; and there might not be any conflicts among tests, i.e. the tests can be scheduled concurrently. However, in the general case, conflicts among tests may occur. Garg et al. proposed a test scheduling technique where test time is minimized for systems with test conflicts [11] and for core-based systems a test scheduling technique is proposed by Chakrabarty [4, 7]. Chou et al. proposed an analytic test scheduling technique where test conflicts and power constraints are considered [9]. A resource graph is used to model the system where an arc between a test and a resource indicate that the resource is required for the test, Fig. 1. From the resource graph, a test compatibility graph (TCG) is generated (Fig. 2) where each test is a node and an arc between two nodes indicates that the tests can be scheduled concurrently. For instance t_1 and t_4 can be scheduled at the same time. Each test is attached with its test time and its power consumption and the maximal allowed power consumption is 10. The tests t_1 and t_5 are compatible, however, due to the power limit they cannot be scheduled at the same time.

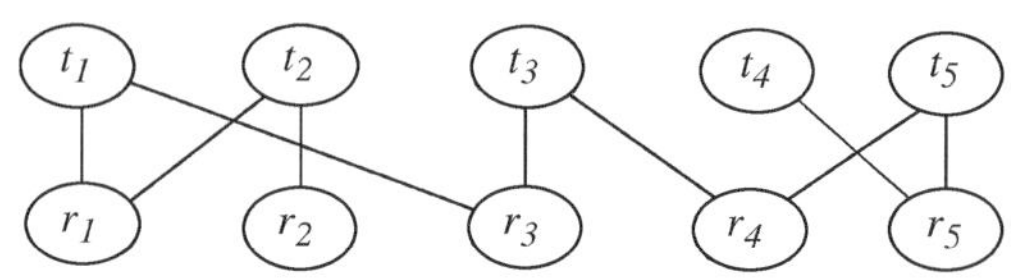

Fig. 1. Resource graph of an example system.

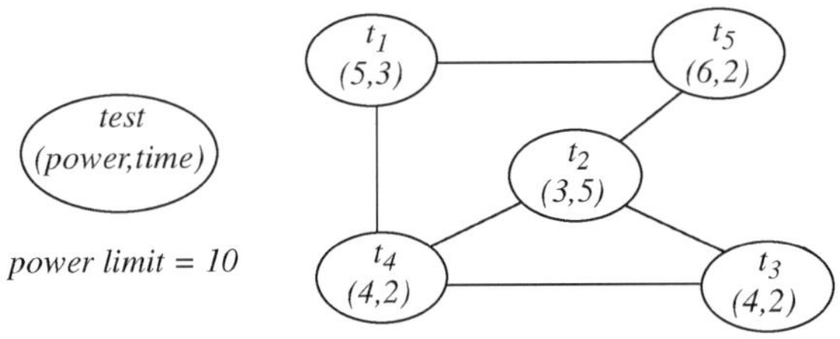

Fig. 2. Test compatibility graph (TCG) of the example system (Fig. 1).

Another test scheduling approach is proposed by Muresan et al. where constraints among tests and on power consumption are considered [23]. Favour is given to reduce the test time by allowing new tests to start even if all tests in a session are not completed. Iyengar and Chakrabarty proposed a pre-emptive test scheduling technique where the test for a testable unit may be interrupted and resumed later, i.e. the test set is partitioned into several test sets [15]. Using a scheme by Craig et al. the above discussed scheduling techniques can be grouped in [10]:

- nonpartitioned testing,
- partitioned testing with run to completion, and
- partitioned testing.

The differences among the grouping are illustrated with five tests ($t_1, \ldots, t_5$) in Fig. 3. The scheduling strategies proposed by Zorian [27] and Chou et al. [9] are nonpartitioned (Fig. 3(a)), the strategy proposed by Muresan et al. is partitioned testing with run to completion (Fig. 3(b)) and the approach proposed by Iyengar and Chakrabarty [15] is partitioned testing (Fig. 3(c)).

A test infrastructure is responsible for the transportation of test vectors from test sources to cores under test and test responses from cores under test to test sinks. It consists of two parts; one for the test data transportation and one for the control of the transportation. In the approach for fully BISTed systems proposed by Zorian [27], tests at cores placed physically close to each other are grouped in the same test session. The main advantage is that the same control structure can be used for all tests in the session, which minimizes the routing of control wires. In general, systems are not tested with a BIST structure only for each testable unit and therefore a TAM is required [5, 6, 8, 24]. Chakrabarty proposed an integer linear programming (ILP) for the allocation of TAM width [5, 6, 8] and the effect on test time for systems using various design styles for test access with the TestShell wrapper is analysed by Aertes and Marinissen [1].

To ease test access the cores can be placed in wrappers such as Boundary scan [3], TestShell [20] or P1500 [14, 22]. The Boundary scan technique, developed for PCB designs, suffers from long testing time due to the shifting process, and it becomes even worse for SOC designs since the amount of test data to be transported increases. An alternative is to use an architectures as proposed by Aertes and Marinissen (Fig. 4(a–c)) where the test time only depends on the number of flip-flops within the cores and the number of test vectors, i.e. the transportation to and from a core is free [1]. The daisy-chained architecture uses a clocked bypass, however, Marinissen et al. have also proposed a wrapper design library allowing bypass structures without clock delay [21].

The above scheduling techniques all assume that all testable units have their selected test sets. Sugihara et al. proposed a technique for selecting test sets where each

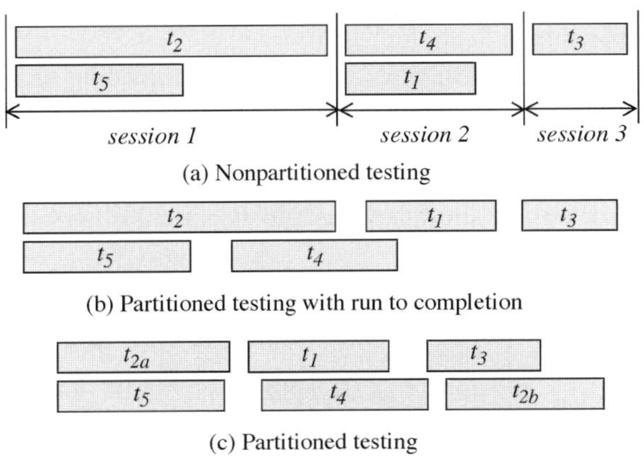

Fig. 3. Scheduling approaches.

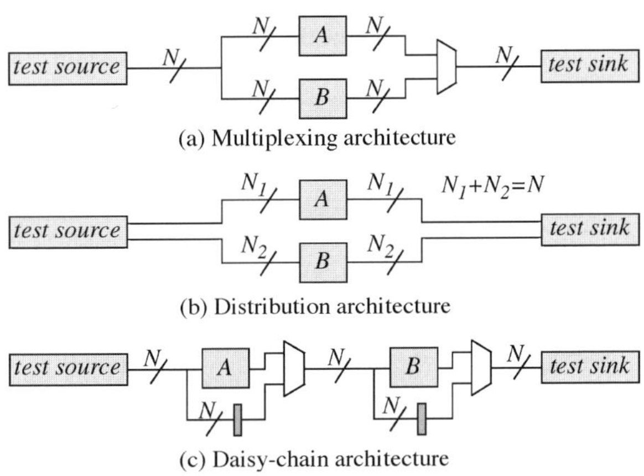

Fig. 4. Multiplexing, distribution and daisy-chain architecture [1].

core may be tested by one test set from an external tester and one test set from a dedicated test generator for the core [25].

3. System Modelling

This section describes the formal notation we use to model the SOC under test. An example of a system under test is given in Fig. 5 where each core is placed in a wrapper to ease test access. A DFT technique is added to each core and in this example all cores are tested using the scan technique. The test access port (*tap*) is the connection to an external tester and for instance the test source, *test source* 1, and test sink, *test sink* 1, are implemented on-chip [17, 18]. Applying several sets of tests, where each test set is produced or stored at a test source and the test response is analysed at a test sink, tests the system.

The system in Fig. 5 can be modelled as a *design with test*, $DT = (C, R_{source}, R_{sink}, p_{max}, T, source, sink, core, constraint, mem, bw)$, where:

$C = \{c_1, c_2, \ldots, c_n\}$ is a finite set of cores where each core $c_i \in C$ is characterized by $p_{idle}(c_i)$: idle power;
$R_{source} = \{r_1, r_2, \ldots, r_p\}$ is a finite set of test sources;
$R_{sink} = \{r_1, r_2, \ldots, r_q\}$ is a finite set of test sinks;
p_{max}: maximal allowed power at any time;
$T = \{t_1, t_2, \ldots, t_o\}$ is a finite set of tests, each consisting of a set of test vectors. Several tests form a *core test* (*CT*) and each core, c_i, can be associated with several core tests, $CT_{ij} (j = 1, 2, \ldots, l)$.
Each test t_i is characterized by:

$\tau_{test}(t_i)$: test time for test t_i,
$p_{test}(t_i)$: test power for test t_i,
$bw(t_i)$: bandwidth required for t_i
$mem(t_i)$: memory required for test pattern storage.

source: $T \to R_{source}$ defines the test sources for the tests;

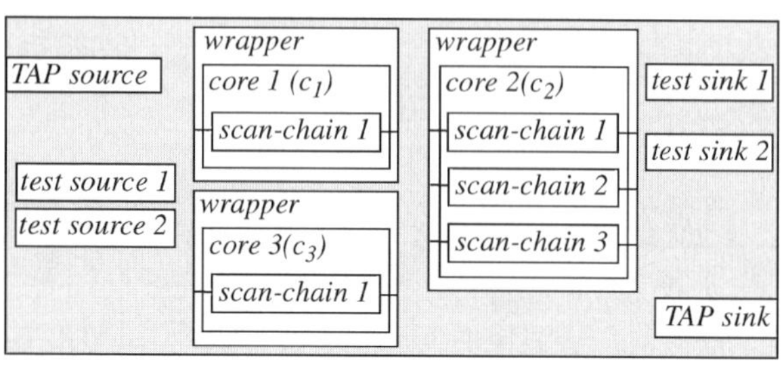

Fig. 5. An example system.

sink: $T \to R_{sink}$ defines the test sinks for the tests;
core: $T \to C$: the core where a test is applied;
constraint: $T \to 2^C$: the set of cores required for a test;
$mem(r_i)$: memory available at test source $r_i \in R_{source}$;
$bw(r_i)$: bandwidth at test source $r_i \in R_{source}$.

For each test, one test sink and one test source are required. In our model, it is possible for every test to combine any type of test source (on-chip or off-chip) with any type of test sink (on-chip or off-chip), which is important to allow high flexibility. To further give the designer a high flexibility to explore different combinations of tests, it is possible to define several set of tests (CT_{ij}) for each core where each such set tests the core.

Given a system as in Fig. 5 where for each of the cores c_1, c_2, c_3, one CT exists. For core c_1, $CT_{11} = \{t_1, t_2\}$, for core c_2, $CT_{21} = \{t_4, t_5\}$ and for c_3, $CT_{31} = \{t_3\}$. The on-chip test source 1 is required by t_1 and t_3 and the on-chip test sink 1 is required by t_3 and t_5. The resource graph in Fig. 1 and the test compatibility graph in Fig. 2 model the system when r_3 (test source 1) is needed by t_1 and t_3 and r_4 (test sink 1) needed by t_3 and t_5.

4. The SOC Test Issues

In this section we discuss the different issues considered by our SOC test framework. We also demonstrate the importance of considering them in an integrated manner.

4.1. Test Scheduling

The test scheduling problem can be seen as placing all tests in a diagram as in Fig. 6 while satisfying all constrains. For instance, t_1 and t_2 cannot be applied at the same time since both tests the same core, c_1 (example in Section 3).

The basic difference between our scheduling technique [17, 18] and the approaches proposed by Zorian [27] and Chou et al. [9] is, besides that we design the TAM while scheduling the tests, that Zorian and Chou et al. do not allow new tests to start until all tests in

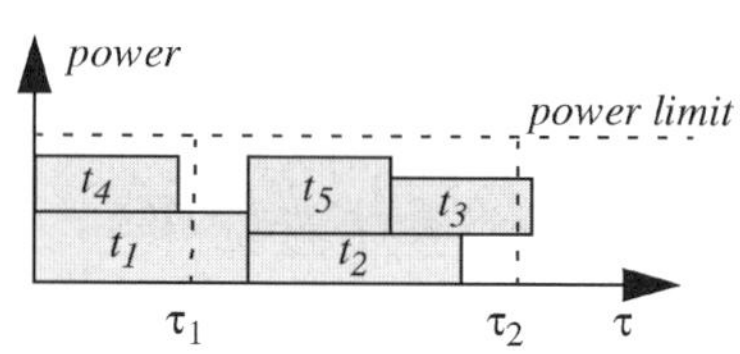

Fig. 6. Example of test scheduling.

a session are completed. It means that if t_3 is about to be scheduled it cannot be scheduled as in Fig. 6. In the approach proposed by Muresan et al. [23], t_3 can be scheduled if it is completed no later than t_2.

In our approach it is optional if tests may start before all tests in a session are completed (nonpartitioned testing) or not (partitioned testing with run to completion). The advantage of using the latter is that it gives more flexibility.

Let a schedule S be an ordered set of tests such that:

$$\{S(t_i) < S(t_j) \mid \tau_{start}(t_i) \leq \tau_{start}(t_j),\ i \neq j,\ \forall t_i \in S,\ \forall t_j \in S\},$$

where $S(t_i)$ defines the position of test t_i in S; $t_{start}(t_i)$ denotes the time when test t_i is scheduled to start, and $\tau_{end}(t_i)$ its completion time: $\tau_{end}(t_i) = \tau_{start}(t_i) + \tau_{test}(t_i)$.

For each test, t_i, the start time and the wires for test data transportation have to be determined before it is inserted into the schedule, S.

Let the Boolean function *scheduled*(t_i, τ_1, τ_2) be true if test t_i is scheduled in such a way that the test time overlaps with the time interval $[\tau_1, \tau_2]$, i.e.,

$$\{t_i \in S \wedge \neg(\tau_{end}(t_i) < \tau_1 \vee \tau_{start}(t_i) > \tau_2)\}.$$

An example to illustrate the function *scheduled* for a set of scheduled tests is shown in Fig. 7.

The Boolean function *scheduled*(r_i, τ_1, τ_2) is true if a source r_i is used by a test t_j between τ_1 and τ_2, i.e.:

$$\{\exists t_j \in S \mid r_i = source(t_j) \wedge scheduled(t_j, \tau_1, \tau_2)\}.$$

A similar definition is used if a sink r_i is scheduled (used by any test) between τ_1 and τ_2. The Boolean function *scheduled* (*constraint*(t_i), τ_1, τ_2) is true if:

$$\{\exists t_j \in S \mid core(t_j) \in constraint(t_i) \wedge scheduled(t_j, \tau_1, \tau_2)\}.$$

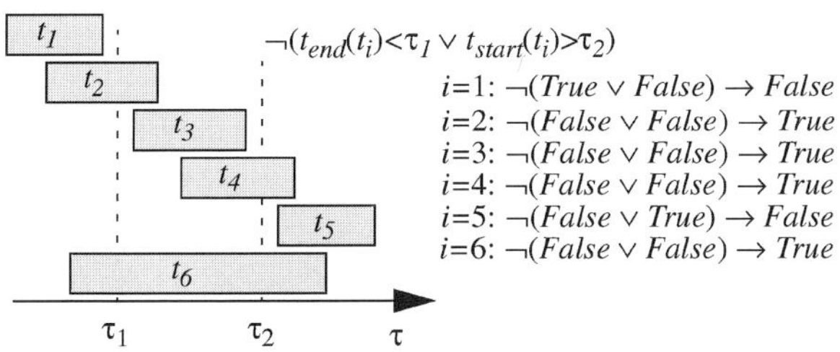

Fig. 7. The function *scheduled*.

4.2. *Power Consumption*

Generally speaking, there are more switching activities during the testing mode of a system than when it is operated under the normal mode. The power consumption of a CMOS circuit is given by a static part and a dynamic part where the latter dominates and can be characterized by:

$$p = C \times V^2 \times f \times \alpha$$

where the capacitance C, the voltage V, and the clock frequency f are fixed for a given design [26]. The switch activity α, on the other hand, depends on the input to the system, which during testing are test vectors and therefore the power dissipation varies depending on the test vectors.

An example illustrating the test power dissipation variation over time τ for two test t_i and t_j is in Fig. 8. Let $p_i(\tau)$ and $p_j(\tau)$ be the instantaneous power dissipation of two compatible tests t_i and t_j, respectively, and $P(t_i)$ and $P(t_j)$ be the corresponding maximal power dissipation.

If $p_i(\tau) + p_j(\tau) < p_{max}$, the two tests can be scheduled at the same time. However, instantaneous power of each test vector is hard to obtain. To simplify the analysis, a fixed value $p_{test}(t_i)$ is usually assigned for all test vectors in a test t_i such that when the test is performed the power dissipation is no more then $p_{test}(t_i)$ at any moment.

The $p_{test}(t_i)$ can be assigned as the average power dissipation over all test vectors in t_i or as the maximum power dissipation over all test vectors in t_i. The former approach could be too optimistic, leading to an undesirable test schedule, which exceeds the test power

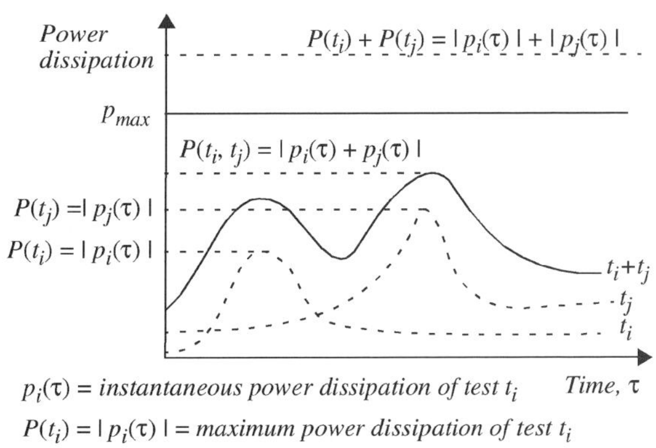

Fig. 8. Power dissipation as a function of time [9].

constraints. The latter could be too pessimistic; however, it guarantees that the power dissipation will satisfy the constraints. Usually, in a test environment the difference between the average and the maximal power dissipation for each test is often small since the objective is to maximize the circuit activity so that it can be tested in the shortest possible time [9]. Therefore, the definition of power dissipation $p_{test}(t_i)$ for a test t_i is usually assigned to the maximal test power dissipation $(P(t_i))$ when test t_i alone is applied to the device. This simplification was introduced by Chou et al. [9] and has been used by Zorian [27] and by Muresan et al. [23].We will use this assumption also in our approach.

Let $p_{sch}(\tau_1, \tau_2)$ denote the peak power between τ_1 to τ_2:

$$max\left\{ \sum_{\forall t_i scheduled(t_i,\tau)} p_{test}(t_i) - p_{idle}(core(t_i)) + \sum_{\forall c_i \in C} p_{idle}(c_i), \tau \in [\tau_1, \tau_2] \right\},$$

where $scheduled(t_i, \tau) = scheduled(t_i, \tau, \tau)$.

As an example, applying the function $p_{sch}(\tau_1, \tau_2)$ on the schedule for a system with 5 tests as in Fig. 6, with τ_1 and τ_2 as indicated in the figure, returns $p_{test}(t_2) + p_{test}(t_5) + p_{idle}(c_3)$, the peak power consumption between τ_1 and τ_2.

In our approach, the maximal power consumption should not exceed the power constraint, p_{max}, for a schedule to be accepted. That is, $p_{sch}(0, \infty) \leq p_{max}$.

4.3. Test Source Limitations

A test source usually has a limited bandwidth. For instance, external tester may only support a limited number of scan chains at a time [13]. There could also be a limitation in the number of available pins. For each test source, this information is given in the attribute bandwidth.

The function $bw_{alloc}(r_i, \tau_1, \tau_2)$ gives the maximal number of wires allocated between τ_1 and τ_2 for a source r_i, i.e.:

$$max\left\{ \sum_{\forall t_i scheduled(t_j,\tau)\wedge r_i = source(t_j)} |t_{wires}(t_j)|, \tau \in [\tau_1, \tau_2] \right\}.$$

For instance, in Fig. 9 a test source r_1 feeds test t_5 using wire w_2 and w_3. In this example, $bw(r_1) = 6$ and

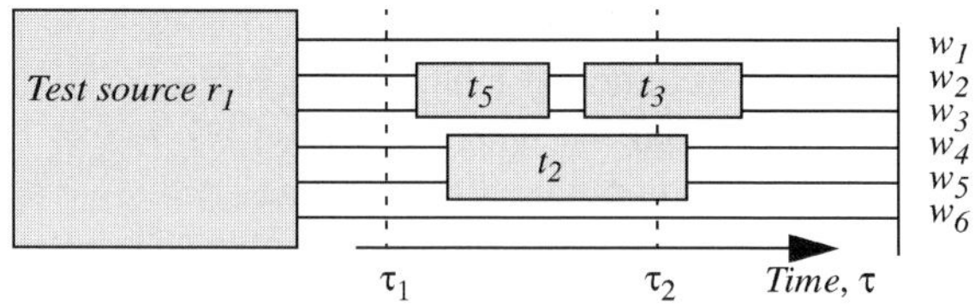

Fig. 9. The TAM allocation.

$bw_{alloc}(r_i, \tau_1, \tau_2) = 4$ since wire w_2 and w_3 are used by t_3 and t_5 and w_5 and w_6 are used by t_2.

A test generator (test source) may use a memory for storing the test patterns. In particular, external test generators use such a memory with a limited size, which may lead to additional constraints on test scheduling [13].

The function $mem_{alloc}(r_i, \tau_1, \tau_2)$ gives the allocated memory between time τ_1 and τ_2 for a given source r_i, i.e.:

$$max\left\{ \sum_{\forall t_j scheduled(t_j,\tau)\wedge r_i = source(t_j)} mem(t_j), \tau \in [\tau_1, \tau_2] \right\}.$$

4.4. Test Set Selection

A test set is attached with its test power consumption, test application time, bandwidth requirement and memory requirement (Section 3). The required test source and the test sink are also defined for each test set. In the approach proposed by Sugihara et al. each testable unit can be tested by two test sets using a dedicated BIST resource and an external tester [25]. In our approach, we assume that an arbitrary number of test sets can be used for each testable unit where each test set can be specified using any type of test resource. For instance, we allow the same test resource to be used by several test sets at different testable units. The major advantage is that the designer can define and explore a wider range of test sets. Due to that the test resources are defined for each test set it is possible to make a comparison of different test sets not only in terms of the number of test vectors but also in respect to test resources.

For each testable unit, the designer defines a core test set where each set is a set of test vectors. For instance, if the following $CT_{11} = \{t_1, t_2\}, CT_{12} = \{t_6\}, CT_{13} = \{t_7, t_8, t_9\}$ is given for core c_1. One of CT_{11}, CT_{12} or CT_{13} must be selected and all tests within the selected CT must be scheduled and no other tests from any other CT for core c_1 should be scheduled.

4.5. *Test Access Mechanism*

A test infrastructure transports and controls the flow of test data in the system under test. The Boundary scan technique can be used for these purposes, however, it suffers from long testing times and due to the amount of test data to be transported in the system, we assume an infrastructure consisting of test wires (lines). It means we add needed number of test wires from test sources to cores and from cores to test sinks. The time to transport a test vector from a test source to a core and the time to transport test response from a core to a test sink is neglected, which means that the test time is determined by the test vectors and design characteristics of each core.

When adding a TAM between a test source and a core or between a core and a test sink, and the test data has to pass another core, c_i, several routing options are possible:

1. through the core c_i using its transparent mode;
2. through an optional bypass structure of core c_i; and
3. around core c_i and not connecting it to the TAM.

The model in Fig. 10(a) of the example system in Fig. 5 illustrates the advantage of alternatives 1 and 2 (Fig. 10(b)) compared to alternative 3 (Fig. 10(c)) since the TAM can be reused in the former two cases. However, a delay may be introduced when the core is in transparent mode or its by-pass structure is used as in the TestShell [22]. On the other hand, Marinissen et al. recently proposed a library of wrapper cells allowing a flexible design where it is possible to design non-clocked bypass structures of TAM width [21]. In the following, we assume that bypass may be solved using such non-delay mechanism.

When designing the TAM, two problems must be solved:

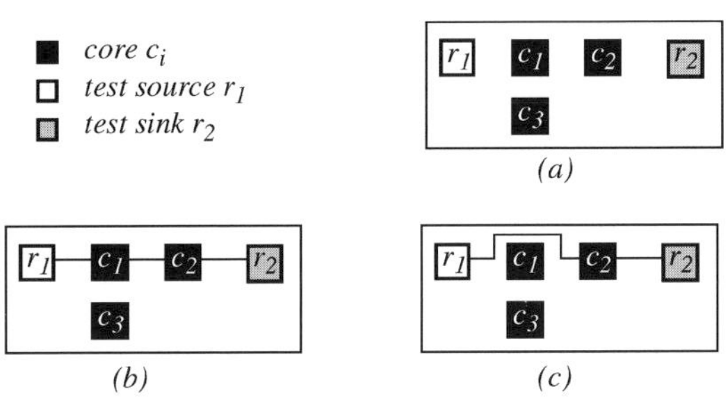

Fig. 10. Illustrating TAM design alternatives using the example in Fig. 5.

Table 1. Test data for the cores in System S.

Core	Index i	External test cycles, e_i	BIST cycles, b_i	Placement x	Placement y
c880	1	377	4096	10	10
c2670	2	15958	64000	20	10
c7552	3	8448	64000	10	30
s953	4	28959	217140	20	30
s5378	5	60698	389214	30	30
s1196	6	778	135200	30	10

- the design and routing of the TAMs, and
- the scheduling of tests using the TAMs.

In order to minimize the routing, few and short wires are desired. However, such approach increases the test time of the system. For instance, consider System S [4] (Table 1) where we added the floor planning (x, y co-ordinates for each core). A minimal TAM would be a single wire starting at the TAP, connecting all cores and ending at the TAP. However, such TAM leads to high test application time since no tests can be applied concurrently, all tests have to be applied in a sequence.

The system (for instance the example system in Fig. 5 or System S) can be modelled as a directed graph, $G = (V, A)$, where V consists of the set of cores (the testable units), C, the set of test sources, R_{source}, and the set of test sinks, R_{sink}, i.e. $V = C \cup R_{source} \cup R_{sink}$ [17, 18]. An arc $a_i \in A$ between two vertices v_i and v_j indicates a TAM (a wire) where it is possible to transport test data from v_i to v_j. Initially no TAM exists in the system, i.e. $A = \emptyset$. However, if the functional infrastructure may be used, it can be included in A initially. A test wire w_i is a path of edges $\{(v_0, v_1), \ldots, (v_{n-1}, v_n)\}$ where $v_0 \in R_{source}$ and $v_n \in R_{sink}$. Let Δy_{ij} be defined as $|y(v_i) - y(v_j)|$ and Δx_{ij} as $|x(v_i) - x(v_j)|$, where $x(v_i)$ and $y(v_i)$ are the *x-placement* respectively the *y-placement* for a vertex v_i and the distance between vertex v_i and v_j is:

$$dist(v_i, v_j) = \sqrt{(\Delta y_{ij})^2 + (\Delta x_{ij})^2}.$$

The Boolean function $scheduled(w_i, \tau_1, \tau_2)$ is true when a wire w_i is used between τ_1 to τ_2:

$$\{\exists t_j \in S \mid w_i \in tam(t_j) \wedge scheduled(t_j, \tau_1, \tau_2)\},$$

where $tam(t_j)$ is the set of wires allocated for test t_j.

The information of the nearest core in four directions, *north, east, south* and *west*, is stored for each

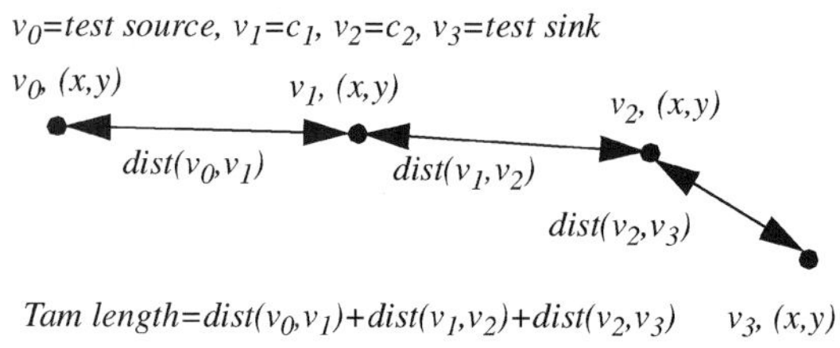

Fig. 11. Computing the TAM length.

vertex. The function $south(v_i)$ gives the closest vertex south of v:

$$south(v_i) = \left\{ v_j \middle| \left(\frac{\Delta y_{ij}}{\Delta x_{ij}} > 1 \vee \frac{\Delta y_{ij}}{\Delta x_{ij}} < -1 \right), \right.$$
$$\left. y(v_j) < y(v_i), i \neq j, \ min \{dist(v_i, v_j)\} \right\}.$$

The functions $north(v_i)$, $east(v_i)$ and $west(v_i)$ are defined in similar ways. The function $insert(v_i, v_j)$ inserts a directed arc from vertex v_i to vertex v_j *if and only if* the following is true:

$$\{south(v_i, v_j) \vee north(v_i, v_j) \vee west(v_i, v_j) \vee east(v_i, v_j)\}.$$

where $south(v_i, v_j)$ is true if $v_i = south(v_i)$.

The function $closest\,(v_i, v_j)$ gives a vertex, v_k, in the neighbourhood of v_i with the shortest distance to v_j. The function $add\,(v_i, v_j)$ adds arcs from v_i to v_j in the following way: (1) find $v_k = closest\,(v_i, v_j)$; (2) add an arc from v_i to v_k; (3) if $v_k = v_j$, terminate otherwise let $v_i = v_k$ and go to (1).

The total length of a path is the sum of the length of all individual edges. An example to illustrate the calculation of the length of a test wire on the example system (Fig. 5) defined as a path is in Fig. 11 (in this path core c_3 is not included).

4.6. *Test Floor-planning*

In the approach proposed by Sugihara et al. each testable unit is tested by a test sets using a dedicated BIST resource and an external tester [25]. However, we allow the sharing of test resources and if a BIST resource is shared among several cores, the floor-planning is not trivial.

For instance, if two cores in the example design (Fig. 5) both use a single on-chip test source and test sink, it is most feasible to place the test source at one core while the test sink is placed at the other core.

Initially, the test resources may not be placed. Our algorithm, described in the next section, determines their placement in this case.

5. The Heuristic Algorithm

In this section the issues discussed above are combined into an algorithm. The algorithm assumes that the tests are initially sorted according to a key k, which characterizes $power(p)$, $testtime(t)$ or $power \times test\ time(p \times t)$.

Let T be the ordered set of the tests, which are ordered based on the key k. If the scheduling approach is partitioned testing with run to completion the function $nexttime(\tau_{old})$ gives the next time where it is possible to schedule a test:

$$\{\tau_{end}(t_i) \mid min\,(\tau_{end}(t_i)), \tau_{old} < \tau_{end}(t_i), \forall t_i \in S\},$$

otherwise if nonpartitioned testing is used the function $nexttime(\tau_{old})$ is defined as:

$$\{\tau_{end}(t_i) \mid max(\tau_{end}(t_i)), \tau_{old} < \tau_{end}(t_i), \forall t_i \in S\}.$$

The algorithm depicted in Fig. 12 can basically be divided into four parts for:

- constraint checking,
- test resource placement,
- test access mechanism design and routing, and
- test scheduling.

A main loop is terminated when there exists a core test (CT) for each core such that all tests within the selected CT are scheduled. In each iteration at the loop over the tests in T a test *cur* is checked. A check is also made to determine if all constraints are fulfilled, i.e. it is possible to schedule test *cur* with a start at τ and an end time at $\tau_{end} = \tau + \tau_{test}$:

- $\neg\exists t_f (t_f \in CT_{ij} \wedge t_f \in S \wedge cur \notin CT_{ij})$ checks that another core test set for current core is not used in the schedule,
- $p_{sch}(\tau, \tau_{end}) + p_{test}(cur) < p_{max}$ checks that the power constraint is not violated,
- $\neg scheduled(v_a, \tau, \tau_{end})$ checks that the test source is not scheduled during τ to τ_{end},
- $\neg scheduled(v_c, \tau, \tau_{end})$ checks that the test sink is not scheduled during τ to τ_{end},
- $\neg scheduled(constraint(cur), \tau, \tau_{end})$ checks that all cores required for *cur* are available during τ to τ_{end},

```
Sort T according to the key (p, t or p×t);
S=∅; τ=0;
until ∀b_pq∃CT_pq∀t_s∈ S do
  for all cur in T do
    v_a=source(cur);
    v_b=core(cur);
    v_c=sink(cur);
    τ_end=τ+t_test(cur);
    if all constraints are satisfied then begin
      ¬scheduled(v_a, 0, t_end) floor-plan v_a at v_b;
      ¬scheduled(v_c, 0, t_end) floor-plan v_c at v_b;
     for all required test resources begin
       new=length of a new wire w_j connecting v_a, v_b and v_c;
       u=number of wires connecting v_a, v_b and v_c not
          scheduled from τ to τ_end;
       v=number of wires connecting v_a, v_b and v_c;
       for all min(v-u, bw(cur))wires w_j
          extend=extend+length of an available wire(w_j);
       if (bw(cur)>u)
          extend=extend+new×(par-u);
          move=par(v_a) × min{dist(v_a, v_b),dist(v_b, v_c)};
          if (move≤min{extend, new × bw(cur)})
            v_x, v_y=min{dist(v_a, v_b), dist(v_b, v_c)}, dist(v_a, v_b)>0,
                 dist(v_b,v_c)>0
            add par(v_a) wires between v_x and v_y;
            if (v_x=source(cur)) then floorplan v_a at v_b;
            if (v_y = sink(cur)) then floorplan v_c at v_b;
     end
     for r = 1 to bw(cur)
       if there exists a not scheduled wire during τ to τ_end
          connecting v_a, v_b and v_c it is selected
       else
          if (length of a new wire < length of extending a wire w_j)
            w_j=add(v_a, v_b)+add(v_b, v_c);
          else
            extend wire;
     schedule cur and remove cur from T;
   end;
 τ = nexttime(τ).
```

Fig. 12. The system test algorithm.

- the available bandwidth at test source v_a is checked to see if: $bw(v_a) > bw(cur) + bw_{alloc}(v_a, \tau, \tau_{end})$ and
- the available memory test source v_a is checked to see if: $mem(v_a) > mem(cur) + mem_{alloc}(v_a, \tau, t_{end})$.

Then the placement of the test resources is checked. If the test resources are on-chip resources and not placed in the system, they are placed at core c_i. If they are floor-planned, a check is performed to determine if they are to be moved.

When the placement of the test resources for the selected test is determined, the test access mechanism is designed and routed. The basic question is if existing wires can be used or new wires must be added. If no routed connection is available connecting all required cores, the distance for adding a completely new connection is re-calculated due to a possible moving of test resources.

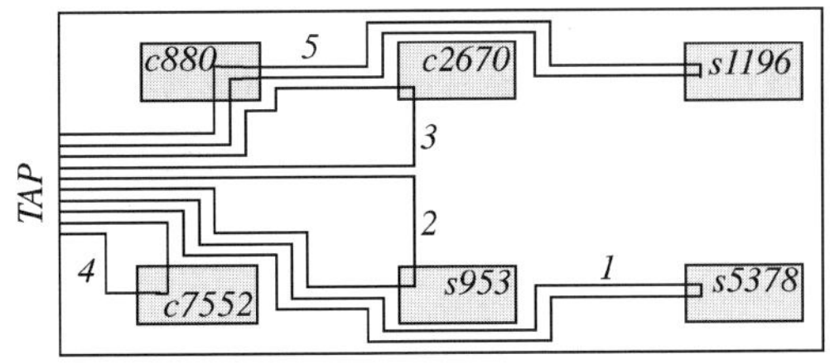

Fig. 13. TAM design using our heuristic on System S.

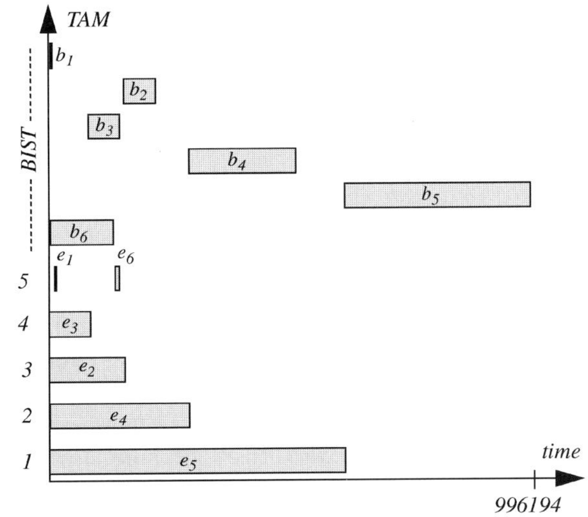

Fig. 14. TAM schedule on System S using our heuristic.

Examples of the produced results from the algorithm using System S [4] (Table 1) are the TAM design as in Fig. 13 and the test schedule as in Fig. 14. The TAM wires 1 to 5 in Fig. 13 correspond to the TAM 1 to 5 in Fig. 14. For instance, b_5 is the BIST test of core indexed 5 (s5378) and e_5 is the external test of s5378 (note that the BIST tests b_i do not require a TAM but they are placed in Fig. 14 to illustrate when they are scheduled).

The computational complexity for the above algorithm, where the test access mechanism design is excluded in order to make it comparable with other approaches, comes mainly from sorting the tests and the two loops. The sorting can be performed using a sorting algorithm at $O(n \times \log n)$. The worst case occurs when in each loop iteration for the loops only one test is scheduled giving a complexity:

$$\sum_{i=0}^{|T|-1} (|T| - i) = \frac{n^2}{2} + \frac{n}{2}$$

The total worst case execution time is $n \times \log n + n^2/2 + n/2$, which is of $O(n^2)$. For instance, the approach by Garg et al. [11] and by Chakrabarty [4] both has a worst case complexity of $O(n^3)$.

6. Simulated Annealing

In this section we outline the Simulated Annealing (SA) technique and describe its adoption to be used for scheduling and TAM design. The technique proposed by Kirkpatrick et al. [16] uses a hill-climbing mechanism to avoid getting stuck in at local optimum.

6.1. *The Simulated Annealing Algorithm*

The SA algorithm (Fig. 15) starts with an initial solution and a minor modification creates a neighbouring solution. The cost of the new solution is evaluated and if it is better than the previous, the new solution is kept. A worse solution can be accepted at a certain probability, which is controlled by a parameter referred to as temperature.

The temperature is decreased during the optimization process, and the probability of accepting a worse solution decreases with the reduction of the temperature value. The optimization terminates when the temperature value is approximately zero.

6.2. *Initial Solution and Parameter Selection*

We use our heuristic described in Section 5 with an initial sorting of the tests based on *power* (using *time* and *power* × *time* results after optimization in the same cost)

```
Construct initial solution, x^now;
Initial Temperature: T=TI;
while stop criteria not met do begin
    for i = 1 to TL do begin
        Generate randomly a neighboring solution
          x'∈ N(x^now);
        Compute change of cost function
          ΔC=C(x')-C(x^now);
        if ΔC≤0 then x^now=x'
            else begin
                Generate q = random(0, 1);
                if q<e^(-ΔC/T) then x^now=x'
            end;
    end;
    Set new temperature T=α×T;
end;
Return solution corresponding to the minimum
    cost function;
```

Fig. 15. Simulated Annealing algorithm.

within the integrated test framework (Section 5) to create the initial solution [17, 18]. An example of an initial solution produced for System S is in Fig. 13 and Fig. 14.

The parameters, initial temperature TI, the temperature length TL and the temperature reduction factor $\alpha (0 < \alpha < 1)$ are all determined based on experiments.

6.3. *Neighbouring Solution in Test Scheduling*

In the case when only test scheduling is considered, i.e. the TAM is not considered or it is fixed and seen as a resource, we create a neighbouring solution by randomly selecting a test from an existing schedule and schedule it as soon as possible but not at the same place as it was in the original schedule. For instance, creating a neighbouring solution given a test schedule as in Fig. 6 with resource graph as in Fig. 1 and test compatibility graph as in Fig. 2, we randomly select a test, let say t_1. We try to schedule t_1 as soon as possible but not with the same starting time as it had while fulfilling all constraints. Test t_1 was scheduled to start at time 0 and no new starting point exists where constraints are fulfilled until end of t_3 where t_1 now is scheduled. In this case, the test time increases after the modification (getting out of a possible local minimum), however, only temporarily since in the next iteration a test may be scheduled at time 0 (where t_1 used to be).

6.4. *Neighbouring Solution in Test Scheduling and TAM Design*

When both the test time and the TAM design are to be minimized, a neighbouring solution is created by randomly adding or deleting a wire and then the tests are scheduled on the modified TAM.

If the random choice is to add a wire, a test is randomly selected and a wire is added from its required test source to the core where the test is applied and from the core to the test sink for the test. For instance, if e_3 in System S (Table 1) is selected, a wire is added from the TAP to core c7552 and from core c7552 to the TAP.

If the random choice is to delete a wire, a similar approach is applied. However, a check is performed to make sure that all tests can be applied.

6.5. *Cost Function*

The cost function of a test schedule, S, and the TAM, A, is:

$$C(S, A) = \beta_1 \times T(S) + \beta_2 \times L(A)$$

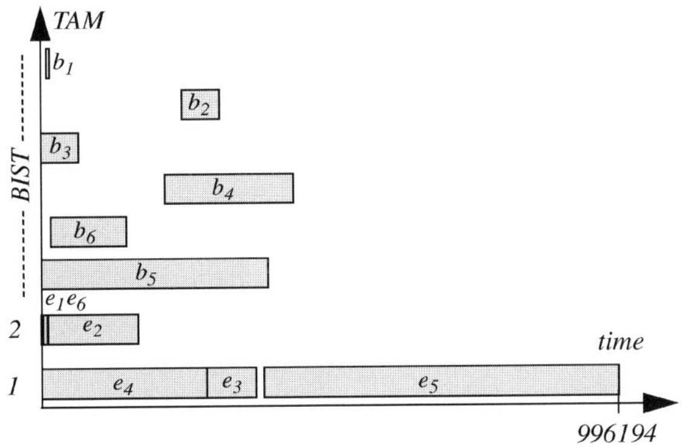

Fig. 16. Test schedule on System S using SA.

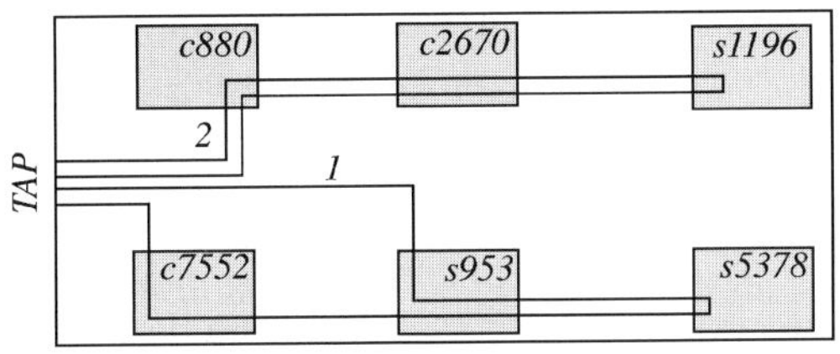

Fig. 17. TAM design using SA on System S.

where: $T(S)$ is the test application time for a sequence of tests, S, $L(A)$ is the total length of the TAM, β_1, β_2 are two designer-specified constants used to determine the importance of the test time and the TAM.

The test application time, $T(S)$, for a schedule, S, is:

$$T(S) = \{max\{t_{end}(t_i) \mid \forall t_i(max\{t_{end}(t_i)\}), t_i \in S\}\}$$

and the length, $L(A)$, of the TAM, A, is given by:

$$\sum_{w_j \in A} \sum_{j=0}^{|w_i|-1} dist(v_j, v_{j+1}), v_j, v_{j+1} \in w_i$$

For the test schedule, S, produced by SA for System S (Fig. 16) the test time, T(S) is 996194 (the end time of test e_5) and the length of the TAM (Fig. 17) is 160. Comparing this to the results produced by our heuristic [18] shows that test time is the same while the TAM is reduced from 320 (Fig. 13) to 160 (Fig. 17).

7. Experimental Results

7.1. *Benchmarks*

We have used the System S [4], which has test conflicts (Table 1) while all other benchmarks and designs have test conflicts and power constraints like the ASIC Z

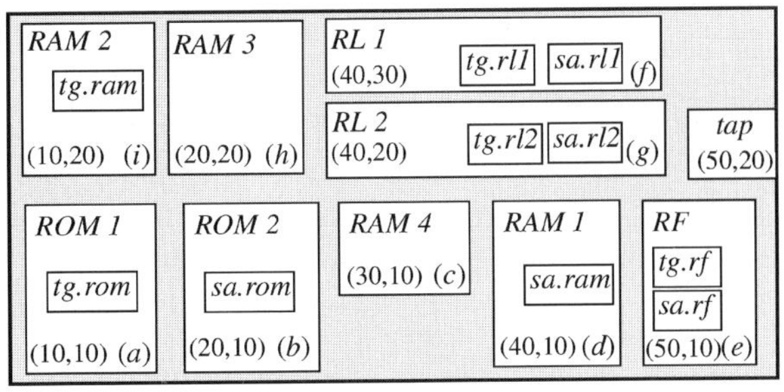

Fig. 18. ASIC Z floor-plan.

design presented by Zorian [27] with added data made by Chou et al. [9] (see the floor-plan in Fig. 18) and we added the placement (x, y) coordinates [18]. For instance, RAM2 is placed at (10, 20), which means that the centre of RAM2 has x-coordinate 10 and y-coordinate 20. The design is fully BISTed with a power dissipation limit at 900 mW.

We have also used an extended ASIC Z design where each core is tested by two test sets (one external test and one BIST) and an interconnection test to a neighbouring core [17]; in total 27 tests.

We have also used a design with 10 tests presented by Muresan et al. [23] and an industrial design with characteristics given in Table 2. The power limitation for the industrial design example is 1200 mW and only one test may use the test bus or the functional pins (fp) at a time. Furthermore block-level tests may not be scheduled concurrently with top-level tests.

Table 2. Characteristics for the industrial design.

Test	Block	Test	Test time	Test power	Test port
Block-level tests	A	Test A	515	379	Scan
	B	Test B	160	205	Testbus
	C	Test C	110	23	Testbus
	E	Test E	61	57	Testbus
	F	Test F	38	27	Testbus
	I	Test I	29	120	Testbus
	J	Test J	6	13	Testbus
	K	Test K	3	9	Testbus
	L	Test L	3	9	Testbus
	M	Test M	218	5	Testbus
Top-level tests	A	Test N	232	379	fp
	N	Test O	41	50	fp
	B	Test P	72	205	fp
	D	Test Q	104	39	fp

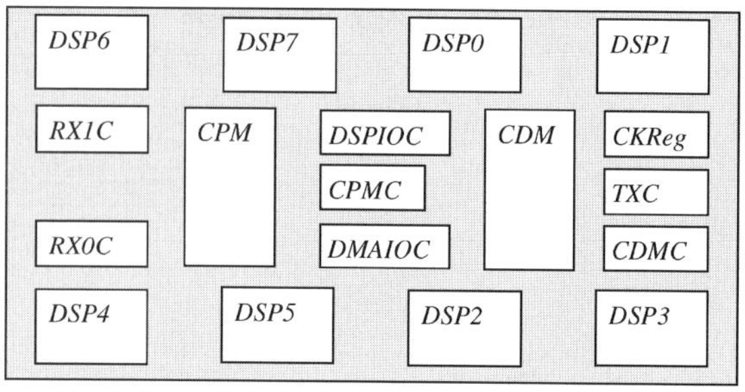

Fig. 19. The Ericsson design.

The largest design is the Ericsson design [17, 19] consisting of 8 DSP cores plus additional logic cores and memory banks, see Fig. 19. Design characteristics are in Table 3 where the following notations are used:

- n: DSP core ($0 \leq n \leq 7$),
- i: common program memory (CPM) bank ($0 \leq i \leq 7$),
- j: common data memory (CDM) bank ($0 \leq j \leq 9$),
- l: local data memory (LDM) at a DSP core ($0 \leq l \leq 3$),
- m: local memory (LZM) bank at a DSP core ($0 \leq m \leq 1$).

All logic blocks in the Ericsson design are tested by two test sets, one using external tester and one using on-chip resources, while memories are tested with one test set; in total 170 tests.The test access port may be used by more than one test concurrently. However, the other test resources cannot be used concurrently. Furthermore, only one test set may be applied concurrently to each block and maximal power dissipation is 5125 mW. For the implementation, we have simplified our system model (Section 3) regarding the TAM wire design and only allowing a single wire per test. When discussing about our algorithm we use *our*1, *our*2 and *our*3, which corresponds to the initial sorting based on *test power*(p), *test time*(t) and *test power* × *test time*($p \times t$).

Unless stated, we use partitioned testing with run to completion, for the cost function (Section 6.5) $\beta_1 = \beta_2 = 1$ and we have used a Sun Ultra10, 450 MHz CPU, 256 MB RAM.

7.2. *Test Scheduling*

We have compared our algorithm (nonpartitioned testing) with the nonpartitioned testing approaches proposed by Zorian [27] and Chou et al. [9]. We have used

Table 3. The Ericsson design characteristics.

Block	Test number	Test time	Test power	Test source	Test sink
RX0C	1	970	375	TAP	TAP
	2	970	375	TG_0	TRA_0
RX1C	3	970	375	TAP	TAP
	4	970	375	TG_0	TRA_0
DSPIOC	5	1592	710	TAP	TAP
	6	1592	710	TG_0	TRA_0
CPMC	7	480	172	TAP	TAP
	8	480	172	TG_0	TRA_0
DMAIOC	9	3325	207	TAP	TAP
	10	3325	207	TG_0	TRA_0
CKReg	11	505	118	TAP	TAP
	12	505	118	TG_0	TRA_0
CDMC	13	224	86	TAP	TAP
	14	224	86	TG_0	TRA_0
TXC	15	364	140	TAP	TAP
	16	364	140	TG_0	TRA_0
CPM_i	$17 + i$	239	80	TG_1	TRA_1
CDM_j	$25 + j$	369	64	TG_1	TRA_1
DSP_n					
LPM	$35 + 17 \times n$	46	16	$TG_{n,0}$	$TRA_{n,0}$
LDM_l	$36 + 17 \times n + l$	92	8	$TG_{n,0}$	$TRA_{n,0}$
LZM_m	$40 + 17 \times n + m$	23	2	$TG_{n,0}$	$TRA_{n,0}$
$Logic_0$	$17 \times n + 42$	4435	152	TAP	TAP
	$17 \times n + 43$	4435	152	$TG_{n,1}$	$TRA_{n,1}$
$Logic_1$	$17 \times n + 44$	4435	152	TAP	TAP
	$17 \times n + 45$	4435	152	$TG_{n,1}$	$TRA_{n,1}$
$Logic_2$	$17 \times n + 46$	7009	230	TAP	TAP
	$17 \times n + 47$	7009	230	$TG_{n,1}$	$TRA_{n,1}$
$Logic_3$	$17 \times n + 48$	7224	250	TAP	TAP
	$17 \times n + 49$	7224	250	$TG_{n,1}$	$TRA_{n,1}$
$Logic_4$	$17 \times n + 50$	7796	270	TAP	TAP
	$17 \times n + 51$	7796	270	$TG_{n,1}$	$TRA_{n,1}$

the same assumptions as Chou et al. and the results are in Table 4. Our approaches (our1, our2, and our3) results, in all cases, in a test schedule with three test sessions (*ts*) at a test time of 300 time units, which is 23% better than Zorian's approach and 9% better than the approach by Chou et al.

In System S, no power constraints are given and therefore only our2 can be used. Our approach finds using partitioned testing with run to completion after 1 second the optimal solution; see Table 5 (first group).

Table 4. ASIC Z test scheduling.

	Zorian		Chou et al.		Our1, Our2, Our3	
ts	Cores	Time	Cores	Time	Cores	Time
1	RAM1, RAM4, RF	69	RAM1, RAM3, RAM4, RF	69	RL2, RL1, RAM2	160
2	RL1, RL2	160	RL1, RL2	160	RAM1, ROM1, ROM2	102
3	RAM2, RAM3	61	ROM1, ROM2, RAM2	102	RAM3, RAM4, RF	38
4	ROM1, ROM2	102				
Test time		392		331		300

The results on the industrial design are in Table 5 (second group) where the industrial designer's solution is 1592 time units while our test scheduling achieve a test time of 1077 time units, the optimum, in all cases (our1, our2, and our3), which is 32.3% better than the designer's solution.

Table 5. Test scheduling results.

Design	Approach	Test time	Difference to SA/optimum (%)	CPU (sec)
System S [4]	Optimal solution	1152180	–	–
	Chakrabarty	1204630	4.5	
	Our2 (t)	1152180	0	1
Industrial design	Optimal solution	1077	–	–
	Designer	1592	47.8	–
	Our1 (p)	1077	0	1
	Our2 (t)	1077	0	1
	Our3 $(p \times t)$	1077	0	1
Muresan [23]	SA	25	–	90
	Muresan [23]	29	16	–
	Our1 (p) [18]	28	12	1
	Our2 (t) [18]	28	12	1
	Our3 $(p \times t)$ [18]	26	4	1
ASIC Z (1)	SA	262	–	74
	Our1 (p)	262	0	1
	Our2 (t)	262	0	1
	Our3 $(p \times t)$	262	0	1
ASIC Z (2)	SA	274	–	223
	Our1 (p) [18]	300	10	1
	Our2 (t) [18]	290	6	1
	Our3 $(p \times t)$ [18]	290	6	1
Extended ASIC Z (3)	SA	264	–	132
	Our1 (p)	313	18	1
	Our2 (t)	287	9	1
	Our3 $(p \times t)$	287	9	1
Ericsson	SA	30899	–	3260
	Our1 (p)	37336	20	3
	Our2 (t)	34762	12	3
	Our3 $(p \times t)$	34762	12	3

The results on design Muresan are in the third group of Table 5. The test time using the approach by Muresan et al. is 29 time units and the results using our approaches our1, our2, and our3 are 28, 28 and 26, respectively, all produced within 1 sec. Our SA ($TI = 400$, $TL = 400$, $\alpha = 0.97$) improves to 25 time units using 90 seconds.

When not considering idle power on ASIC Z, the test schedules using our1, our2, and our3 (fourth group in Table 5) all result in a test time of 262. The SA ($TI = 400$, $TL = 400$ and $\alpha = 0.97$) did not find a better solution.

In the experiments considering idle power (fifth group of Table 5), our heuristic approaches our1, our2, and our3 resulted in a solution of 300, 290 and 290, respectively, each produced within 1 second. The SA ($TI = 400$, $TL = 400$ and $\alpha = 0.99$) produced a solution of 274 requiring 223 sec., i.e. a cost improvement in the range of 6% to 10%.

The results on Extended ASIC Z when not considering idle power are 313 (our1), 287 (our2) and 287 (our3) (sixth group in Table 5), which were produced by our heuristic after 1 second. The SA optimization ($TI = TL = 400$, $\alpha = 0.97$) produced a solution at a cost of 264 running for 132 seconds, i.e. a cost improvement in the range of 9% to 18%.

The results on the Ericsson design (seventh group of Table 5) are 37226, 34762, 34762 produced by our heuristic our1, our2, and our3 (within 3 sec.). The SA algorithm ($TI = 200$, $TL = 200$, $\alpha = 0.95$) produced a solution at 30899 after 3260 seconds.

7.3. *Test Resource Placement*

In the ASIC Z design all cores have their own dedicated BIST structure. Let us assume that all ROM cores share one BIST structure and all RAM memories share another BIST structure; the rest of the cores have their own dedicated BIST structure. Using our placement strategy the test resources in ASIC Z will be placed as in Fig. 18.

7.4. *Test Access Mechanism Design*

Assume for ASIC Z (Fig. 18) that all tests are scan-based (1 scan-chain per core) applied with an external tester allowing a maximum of 8 scan chains to operate concurrently. The results using our1, our2, and our3 considering idle power are collected in Table 6. The test schedule and the TAM schedule achieved with heuristic our3 are in Fig. 20. The total length of the test access mechanism is 360 units and it is routed as in Fig. 21. All solutions were produced within 1 second.

Table 6. Results on ASIC Z.

Approach	Test time	TAM cost
Our1 (p)	300	360
Our2 (t)	290	360
Our3 ($p \times t$)	290	360

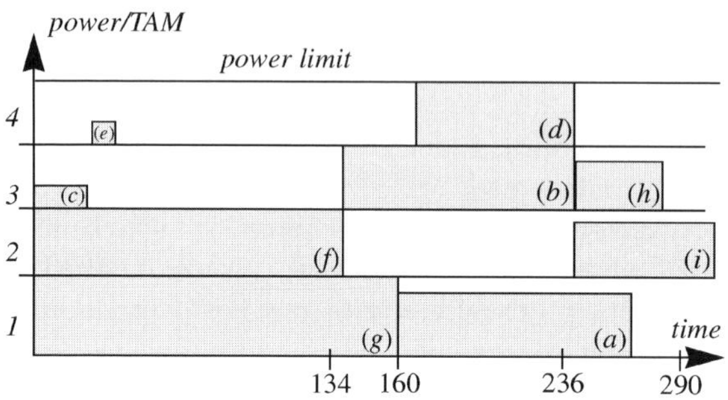

Fig. 20. Test schedule for ASIC Z.

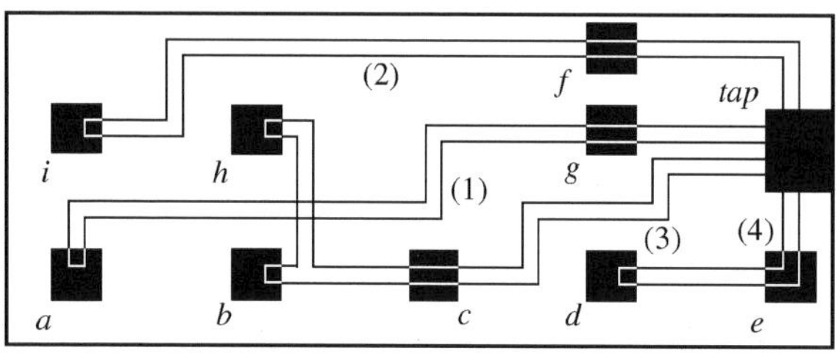

Fig. 21. ASIC Z with test data access mechanism.

Table 7. TAM design and test scheduling.

	System S			ASIC Z		
Approach	Time	TAM	Total	Time	TAM	Total
Naive 1	1541394	100	1641	699	110	809
Naive 2	996194	360	1356	300	500	800
Our1 (p)	–	–	–	300	360	660
Our2 (t)	996194	320	1316	290	360	650
Our3 ($p \times t$)	–	–	–	290	360	650
SA	996194	160	1156	334	180	514

7.5. *Test Scheduling and TAM Design*

We have made experiments using System S and ASIC Z to demonstrate the importance of integrating test scheduling and TAM design. The ASIC Z is fully BISTed; however, in this experiment we assume all tests are applied using an external tester able of supporting several tests concurrently. We have used two naive approaches, Naive1 and Naive2. Naive1 uses a minimal TAM design connecting all cores in a ring and Naive 2 uses an extensive TAM where each core gets its own dedicated TAM.

For the cost function, we have for ASIC Z used $\beta_1 = \beta_2 = 1$ and for System S $\beta_1 = 1/1000$ and $\beta_2 = 1$ and the results are collected in Table 7. Naive 1 produces a low cost TAM design but the test time is long leading to high total cost due to that all tests have to be scheduled in a sequence. The total cost of using the approach Naive 2 is also high due to the extensive TAM design. Such extensive TAM gives more flexibility, however, other constraints in the design limits its use. Our approaches (our1, our2, and our3) produces better results indicating the importance of considering TAM design and test scheduling in an integrated manner. The TAM design (Fig. 13) and the test schedule (Fig. 14) achieved using our heuristic for System S (Table 1) have a test time of 996194 and a TAM length of 320 computed within 1 second. The SA (TI = TL = 100, $\alpha = 0.99$) was running for 1004 seconds producing a test schedule (Fig. 16) and a TAM design (Fig. 17) with a test time of 996194 and a TAM length of 160, a TAM improvement of 50%. For ASIC Z, the SA (TI = TL = 300, $\alpha = 0.97$) produced after 855 seconds a solution with a cost of 514 (334 for test time and 180 for TAM).

The results comparing our heuristic with SA are collected in Table 8 where, for instance, our heuristic our1 produced a solution for ASIC Z with a total cost of 660 (a test time of 290 and a TAM cost of 360)

Table 8. TAM and scheduling results.

Approach	SA	Our1	Our2	Our 3
ASIC Z				
Test time	334	300	290	290
Diff to SA	–	−10%	−13%	−13%
TAM cost	180	360	360	360
Diff to SA	–	100%	100%	100%
Total cost	514	660	650	650
Diff to SA	–	28%	21%	21%
Comp. cost	855 sec.	1 sec.	1 sec.	1 sec.
Diff to SA	–	−85400%	−85400%	−85400%
Extended ASIC Z				
Test time	270	313	287	287
Diff to SA	–	16%	6%	6%
TAM cost	560	720	660	660
Diff to SA	–	29%	18%	18%
Total cost	830	1033	947	947
Diff to SA	–	24%	14%	14%
Comp. cost	4549 sec.	1 sec.	1 sec.	1 sec.
Diff to SA		−454800%	−454800%	−454800%
Ericsson				
Test time	33082	37336	34762	34762
Diff to SA	–	11%	5%	5%
TAM	6910	8245	9350	8520
Diff to SA	–	19%	35%	23%
Total cost	46902	53826	53462	51802
Diff to SA	–	15%	14%	10%
Comp. cost	15 h	81 sec.	79 sec.	62 sec.
Diff to SA		−66567%	−68254%	−86996%

after 1 second. The test time results are in the range of 10% to 13% better using our fast heuristic (in all cases) compared to the SA optimization. However, the TAM results are much worse using our heuristics and the total cost improvements by the SA are in the range from 21% to 28%.

The results from experiments on Extended ASIC Z are in Table 8 (second group). Our heuristic our1 produced a solution after 1 second with a test time of 313 and a TAM cost of 720, resulting in a total cost of 1033. The solution produced after 4549 seconds by our SA (TI = TL = 200, $\alpha = 0.97$) optimization has a test time of 270 and a TAM cost of 560. In this experiment, SA produced a better total cost (range 14% to 24%) as well as better cost regarding test time (range 6% to 16%) and TAM cost (range 18% to 29%). The results on the Ericsson design are collected in Table 8 (third group). For instance, our heuristic our1 results in a solution with a test time of 37336 and a TAM cost of 8245, which took 81 seconds to produce. The total cost is 53826 when using $\beta_1 = 1$ and $\beta_2 = 2$. The SA (TI = TL = 200, $\alpha = 0.95$) optimization produced a solution with a test time of 33082 and a TAM cost of 6910 after 15 hours. In all cases, the SA produces better results. Regarding test time the SA improvement is in the range from 5% to 11%, for the TAM cost in the range from 19% to 35% and the total cost in the range from 10% to 15%.

For all experiments with the SA, the computational cost is extremely higher compared to our heuristics. A finer tuning of the SA parameters could reduce it, however, the extensive optimization is only used for the final design and therefore a high computational cost can be accepted.

8. Conclusions

For complex systems such as SOCs, it is a difficult problem for the test designer to develop an efficient test solution due to the large number of factors involved. The work-flow consists of two consecutive parts: an early design space exploration and an extensive optimization for the final solution. In this paper we have proposed a framework suitable for these two parts where test scheduling, test access mechanism design, test set selection and test resource placement are considered in an integrated manner minimizing the test time and the size of the test access mechanism while satisfying test conflicts and test power constraint. For the early design space exploration, we have implemented an algorithm running at a low computational cost, which is suitable to be used iteratively many times. For the final solution, a more extensive optimization can be justified and we have proposed and implemented a technique based on Simulated Annealing for test scheduling and test access mechanism design.

We have performed several experiments on benchmarks and industrial designs to show the efficiency and usefulness of our approach.

References

1. J. Aerts and E.J. Marinissen, "Scan Chain Design for Test Time Reduction in Core-Based ICs," in *Proceedings of IEEE International Test Conference (ITC)*, Washington, DC, Oct. 1998, pp. 448–457.

2. A. Benso, S. Cataldo, S. Chiusano, P. Prinetto, and Y. Zorian, "A High-Level EDA Environment for the Automatic Insertion of HD-BIST Structures," *Journal of Electronic Testing; Theory and Applications (JETTA)*, vol.16, no. 3, pp. 179–184, June 2000.
3. H. Bleeker, P. Van Den Eijnden, and F. De Jong, *Boundary-Scan Test: A Practical Approach*, Boston, MA: Kluwer Academic Publishers, 1993.
4. K. Chakrabarty, "Test Scheduling for Core-Based Systems," in *Proceedings of IEEE/ACM International Conference on Computer Aided Design (ICCAD)*, San Jose, CA, Nov. 1999, pp. 391–394.
5. K. Chakrabarty, "Design of System-on-a-Chip Test Access Architectures Using Integer Linear Programming," in *Proceedings of IEEE VLSI Test Symposium (VTS)*, Montreal, Canada, April 2000, pp. 127–134.
6. K. Chakrabarty, "Design of System-on-a-Chip Test Access Architecture under Place-and-Route and Power Constraints," in *Proceedings of ACM/IEEE Design Automation Conference. (DAC)*, Los Angeles, CA, June 2000, pp. 432–437.
7. K. Chakrabarty, "Test Scheduling for Core-Based Systems Using Mixed-Integer Linear Programming," *IEEE Transactions on Computer-Aided Design of Integrated Circuits and Systems*, vol. 19, no. 10, pp. 1163–1174, Oct. 2000.
8. K. Chakrabarty, "Optimal Test Access Architectures for System-on-a-Chip," *ACM Trans. on Design Automation of Electronic Systems*, vol. 6, pp. 26–49, Jan. 2001.
9. R. Chou, K. Saluja, and V. Agrawal, "Scheduling Tests for VLSI Systems Under Power Constraints," *IEEE Transactions on VLSI Systems*, vol. 5, no. 2, pp. 175–185, June 1997.
10. G.L. Craig, C.R. Kime, and K.K. Saluja, "Test Scheduling and Control for VLSI built-in-self-test," *IEEE Transactions on Computers*, vol. 37, no. 9, pp. 1099–1109, Sept. 1988.
11. M. Garg, A. Basu, T.C. Wilson, D.K. Banerji, and J.C. Majithia, "A New Test Scheduling Algorithm for VLSI Systems," in *Proceedings of the CSI/IEEE Symposium on VLSI Design*, New Delhi, India, Jan. 1991, pp. 148–153.
12. S. Gerstendorfer and H.-J. Wunderlich, "Minimized Power Consumption for Scan-Based BIST," in *Proceedings of IEEE International Test Conference (ITC)*, Atlantic City, NJ, Sept. 1999, pp. 77–84.
13. G. Hetherington, T. Fryars, N. Tamarapalli, M. Kassab, A. Hassan, and J. Rajski, "Logic BIST for Large Industrial Designs: Real Issues and Case Studies," in *Proceedings of IEEE International Test Conference (ITC)*, Atlantic City, NJ, September 1999, pp. 358–367.
14. IEEE P1500 Web site. http://grouper.ieee.org/groups/1500.
15. V. Iyengar and K. Chakrabarty, "Precedence-Based, Preemptive, and Power-Constrained Test Scheduling for System-on-a-Chip," in *Proceedings of IEEE VLSI Test Symposium (VTS)*, Marina Del Rey, CA, April 2001, pp. 368–374.
16. S. Kirkpatrick, C. Gelatt, and M. Vecchi, "Optimisation by Simulated Annealing," *Science*, vol. 220, no. 4598, pp. 671–680, 1983.
17. E. Larsson, "An Integrated System-Level Design for Testability Methodology." Ph. D. Thesis no. 660, Linköpings Universitet, Sweden 2000.
18. E. Larsson and Z. Peng, "An Integrated System-On-Chip Test Framework," in *Proceedings of the Design, Automation and Test in Europe Conference (DATE)*, Munchen, Germany, March 2001, pp. 138–144.
19. E. Larsson, Z. Peng, and G. Carlsson, "The Design and Optimization of SOC Test Solutions," in *Proceedings of IEEE/ACM International Conference on Computer-Aided Design (ICCAD)*, San Jose, CA, Nov. 2001, pp. 523–530.
20. E.J. Marinissen, R. Arendsen, G. Bos, H. Dingemanse, M. Lousberg, and C. Wouters, "A Structured and Scalable Mechanism for Test Access to Embedded Reusable Cores," in *Proceedings of IEEE International Test Conference (ITC)*, Washington, DC, Oct. 1998, pp. 284–293.
21. E.J. Marinissen, S.K. Goel, and M. Lousberg, "Wrapper Design for Embedded Core Test," in *Proceedings of IEEE International Test Conference (ITC)*, Atlantic City, NJ, Oct. 2000, pp. 911–920.
22. E.J. Marinissen, Y. Zorian, R. Kapur, T. Taylor, and L. Whetsel, "Towards a Standard for Embedded Core Test: An Example," in *Proceedings of IEEE International Test Conference (ITC)*, Atlantic City, NJ, Sept. 1999, pp. 616–627.
23. V. Muresan, X. Wang, V. Muresan, and M. Vladutiu, "A Comparison of Classical Scheduling Approaches in Power-Constrained Block-Test Scheduling," in *Proceedings of IEEE International Test Conference (ITC)*, Atlantic City, NJ, Oct. 2000, pp. 882–891.
24. M. Nourani and C. Papachristou, "An ILP Formulation to Optimize Test Access Mechanisms in System-on-a-Chip Testing," in *Proceedings of IEEE International Test Conference (ITC)*, Atlantic City, NJ, Oct. 2000, pp. 902–910.
25. M. Sugihara, H. Date, and H. Yasuura, "A Test Methodology for Core-Based System LSIs," *IEICE Transactions on Fundamentals*, vol. E81-A, no. 12, Dec. 1998, pp. 2640–2645.
26. N. Weste and K. Eshraghian, *Principles of CMOS VLSI Design*, Addison-Wesley, 1993.
27. Y. Zorian, "A Distributed BIST Control Scheme for Complex VLSI Devices," in *Proceedings of the IEEE VLSI Test Symposium (VTS)*, Atlantic City, NJ, April 1993, pp. 4–9.

Erik Larsson received his M.S. and Ph.D. degrees in Computer Systems from Linköping University in 1994 and 2000, respectively. Since October 2001, he is on leave from his Assistant Professor position at the Computer Science department at Linköping University for a Research Fellow position at Fujiwara Laboratory, Nara Institute of Science and Technology, Japan, funded by JSPS (Japan Society for the Promotion of Science). His main research interests are system-level design for testability and design of SOC test solutions.

Zebo Peng is Professor of the chair in Computer Systems, Director of the Embedded Systems Laboratory, and Chairman of the Division for Software and Systems at Linköping University. He received his Ph.D. degree in Computer Science from Linköping University in 1987.

Prof. Peng's current research interests include design and test of embedded systems, electronic design automation, design for testability, hardware/software co-design, and real-time systems. He has published over 120 technical papers in these areas and coauthored the book "System Synthesis with VHDL" (Kluwer Academic, 1997). He was corecipient of two best paper awards at the European Design Automation Conferences (EURO-DAC) in 1992 and 1994. He has served on the program committee of several IEEE/ACM conferences and workshops and was the General Chair of the 6th IEEE European Test Workshop.

JOURNAL OF ELECTRONIC TESTING: Theory and Applications 18, 401–414, 2002

On Concurrent Test of Core-Based SOC Design

YU HUANG*, WU-TUNG CHENG, CHIEN-CHUNG TSAI, NILANJAN MUKHERJEE, OMER SAMMAN AND YAHYA ZAIDAN
Mentor Graphics Corporation, 8005 S.W. Boeckman Rd., Wilsonville, OR 97070, USA
Yu_Huang@mentor.com
Wu-Tung_Cheng@mentor.com
Chien-Chung_Tsai@mentor.com
Nilanjan_Mukherjee@mentor.com
Omer_Samman@mentor.com
Yahya_Zaidan@mentor.com

SUDHAKAR M. REDDY
Department of Electrical & Computer Engineering, University of Iowa, Iowa City, IA 52242, USA
Reddy@engineering.uiowa.edu

Received August 29, 2001; Revised January 28, 2002

Editor: Krishnendu Chakrabarty

Abstract. In this paper, a method to solve the resource allocation and test scheduling problems together in order to achieve concurrent test for core-based System-On-Chip (SOC) designs is presented. The primary objective for concurrent SOC test is to reduce test application time under the constraints of SOC pins and peak power consumption. The methodology used in this paper is not limited to any specific Test Access Mechanism (TAM). Additionally, it can also be applied to SOC budgeting at design phase to predict a tradeoff between test application time and SOC pins needed. The contribution of this paper is the formulation of the problem as a well-known 2-dimensional bin-packing problem. A best-fit heuristic algorithm is adopted to achieve optimal solution.

Keywords: concurrent SOC test, test scheduling, pin mapping, 2-dimensional bin-packing

1. Introduction

Core-based SOCs have become an attractive design strategy because designing a multi-million gate system with pre-designed reusable cores is the most effective way to reduce design time and cost. The Semiconductor Industry Association's Technology Roadmap [16] predicts the percentage of reusable cores in SOC to be rising to 80% in 2006. However, with the increasing complexity and reduction of design cycle, the test application time of SOC is becoming a major bottleneck for time-to-market. In this paper, the test time problem is addressed at the system level rather than core level by proposing a methodology for concurrent SOC test. An algorithm is presented to optimize the test resource allocation and test scheduling problems under the concurrent SOC test paradigm.

*The research was conducted when the author was with Mentor Graphics as an intern from Department of Electrical & Computer Engineering, University of Iowa.

1.1. Terminology Definitions

The terminology used in this paper is given as follows.

SOC: The integration of a complete system, which until recently consisted of multiple ICs on a PCB, onto one chip. An SOC often includes multiple types of design blocks and intellectual property on a chip, such as digital logic, memories, and analog circuitry. Typically, SOCs are designed using embedded reusable cores [23].

Core: In SOC designs, the design process involves an IC that is often made up of large pre-defined and pre-verified reusable building blocks. The IC building blocks are called embedded cores.

Parallel Access: This technique is based on parallel access to the core terminals by either using dedicated SOC pins or by sharing SOC pins through multiplexers [22].

Serial Access: This technique for peripheral access is based on serialization of test patterns. A scan chain around the core allows indirect but full access to all the core terminals [22].

Sequential SOC Test: In the entire SOC test process, one core is tested at a time.

Concurrent SOC Test: In order to reduce the total test application time, multiple cores are tested concurrently, under the constraints of available SOC pins and peak power consumption.

Test Resource: In this paper, the total available SOC pins or the total test bus bandwidth is considered as test resource.

Test Scheduling: Test scheduling for core-based SOCs focuses on scheduling the core tests in such a way that (1) no resource conflict occur with respect to the test access infrastructure between SOC pins and core terminals, and (2) the overall SOC test time is minimized [10].

1.2. Earlier Works

The complexity of SOC makes test much more difficult than before. Many new DFT techniques have been exploited to address this problem. The research in this field can be placed into two categories. One is applicable to core providers while the other is for core integrators. The proposed method in this paper addresses the point of view of core integrators. The cores are assumed to be given as black boxes with known interfaces, test protocols, and test vector files. This reflects the most common situation when using reusable IP cores.

Most of the earlier research have focused on two aspects: Test Access Mechanism (TAM) [7–9, 11–13, 18–21] and test scheduling for core-based SOC [2, 14, 15, 17].

There are several major techniques for embedded core peripheral access, such as parallel access (multiplexed access) [9, 12], serial access [18, 20], and functional access [7, 8]. In recent years, several new TAMs have been proposed such as Test Bus [19], TestRAIL [11], and Addressable Test Ports [21].

Parallel access to cores [9, 12] is realized by either using dedicated SOC pins or sharing SOC pins to access cores through MUXes. The advantage of this method is that the core can be tested as if it is a stand-alone device. Hence previously created tests for the core can be applied directly. The disadvantage of this method is that the area cost can be high due to routing and multiplexing overheads. Serial access to a core [18, 20] is allowed through scan chains around it. This method is less restricted by any pin limitation, but suffers from increased test application time due to scan chain shifting. Functional access [7, 8] is based on core transparency characteristics and is not considered in this paper. In addition to the traditional TAM techniques, Varma and Bhatia [19] propose to use dedicated on-chip variable-width buses for propagating test data signals. The control signals are generated from an on-chip JTAG controller to control the configuration of test collars (or wrappers). Test collars around the cores provide paths between test buses and core terminals. The cores-under-test are the only one connected to the test buses, while all the other cores on the SOC are disconnected from them. Marinissen et al. [11] build a shell around each core. Each Test shell contains a local n-bit wide test rail. Test shells (or wrappers) could be set to function mode, IP test mode, or bypass mode based on the control signals. At the SOC level, global test rails are built to connect different local test rails in different cores or to connect local test rails to SOC pins to obtain a flexible and scalable test bus architecture. Whetsel [21] proposes an addressable test access port architecture to directly address the core-under-test. Serial and parallel communication paths between each core and SOC pins are provided. Nourani and Papachristou [13] formulate the TAM optimization problem as an Integer Linear Programming (ILP) problem. They assume that there are many bridges existing in the SOC design, such as MUXes, controllable buffers, bypass routes, tri-state ports etc. An ILP method is applied to select the best subset of bridges to provide access to all the cores, while minimizing the cost function.

In all the above techniques, no concurrent core test methodology was proposed although some TAMs can potentially offer this function. In this paper, our primary objective is to implement SOC concurrent testing to reduce total test application time. The proposed method can be applied to any TAM that provides access to cores. Note that a serial access mechanism is automatically considered when complete parallel access is impossible due to test resource limitation.

For test scheduling of core-based systems, most of the previous research concentrate on viewing it as a task after the TAM is fixed. Sugihara et al. [17] address the problem by selecting a test set for each core from a set of test sets provided by the core providers, and then scheduling the test sets in order to minimize test time. The authors assume: (1) each core has its own BIST logic, (2) external test can be carried out for only one core at a time, and (3) core providers provide multiple test sets comprising of both BIST and external test patterns for each core. The problem is modeled as a combinatorial optimization problem and solved by using a heuristic method. Chakrabarty [2] generalizes the test scheduling problem of [17]. He assumes: (1) TAM has one external test bus and multiple BIST resources, and (2) the cores have been assigned to the test bus. The problem is formulated as an m-processor open shop scheduling problem, and is solved by using a Mixed-Integer Linear Programming (MILP) model to minimize the test time. Ravikumar et al. [14, 15] propose a method to solve the test scheduling problem under the constraint of power consumption. The assumption in this case is that BIST is the only methodology for testing individual cores.

In all the above techniques, BIST scheduling is their focus. External test can not be carried out for multiple cores concurrently. In our method, we will focus on external test scheduling for multiple cores. However, the proposed method can be applied to BIST scheduling as well.

Recently, there have been a few papers considering test scheduling problem together with the test resource allocation problem [1, 3, 4]. Chakrabarty [3, 4] assumes a test bus architecture and formulates the problem with an ILP model to find the minimum total test time by allocating N test lines to a fixed number of test buses, and assigning each core to only one test bus. The place-and-route and power constraints are also considered in [4]. Bagchi et al. [1] address the same problem as in [3], but considered clustering some cores into a module, and scheduled testing of modules rather than individual cores in order to reduce the total test time.

The above techniques and the method proposed in this paper try to solve the same problem but in different ways. It is shown in Section 4 that test application time can be reduced by using the proposed algorithm as compared to the above mentioned methodologies [3, 4].

1.3. Characteristic of the Proposed Work

The characteristics of the proposed method are as follows.

(i) It is not restricted to one specific TAM.
(ii) It provides a clear picture of the tradeoffs between test resources (number of SOC pins or bandwidth of test buses) and test application time.
(iii) Given a fixed number of test resources (total SOC pin count or total SOC bus bandwidth), it allows for optimal resource allocation (pin mapping or bus assignment) to achieve the shortest test application time.
(iv) Given a fixed test application time, it can predict the minimum resource requirement and allocation, which can be used as a guideline during SOC integration.
(v) Functional test patterns can be applied directly from SOC pins to terminals of a core as long as there are enough test resources, and it provides a solution for testing legacy IP cores without scan chains.
(vi) For cores that cannot have complete parallel access, they can still be tested with partial parallel and partial serial access.
(vii) Power constraint is considered.
(viii) The problem is formulated as a 2-dimensional bin-packing problem, and a known heuristic is improved to solve the problem.

The paper is organized in the following manner. In Section 2, the problem is formulated as a 2 dimensional bin packing problem and a known heuristic algorithm is adopted to solve the problem of test resource allocation and test scheduling to achieve optimized concurrent SOC test. SOC design budgeting and TAM synthesis are introduced in Section 3. Experimental results are presented in Section 4, followed by the conclusions section.

2. Proposed Method

2.1. *Problem Formulation*

The problem considered in this paper is formally stated as follows.

Given N SOC pins and K cores, for each core C_i $(0 < i \leq K)$, its test application time T_i and peak power consumption P_i during test are known. Suppose core C_i has W_i terminals. Ω is the maximum peak power that should not be exceeded at any time during the test of the SOC. The problem is to determine a mapping from terminals of each core C_i to SOC pins and a test schedule for each core C_i. The objectives of the concurrent SOC test scheduling are as follows:

1. The total test application time is minimized.
2. There is no test resource conflict (Pin Constraint).
3. At any time, the sum of peak power of a set of cores that are tested simultaneously should not exceed Ω (Power Constraint).

The problem can be transformed to a well-known 2-dimensional bin-packing problem [5], which is often referred to as the rectangle packing problem, as follows. The conceptual view is illustrated in Fig. 1.

Each core C_i is represented by a rectangle that has height T_i, width W_i, and weight P_i. An SOC with N pins is represented by a rectangular bin with width N. The objective is to pack total of K rectangles representing the cores into the bin representing the SOC such that:

1. The height of the bin is minimized.
2. There is no overlap between rectangles packed in the bin (Pin Constraint).
3. If any horizontal line drawn on the bin cuts through a set of rectangles, the total weight of this set of rectangles has to be less than a fixed threshold Ω (Power Constraint).

2.2. *Modified Best-fit Algorithm*

Rectangle packing is a well studied NP-hard problem [5], and therefore, a known heuristic algorithm, called "Best-fit Algorithm," proposed by Fenrich et al. [6] has been adopted and modified as illustrated in Fig. 2.

Fig. 2 shows the flow chart of the proposed algorithm. Note that the shapes of the rectangles can be transformed in order to fit in the bin to get optimized solution. This is referred to as rectangle transformation in the algorithm description and will be explained in more details later.

Step 1. Pre-processing transforms the unpackable rectangles to packable ones (see Algorithm 1), and sorts all the rectangles in a descending order of their height to build an ordered list L.

Step 2. If R (the current rectangle being processed) can be packed to some existing levels without transformation, then pack R on level k, which has the minimum unused_width(k) after R is packed, and the weight constraint is satisfied.

Step 3. Do rectangle transformation for R on every available level assuming that all the unused width is allocated to R. Then pick the level where R's new height is lower than the current height of this level. If multiple levels meet the above requirement, choose level k with minimum unused_width(k). Note that the weight constraint is still considered in this step.

Step 4. Post-processing combines two adjacent test sessions (levels) to reduce the total test application time. It takes the upper level, reverses all the

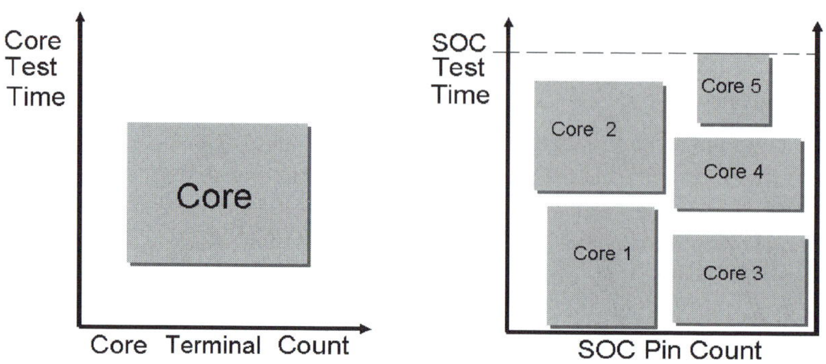

Fig. 1. Conceptual view of 2-D bin packing formulation.

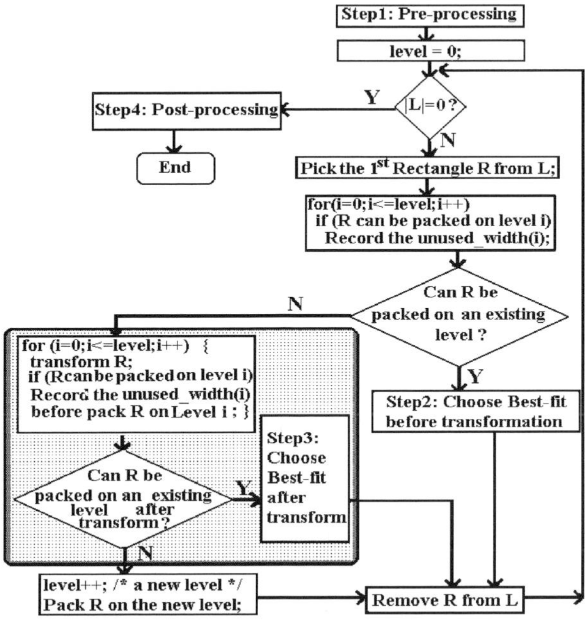

Fig. 2. Flow chart of the proposed algorithm.

rectangles on this level (i.e. the leftmost rectangle will be the rightmost), and pushes down all the newly arranged rectangles on the top of the lower level as far as possible. The weight constraint is still maintained during the process.

2.3. *Rectangle Transformation*

As discussed above there are two situations when rectangle transformation is performed by the proposed procedure. In both situations the width of a rectangle is reduced and its height is increased. The first situation occurs if the number of terminals on some cores are more than the number of SOC pins, i.e. test resource is limited and complete parallel access becomes impossible. In other words, there are some rectangles with their widths larger than N. In this case, the shapes of the rectangles have to be transformed in order to fit in the bin. This is done in our procedure in Step 1. The second situation occurs in Step 3 of the proposed procedure. Here the procedure attempts to pack the next rectangle from the ordered list in to an existing level. In order to achieve this it may be necessary to transform the rectangle by reducing its width and increasing its height. Reducing the width of a rectangle corresponds to reducing the number of pins for parallel access of the core pins. We describe next how the rectangle transformation is done in our procedure.

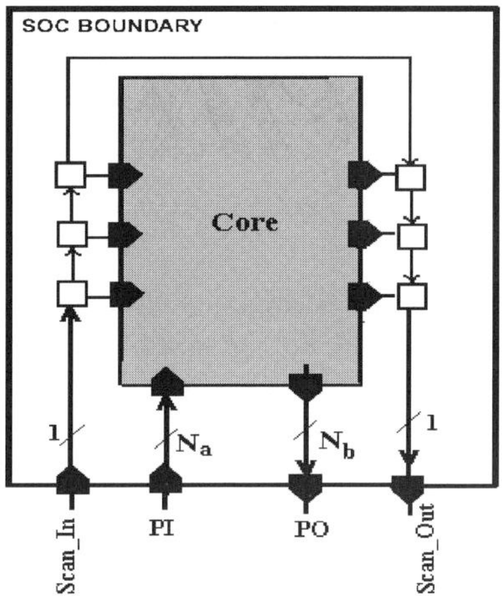

Fig. 3. Partial parallel and partial serial core access.

Rectangle transformation can be realized by accessing a subset of core terminals serially through one or more scan chains outside the core, which are called wrapper scan chains. The rest of the core terminals are accessed in parallel from the SOC pins. In other words, after rectangle transformation, a core can have partial parallel and partial serial access as illustrated in Fig. 3. In the experimental results presented later we assume a single wrapper scan chain in order to compare our results with those of earlier works. It should be noted that with a single wrapper scan chain, after rectangle transformation, the width of a rectangle changes to $N_j + 2$. N_j SOC pins are allocated for parallel access and two pins are necessary for scan input and scan output to serially access all other pins that can not have parallel access. The test application time increases proportionately due to serial access of some of the core pins. The increased test application time is accommodated by increasing the height of the rectangle obtained after the transformation.

In Step 1 of the procedure we use a heuristic to transform the rectangles corresponding to the cores with number of pins larger than N. This is described next. Assume that N_j SOC pins are allocated to a core C_j for parallel access. The procedure we used is given below as Algorithm 1. The heuristic we used allocates more SOC pins for parallel access to the cores that need longer test application time and more test resources (pins). Note that after transformation, these cores will be processed together with the other cores. They might be transformed once more in Step 3.

Let $C_j, (j = 1 \ldots t)$, be the cores that have more terminals than the number of SOC pins;
Let $Area_j = T_j * W_j$;
Then, $N_j = \lceil N * (Area_j / \sum_{i=1}^{t}(Area_i))\rceil$ *for* $j = 1 \ldots t$;
Assign N_j *SOC pins to Core*$_j$ *for parallel access;*

Algorithm 1. Allocate N_j SOC pins to Core$_j$ for parallel access.

After rectangle transformation, the height of the rectangle changes as well. For the sake of simplicity of presentation and the reasons given next we do not consider internal scan chains of cores.

(1) The primary purpose of this paper is to provide a solution to the concurrent SOC test scheduling problem by a 2-D bin packing model that is the same for both non-scan cores and scan cores.
(2) The concept of rectangle transformation used in the proposed 2-D bin packing algorithm is exactly same for both non-scan cores and scan cores. Whereas the calculation of the new height after transformation is different for these two scenarios. For scan cores, the new height after rectangle transformation depends on multiple parameters such as the number of internal scan chains, the length of each scan chain and the number of core functional terminals etc. Below we give the method used to compute the height of a transformed rectangle for non-scan cores. Similar formulae can be used for cores with internal scan chains.
(3) In all industrial SOC designs used in our experimentation, all test vectors are completely serialized and no internal scan information is available.

For two of the hypothetical SOC designs (S1 and S2) we used in our study, methods in [3] and [1] also used the same assumptions and hence we can do a fair comparison with the results in [3] and [1].

Now suppose it is known that there are N_j SOC pins allocated for Core C_j that has M_j^{in} inputs and M_j^{out} outputs. Suppose N_a inputs and N_b outputs of the core are mapped to N_j SOC pins, then the following conditions must be satisfied:

$$N_a + N_b = N_j \tag{1}$$

$$Tj\_new = \left[1 + \mathrm{Max}\left(M_j^{in} - N_a, M_j^{out} - N_b\right)\right] * T_j \tag{2}$$

In (2), T_{j_new} is the new time for testing core C_j after performing rectangle transformation described in Fig. 4. $(M_j^{in} - N_a)$ and $(M_j^{out} - N_b)$ gives the number of inputs and outputs of the core that needs serial access. The number of additional shift cycles is therefore controlled by the maximum of $(M_j^{in} - N_a)$ and $(M_j^{out} - N_b)$ plus an extra cycle to load / unload the patterns. In other words, given that T_j denotes the test application time for the original core test, the test application time for the transformed core (T_{j_new}) will scale up by a factor dependent on the number of serial access pins.

The numbers of N_a and N_b are determined such that the value of $\mathrm{Max}(M_j^{in} - N_a, M_j^{out} - N_b)$is minimized. A solution is given below as Algorithm 2 that will result in a minimal increase in test application time for the core.

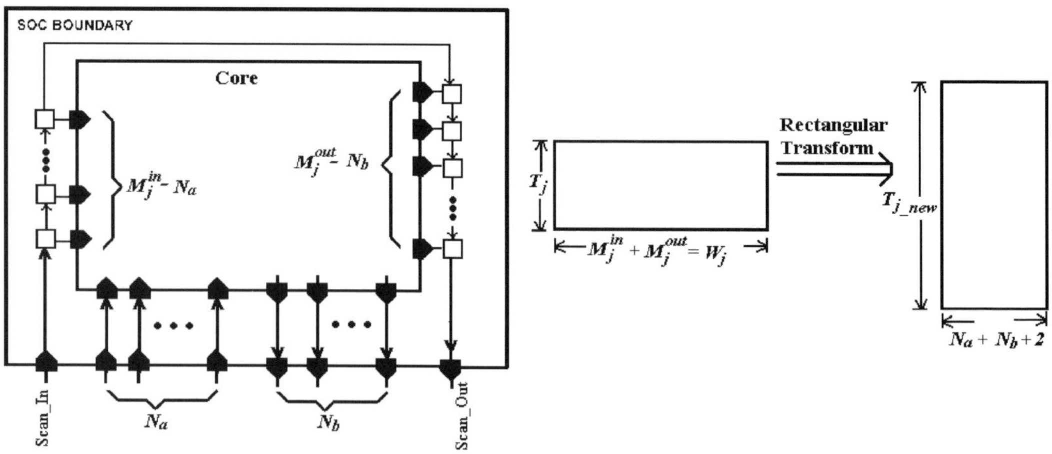

Fig. 4. Rectangle transformation.

$$
\begin{aligned}
&\textit{if}\,(M_j^{in} > M_j^{out} + N_j)\,\{ && N_a = N_j;\ N_b = 0;\\
&\}\ \text{elsif}\ (M_j^{out} > M_j^{in} + N_j)\,\{ && N_a = 0;\ N_b = N_j;\\
&\}\ \text{else}\ \{ && N_a = \lceil (N_j + M_j^{in} - M_j^{out})/2 \rceil;\\
& && N_b = \lceil (N_j - M_j^{in} + M_j^{out})/2 \rceil;\\
&\}
\end{aligned}
$$

Algorithm 2. Map N_j SOC pins to N_a inputs and N_b outputs of a transformed core.

Some practical considerations are discussed below when applying the rectangle transformation algorithm.

1. In practice, some designs are not allowed to mix parallel and serial access. In this case, the rectangle transformation will not be performed on specific cores. If the core has more terminals than SOC pins, then it will be completely serially accessed and scheduled for test independently.
2. One assumption used in the rectangle transformation is that the peak power consumption of the core remains the same after rectangle transformation. This assumptions needs to be verified in practice.

3. Related Issues

3.1. SOC Design Budgeting

One main aspect of designing a SOC is to determine the minimum resource requirement given an upper bound on the test application time. The proposed algorithm can be applied to this problem by switching width and length of each core and the SOC in the formulation.

3.2. TAM Synthesis

After the test resource allocation and test scheduling results are obtained, the TAM is inserted to implement the concurrent SOC test based on the results. In this subsection, the synthesis of TAM is illustrated and an example is shown in Fig. 5.

As an example, a core test controller is synthesized from a VHDL behavioral description in Fig. 5 to control the concurrent testing of SOC by activating/de-activating certain core_test signals. The reset signal typically exists on the chip. A rising edge in the "control" signal indicates a new test session, during which a new subset of cores are tested by activating their core_test signals simultaneously, and de-activating the core_test signals for all the other cores. The waveform on the control pin is based on the scheduling information obtained earlier. "Test_mode" is another global control signal in SOC. When it is not active, all cores are in their functional mode. When it is active, only cores with their core_test pins active are in their test mode, the other cores are in an isolation mode.

```
entity Controller is
port(      Reset, control: in bit;
           core_test: out bit_vector (31 downto 0) );
end Controller;
architecture BEHAV of Controller is
begin
  process (Reset, control)
  begin
  if Reset = '1' then
           core_test(1) <='0'; core_test(2) <='0';
           ...............
           core_test(K-1) <='0'; core_select(K) <='0';
           decoder := 0;
  else if control's event and control = '1' then
           decoder := decoder + 1;
           switch (decode)
           case 0:    -- Initialization
           case 1:    -- Test Core2, Core3, Core 5....
             core_test(1) <= '0'; core_test(2) <= '1';
             core_test(3) <= '1'; core_test(4) <= '0';
             core_test(5) <= '1';              ........;
           case 2:    .......
           case r:    .......
           End switch;
  endif;
  end process;
end BEHAV;
```

Fig. 5. RTL design of core test controller.

Fig. 6 illustrates the structures necessary to allow access to the terminals of cores from the SOC pins. Core terminals that are directly connected to SOC pins do not need any additional logic.

In Fig. 6, it can be seen that one input of $Core_1$ and one input of $Core_2$ share a SOC input PI_1. The outputs from $Core_1$, $Core_2$ and $Core_3$ share an output PO_1. Note that, since these 3 cores share the same resources, they must be scheduled for test at different time, i.e. signals S1, S2 and S3 are mutually exclusive in time domain. $Core_4$ doesn't share any resources with other cores, therefore, it can be tested concurrently with either $Core_1$, $Core_2$, or $Core_3$.

In order to reduce the required number of bi-directional SOC pins, the following method is used. When mapping a core input (output), first an SOC input (output) is searched. If it is not available, then find an SOC Bi-Directional (BiDi) pin. If no SOC Bidi is

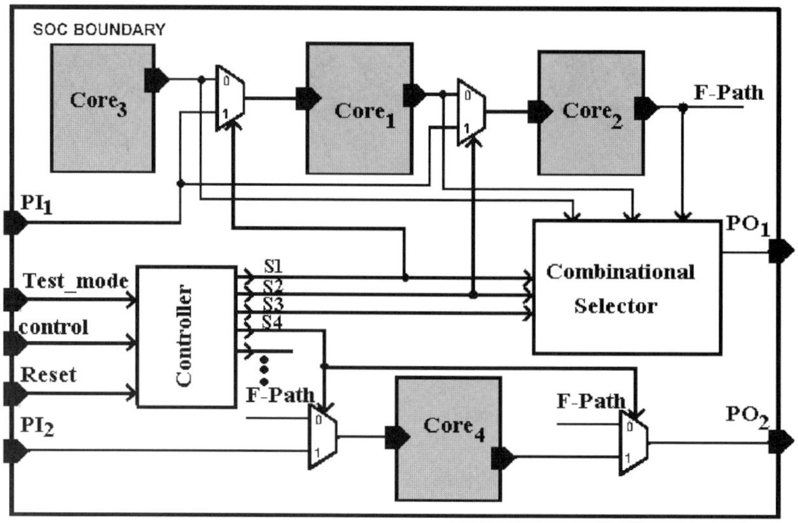

Fig. 6. TAM synthesis without considering BiDis.

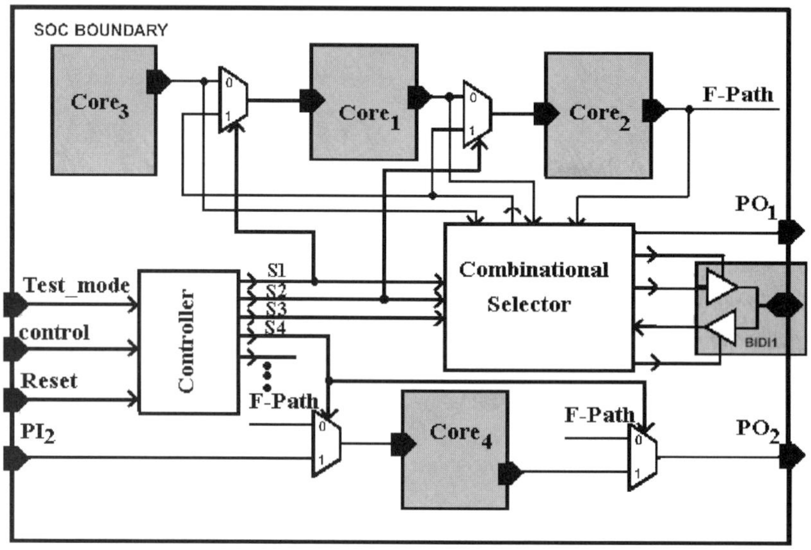

Fig. 7. TAM synthesis with transformed BiDi.

available, an SOC output (input) is transformed to a BiDi.

Fig. 7 shows an example of mapping an input of $Core_1$ and $Core_2$ and an output of $Core_3$ to a transformed SOC BIDI1. The output of $Core_1$ and $Core_2$ will map to another SOC output PO_1. A simple combinational logic is synthesized to control the enable pins of the two tri-state gates to configure the BIDI1 as input or output at appropriate test session.

4. Experimental Results

The proposed algorithm is run on 6 hypothetical but nontrivial SOCs (S1 to S6) composed from some ISCAS85 and ISCAS89 benchmark circuits, and 4 industry designs (IND1-IND4). Since S1 and S2 are the only benchmarks used in both [3] and [1], we also use these 2 SOCs to compare with the results in these works. The information of the cores used in the hypothetical and industrial SOCs are listed in Table 1 and Table 3

Table 1. Information of cores used in SOCs S1 and S2.

Core	No. of inputs	No. of outputs	No. of test cycles	Core	No. of inputs	No. of outputs	No. of test cycles
C432	36	7	27	S838	35	2	2507
C499	41	32	52	S5378	39	53	4507
C880	60	26	16	S9234	23	26	6147
C1355	41	32	84	S13207	47	137	10061
C3540	50	22	84	S15850	30	103	3743
C6288	32	32	12	S35932	67	352	714
C7552	207	108	73	S38417	60	138	3656
				S38584	44	310	5216

respectively. S3 was made from 2 S1, and S4 was made from 3 S2 in order to create more complex SOCs. S5 and S6 are constituted with randomly picked cores. The compositions of the SOCs (S1 to S6) are shown in Table 2.

The comparison between the proposed method and the ILP method in [3] is shown in Table 4. The first column is the total number of SOC pins. The results for S1 and S2 are in Columns 2, 3, and 4 and Columns 5, 6, and 7, respectively. The test application time necessary for the proposed mapping and scheduling algorithms are shown in columns 2 and 5, followed by the test time reported in [3] in columns 3 and 6. The percentage of test application time reduced by using the proposed method for SOCs S1 and S2 are listed in columns 4 and 7, respectively. The complexity of the algorithm is $O(K^2)$, where K is the total number of cores. The algorithm was implemented in C running on a HP9000 workstation. The CPU time is less than 1 second for each benchmark used for experimentation. Therefore the CPU time is not included in the results.

Table 2. Composition of S1–S6.

Core	No. in S1	No. in S2	No. in S3	No. in S4	No. in S5	No. in S6
C432	1	0	2	0	3	6
C499	1	0	2	0	2	1
C880	1	0	2	0	1	5
C1355	1	0	2	0	4	2
C3540	1	0	2	0	1	3
C6288	1	1	2	3	3	5
C7552	1	1	2	3	2	4
S838	0	1	0	3	5	2
S5378	1	1	2	3	4	3
S9234	0	1	0	3	1	4
S13207	0	1	0	3	3	5
S15850	0	1	0	3	2	1
S35932	1	1	2	3	4	3
S38417	1	1	2	3	3	2
S38584	0	1	0	3	2	4

From the results, it is evident that:

1. The proposed algorithm mostly results in reduced test application time.
2. For a fixed test application time, the proposed method uses much less test resources. For example, 430403 test cycles are necessary when the SOC has 48 pins, whereas, in [3], similar test time can be obtained by using 72 SOC pins.
3. In the proposed method, the test application time is only bounded by the maximum test time of individual cores if test resources are not limited. However, in [3], there is a theoretical lower bound on test application time even if there is unlimited resource. For example, in the case of S1, the method presented in this paper could achieve the lowest test application time reported in [3] with fewer SOC pins (72 vs. 112). There is continued reduction with increasing resources. One possible explanation for the difference is that in [3] it was assumed that the number of test buses is fixed. Whereas, it is optimized dynamically in the concurrent SOC test scheduling process in the proposed method.

In Table 5, the proposed method is compared with the method in [1]. In [1], several cores are clustered into a module, and modules are processed instead of cores. We use the clustering result from [1] and compare the test application times. In [1], the lower bound on the test application time for SOC is determined by the longest test time for an individual core assuming the test resource is enough. This lower bound is also applicable to the proposed method. The total number of

Table 3. Information of cores used in IND1–IND4.

Core	No. of inputs	No. of outputs	No. of BiDis	No. of test application time (10 ns)	
				Test set I	Test set II
1	58	95	16	100000	100000
2	34	4	0	999999	999999
3	14	2	0	999999	999999
4	140	83	16	999999	999999
5	5	16	33	7600	333700
6	262	318	131	151100	200000
7	12	90	67	58000	200000
8	51	57	16	100000	100000
9	106	76	16	80000	80000
10	105	87	16	60000	60000
11	10	42	189	3400	246000
12	11	36	12	25605	25605
13	93	33	0	999999	999999
14	35	43	25	1300	32800
15	4	16	33	1300	38400
16	103	30	16	60000	60000
17	4	6	31	1300	52400
18	123	117	29	41900	200000
19	3	216	54	3400	278800
20	82	17	19	1300	172100
21	7	136	37	1300	62900
22	21	2	21	2000	38000
23	7	40	106	26500	200000

Table 4. Comparison of the proposed concurrent SOC test method with the method in [3].

Total number of SOC pins	SOC S1			SOC S2		
	Optimal testing time of the proposed method	Optimum testing time reported in [3]	Time saved (%)	Optimal testing time of the proposed method	Optimum testing time reported in [3]	Time saved (%)
16	487272	497200	2.00	1601312	–	–
24	470152	487940	3.65	1585664	–	–
32	455080	478936	4.98	1575232	2423712	35.01
40	440131	470380	6.43	1559980	2363126	33.99
48	430403	461277	6.69	1549548	2278443	31.99
56	424468	452781	6.25	1528288	–	–
64	419247	443620	5.49	1518108	2202286	31.07
72	380347	435042	12.57	1502412	2174501	30.91
80	365859	426043	14.13	1489067	2149720	30.73
88	351251	417057	15.78	1476236	2123437	30.48
96	271586	408077	33.45	1457062	2099390	30.60

(*Continued on next page.*)

Table 4. (*Continued*).

	SOC S1			SOC S2		
Total number of SOC pins	Optimal testing time of the proposed method	Optimum testing time reported in [3]	Time saved (%)	Optimal testing time of the proposed method	Optimum testing time reported in [3]	Time saved (%)
104	253306	399290	36.56	1441414	2086542	30.92
112	233880	391190	40.21	1425766	2069738	31.11
120	232437	391190	40.58	1410118	2044346	31.02
128	228266	391190	41.65	1394470	2029753	31.30
192	213622	391190	45.39	1004135	–	–
256	91003	391190	76.74	508528	–	–
512	4607	391190	98.82	19647	–	–

Table 5. Comparison of the proposed concurrent SOC test method with the method in [1].

	SOC S1			SOC S2			
Total number of SOC pins	Optimal testing time of the proposed method	Optimum testing time reported in [1]	Time saved (%)	Total number of SOC pins	Optimal testing time of the proposed method	Optimum testing time reported in [1]	Time saved (%)
704	5346	9225	42.05	704	23159	35290	34.38
980	5237	5346	2.04	1118	15357	18789	18.27
1044	5221	5346	2.34	1146	15357	15357	0
1144	5221	5346	2.34	1210	15357	15357	0
1264	5221	5346	2.34	1624	15357	15357	0
1678	5221	5221	0	2296	15357	15357	0

Table 6. Comparison of the proposed concurrent SOC test method with sequential SOC test.

Benchmarks	Total no. of SOC pins	No. of cores	Max no. of cores tested concurrently	Max no. of cores mapped by one pin	1–(Concurrent test time/ Sequential test time) (%)
S1	48	10	7	2	51.15
S2	48	10	5	3	64.02
S3	56	20	10	3	75.64
S4	64	30	11	7	87.10
S5	72	40	11	10	83.88
S6	96	50	14	22	82.55
IND1	205	23	5	10	54.20
IND2	205	23	6	10	72.85
IND3	196	34	6	16	75.82
IND4	196	34	8	18	81.95
Avg.	118	27	8	10	72.92

SOC pins in Colume1 and 5 of Table 5 are taken from [1]. As can be seen, the proposed method can achieve the test application time equal to the lower bound using fewer SOC pins than [1].

Table 6 presents a comparison of the concurrent SOC testing method with sequential SOC testing. The second column is the SOC pins, followed by the number of cores in the SOC. In the next column, the maximum number of cores that can be scheduled to test at the same time is presented. This gives a measure of the concurrency achieved. Column 5 shows the maximum number of cores in different concurrency group that one SOC pin can drive. This is useful information when considering fan-out capability of a design. In the last column, the percentage of test application time saved by concurrent SOC testing compared with sequential SOC testing is presented. On average, when using the proposed concurrent SOC test method, the test application time can be saved by about 73% compared with the test time used by traditional sequential SOC test techniques. Note that an interesting fact is that IND1 and IND2 are exactly the same industry circuits (so are IND3 and IND4), but the core provider offers two different sets of test vectors. The results show that, for different set of test vectors, the benefit of using the proposed concurrent SOC test is different.

5. Conclusions

A method to allocate test resources and schedule test sets is proposed to achieve concurrent SOC test. Our objective is to minimize the test application time so that time-to-market is reduced, while offering structural and functional tests for different TAMs (Test Access Mechanism) under the constraint of peak power consumption. A 2-dimensional bin-packing model is proposed to solve the problem, and a heuristic best-fit algorithm is implemented. Experimental results show that the proposed method produces better results as compared to other available methods.

References

1. D. Bagchi, D.R.Chowdhury, and J. Mukherjee, "A Novel Strategy to Test Core Based Designs," in *VLSI Design 2001*. pp. 122–127.
2. K. Chakrabarty, "Test Scheduling for Core-Based Systems Using Mixed-Integer Linear Programming," in *IEEE TCAD*, vol. 19, pp. 1163–1174, Oct. 2000.
3. K. Chakrabarty, "Design of System-on-a-Chip Test Access Architectures Using Integer Linear Programming," in *VTS 2000*. pp. 127–134.
4. K. Chakrabarty, "Design of System-on-a-Chip Test Access Architectures Under Place-and-Route and Power Constraints," in *DAC 2000*. pp. 432–437.
5. E.G. Coffman, M.R. Garey, D.S. Johnson, and R.E. Tarjan, "Performance Bounds for Level-Oriented Two-Dimensional Packing Algorithms," *SIAM J. Computer*, pp. 808-826, 1980.
6. R. Fenrich, R. Miller, and Q.F. Stout, "Multi-Tiered Algorithms for 2-Dimensional Bin Packing," in *Proc. of Distributed Memory Computing Conference*, 1990.
7. I. Ghosh, S. Dey, and N.K. Jha, "A Fast and Low-Cost Testing Technique for Core-Based System-Chips," *IEEE TCAD*, vol. 19, pp. 863–877, Aug. 2000.
8. I. Ghosh, N.K. Jha, and S. Dey, "A Low Overhead Design for Testability and Test Generation Technique for Core-Based System-on-a-Chip," *IEEE TCAD*, vol. 18, pp. 1661–1676, Nov. 1999.
9. V. Immaneni and S. Raman, "Direct Access Test Scheme—Design of Block and Core Cells for Embedded ASICs," in *ITC 1990*. pp. 488–492.
10. E.J. Marinissen and J. Aerts, "Test Protocol Scheduling for Embedded Core-Based System ICs," in *IEEE International Workshop of Testing Embedded Core-based System-Chips*, 1998.
11. E.J. Marinissen, R. Arendsen, G. Bos, H. Dingemanse, M. Lousberg, and C. Wouters, "A Structured and Scalable Mechanism for Test Access to Embedded Reusable Cores," in *ITC 1998*. pp. 284–292.
12. J. Monzel, "Low Cost Testing of System-On-Chip," in *Proc. Digest of 1st IEEE Internation TECS*, 1997.
13. M. Nourani and C. Papachristou, "An ILP Formulation to Optimize Test Access Mechanism in System-on-Chip Testing," in *ITC 2000*, pp. 902–910.
14. C.P. Ravikumar, G. Chandra, and A. Verma, "Simultaneous Module Selection and Scheduling for Power-Constrained Testing of Core Based Systems," in *VLSI Design 2000*.
15. C.P. Ravikumar, A. Verma, and G. Chandra, "A Polynomial-Time Algorithm for Power Constrained Testing of Core Based Systems," in *ATS 1999*, pp. 107–112.
16. Semiconductor Industry Association, *The National Technology Roadmap for Semiconductors*, Sematech, Inc. pp. 24, 1997.
17. M. Sugihara, H. Date, and H. Yasuura, "A Novel Test Methodology for Core-Based System LSIs and a Testing Time Minimization Problem," in *ITC 1998*, pp. 465–472.
18. N.A. Touba and B. Pouya, "Testing Embedded Cores Using Partial Isolation Rings," in *VTS 1997*, pp. 10–15.
19. P. Varma and S. Bhatia, "A Structured Test Re-Use Methodology for Core-Based System Chips," in *ITC 1998*, pp. 294–302.
20. L. Whetsel, "An IEEE 1149.1 Based Test Access Architecture for ICs with Embedded IP Cores," in *ITC 1997*, pp. 69–78,
21. L. Whetsel, "Addressable Test Ports An Approach to Testing Embedded Cores," in *ITC 1999*, pp. 1055–1064.
22. Y. Zorian, "Test Requirement for Embedded Core-Based Systems and IEEE P1500," in *ITC 1997*, pp. 191-199.
23. Y. Zorian, E.J. Marinissen and S. Dey, "Testing Embedded Core-Based System Chips," *Computer*, vol. 32, no. 6, pp. 52–60, June 1999.

Yu Huang received a B.S. (hons.) degree in Electronics and Information Systems in 1993 and a master's degree in Semiconductor Devices and Physics in 1996 from Nankai University, Tianjin, China, and a Ph.D. degree in Electrical and Computer Engineering in 2002 from the University of Iowa, Iowa City, Iowa, USA.

Prior to joining the Ph.D. program, he worked at Quantum Data Systems, Beijing, as a principal R&D engineer. Dr. Huang is currently working as a member of the scientific staff in the System-On-Chip Test Group within the DFT division at Mentor Graphics. His primary interests are focused on VLSI testing and CAD for VLSI, in particular, SOC testing, RT-Level DFT and BIST. He has published more than 10 technical papers in leading IEEE journals and international conferences. His contribution at the 2001 Asian Test Symposium won him one of the Best Paper Awards and Teruhiko Yamada Memorial Awards.

Wu-Tung Cheng received his B.S. and M.S. degrees in electrical engineering from National Taiwan University in 1978 and 1982, respectively, and his Ph.D. degree in Computer Science from the University of Illinois at Urbana-Champaign in 1985.

From 1985 to 1990, he was with AT&T Bell Laboratories developing GENTEST, a test pattern generation system for general sequential circuits. In GENTEST, he created the BACK test generation algorithm, the SPLIT circuit model and the Differential fault simulation algorithm. GENTEST supported stuck-at fault model and IDDQ test methods.

In 1989 he held a visiting position at the University of Illinois where he improved the Differential fault simulation algorithm with the PROOF fault simulation algorithm.

From 1990 to 1993, he was the Vice President of Engineering and a cofounder of CheckLogic Systems, Inc. (formerly ADAS Software, Inc.) His responsibilities included:

1. Managing the product development team for FastScan (aka ScanCheck), FlexTest (aka CyCheck) and DFTAdvisor (aka DFT-Maker).
2. Developing software tools of FlexTest and other products in the area of test synthesis, netlist knowledge extraction, design for testability, built-in self test, Boolean verification, and logic minimization.

In 1993, CheckLogic merged with Mentor Graphic Corporation. Since then he has had various technical management positions. His current positions are Chief Scientist and System-on-Chip Test Business Unit Manager.

Dr. Cheng serves as a member of program committees of several DFT related workshops and conferences. He is a member of IEEE P1500 standard committee.

Chien-Chung Tsai received his B.S. degree from National Chengchi University, Taiwan, in 1977 and his M.S. degree from Illinois Institute of Technology, Chicago, in 1981. Both are in applied mathematics. Between August 1977 and August 1979, he served in the army of Taiwan, R.O.C. He became a Ph.D. candidate in mathematics at University of Florida, Gainesville, in 1984 and left school without completing the degree. From August 1984 to December 1988, he was with Unisys Corp, St. Paul, Minnesota. From December 1988 to September 1990, he was with Cadence Design Systems, San Jose, California. His work includes areas of test pattern generation, testability analysis and scan design.

He returned to school in September 1990 and studied at the Department of Electrical and Computer Engineering, University of Califronia, Santa Barbara. He received his Ph.D. degree in January, 1996. He has been a develop engineer since 1996 in the Design-For-Test group of Mentor Graphics Corporation, Wilsonville, Oregon. His research interests include logic synthesis, design for test, testability analysis.

Nilanjan Mukherjee received a BTech (hons.) degree in electronics and electrical communication engineering from Indian Institute of Technology, Kharagpur, India, a master's degree in computer science from the University of Guelph, Canada, and a Ph.D. degree in computer engineering in 1996 from McGill University, Montreal, Canada.

Prior to joining the master's program, he worked at TELCO, India, as a systems engineer. Dr. Mukherjee is presently working as a member of the scientific staff in the Advanced Technology Group within the DFT division at Mentor Graphics. His primary interests are focused on developing next generation test methodologies and DFT tools targeting DSM designs. Prior to joining Mentor Graphics, he worked at Lucent Bell Laboratories in Princeton, New Jersey, where he mainly contributed in the areas of Logic BIST, RTL testability analysis, path-delay testing, and on-line testing. He has coauthored more than 20 technical articles in leading IEEE journals and international conferences. His contribution at the 1995 VLSI Test Symposium won him the Best Paper Award. In addition, he is also the coinventor of seven US patents. He is presently involved with the IEEE P1500 committee working towards standardization of a test methodology for reusable cores. He is a member of the IEEE.

Omer Samman received a B.S.E.E. degree in electrical and computer engineering from the University of Portland, Portland, Oregon, in 1985; an M.S.E.E degree in electrical and computer engineering from the University of Portland, Portland, Oregon in 1987; and was a Ph.D. candidate in the department of electrical and computer engineering at Oregon State University from 1987 to 1989.

Prior to joining the Design-for-Test group at Mentor Graphics Corp. in 1993, he worked at Oce-Oregon Inc., Beaverton, Oregon as a software development engineer, designing and implementing embedded systems software and firmware. Mr. Samman is presently working as a member of the scientific staff in the SoC Group within the DFT division at Mentor Graphics Corp. His primary interests are focused on developing next generation test methodologies and DFT tools. His other interests span many areas of EDA, from physical design to formal verification as well as disciplined software engineering practices. He is a member of the IEEE.

Yahya Zaidan received a B.S. degree in computer science from Yarmouk University, Jordan, M.S. degree in information and computer science from King Fahd University of Petroleum & Minerals, Saudi Arabia.

Presently he is working as software development engineer in System-On-Chip test Group within DFT division at Mentor Graphics. Prior to joining Mentor Graphics, he worked at King Fahd University as a Lecture in Information and Computer Science department. His primary interest is focused on developing DFT tools to test System-On-Chip designs. He has co-authored several technical articles. In addition, he is also the coinventor of some US patents. He is a member of the IEEE.

Sudhakar M. Reddy obtained his undergraduate degree in Electrical and Communication Engineering from Osmania University, M.S. degree from Indian Institute of Science, and Ph.D. degree in Electrical Engineering from the University of Iowa, Iowa City, Iowa.

Dr. Reddy has been active in the areas of testable designs and test generation for logic circuits since 1972. He has an associate editor and twice a guest editor of IEEE Transactions on Computers. He was Program Committee Chair for 1989 International Symposium on Fault-tolerant Computing.

Since 1968 he has been a member of the faculty of the Department of Electrical and Computer Engineering, University of Iowa, where he is currently a professor. In 1990 he was made a University of Iowa Foundation Distinguished Professor. In 1995 he received the Von Humboldt Senior Researcher Fellowship. In January 2000 he received the First Life Time Achievement Award from VLSI Design Conference. He is currently an associate editor of IEEE Transactions on CAD. He is a Fellow of IEEE.

JOURNAL OF ELECTRONIC TESTING: Theory and Applications 18, 415–434, 2002

A Novel Reconfigurable Wrapper for Testing of Embedded Core-Based SOCs and its Associated Scheduling Algorithm

SANDEEP KORANNE*
Tanner Research, Inc., 2650 E. Foothill Bl., Pasadena, CA 91107, USA
sandeep.koranne@tanner.com

Received August 28, 2001; Revised January 3, 2002

Editor: Krishnendu Chakrabarty

Abstract. In this paper a mathematical formulation and an efficient solution, of the embedded core-based system-on-chip (SOC) test scheduling problem (ECTSP) is presented. The ECTSP can be stated as follows; given a chip with N_C cores each having a test T_i; where T_i takes time $\mathcal{F}_T(T_i, w_j)$ to execute on a test access mechanism (TAM) of width w_j, and a constraint W on the number of top-level test pins; calculate the TAM assignment vector π and the schedule Σ for each test T_i, such that the completion time of the full chip test is minimized. All existing approaches have solved the ECTSP by solving the TAM partition and scheduling problem sequentially. In this paper we present an unified approach to solve the ECTSP. We present the first report of a design of reconfigurable core wrapper which allows for a dynamic change in the width of the test access mechanism (TAM) executing a core test. An automatic procedure for the creation of DfT hardware required for reconfiguration using a graph theoretic representation of core wrappers is also presented. For the case of reconfigurable wrappers, efficient algorithms to compute the schedule are presented based upon some recent results in the field of malleable task scheduling. Cases in which the degree of reconfigurability are constrained are considered; the case when only a single core can have reconfigurable wrapper, a schedule with zero TAM idle time can be found in time $O(N_C(N_C + W)\lg W)$, and the case when only 2 different wrapper configurations are allowed can be solved in time $O({N_C}^3)$. Comparison with existing results on benchmark SOCs show that our algorithms outperform state-of-art ILP formulations not only in schedule makespan, but also significantly reduce computation time.

Keywords: embedded core based test scheduling, system-on-chip test, reconfigurable wrapper, parallel scheduling of malleable tasks, VLSI test

1. Introduction

The widening design productivity gap between VLSI system capabilities and design engineering capability, in a limited time to market scenario, has prompted many design groups to adopt a policy of design reuse at the core level [21]. Typical cores include CPUs like MIPS and ARM, network controllers, embedded memories, DSP cores and associated peripherals like the IEEE 1394 Firewire and UARTs. As has been noted in [39], reusability of design alone is not sufficient as the verification and test generation efforts now dominate the typical design time. Reusability of tests is crucial for reducing total design time. This raises the problem of test knowledge transfer and physical test application. The proposed IEEE P1500 (SECT) [15, 38] standard provides facilities for test knowledge transfer using Core Test Language (CTL) [20] and has suggested various core wrapper designs for providing test access to

*This research was conducted when the author was with ED&T/Test, Philips Research Laboratories, Eindhoven, The Netherlands.

embedded cores. In this paper we focus on a related problem, i.e., optimization of TAM in conjunction with efficient scheduling of tests on system level. The complete problem of core test application can be divided into three sub-problems:

1. *Top level TAM partitioning*: as the number of test pins available at the IC level is constrained, an optimal partition of these test bits must be done so as to reduce the total test cost. Test cost in this paper refers to the test application time on the automatic test equipment (ATE). The test application model assumed in this paper is based on a TESTBUS model which is an extension of the model proposed in [26]. In that paper Marinissen et al. propose a structured and scalable method of test access to embedded cores. Their design, the TESTSHELL, separates the problem of test access, which is done through test access mechanism (TAM) from test application to individual scan chains, which is done by the *wrapper*. The same approach is advocated in the upcoming IEEE P1500 [28]. In this paper we propose to design reconfigurable wrappers that can dynamically change the wrapper structure using test control mechanisms (TCMs), such that it is possible to assign a core varying number of TAM bits over the course of the complete SOC test,
2. *Test port partitioning at core level*: the second problem is scan chain (or test ports) partitioning at the core level; given a particular width of the TAM, how should the core level scan chains be connected so as to reduce the length of the longest scan chain for that core,
3. *Test scheduling*: given a set of tests and the test resources like TAM (many of which are shared between many cores), to determine the TAM bit assignment to execute the tests of each core, so as to reduce the total SOC test time.

These problems are referred to in literature as (i) the TAM design problem [6], (ii) the wrapper design problem [27] and (iii) the test scheduling problem [5], respectively. There is a fundamental difference between these problems. Problems (i) and (iii) are encountered by the SOC integrator, when the core are delivered containing the pre-computed tests. Problem (ii) has to be solved prior to the core delivery as the wrapper itself is part of the core and the tests have to be generated keeping in mind the scan chain structure. The SOC integrator would not like to modify the core as that can lead to substantial test (re-)generation time for large cores. By definition SOC integration is on the critical path of any SOC design cycle and hence a practical solution would have to be not only efficient in terms of the quality of the SOC test, but also not consume inordinate amount of time that late in the product realization phase. A sequential solution to these problems may be sub-optimal. In this paper we propose an unified approach that simultaneously solves problem (i), (ii) and (iii).

Notation: Throughout in this paper we refer to the top-level TAM width by W, the number of TAM partitions by B, the width of the TAM assigned to a core by w_i, and the number of cores by N_C. Each core has a test T_i. Test T_i is characterised by its number of patterns p_i and by the bitwidth[1] of the core, ϕ_i. The schedule is denoted by Σ and the TAM assignment vector by π. The schedule contains tuples of the form $(T_i, w_i, t_1 \leftrightarrow t_2)$ to denote the width of the TAM responsible for executing test T_i from time instance t_1 upto time instance t_2. The length of the schedule (makespan) is denoted $|\Sigma|$. The assignment vector π is defined as $\pi = \{w_i\} : \sum_{i=1}^{B} w_i \leq W$. The major technical contributions of this paper are:

Design of reconfigurable core wrappers: All previous work in the area of core wrapper design has focused on static TAM partitions i.e., the number of TAM bits (the TAM width) assigned to each core are not allowed to be changed between core tests. In this paper we present the first report of designing reconfigurable core wrappers with little area and delay overhead. Using our design the width of the TAM connected to a core is allowed to change dynamically using test control blocks. The formation of balanced internal and surround scan chain for more than one TAM width by adding DfT hardware is explained. An automated procedure for arriving at this DfT hardware using graph theoretical representation is also described,

Efficient algorithms for ECTSP: The relation between the number of test bits p_i, bitwidth ϕ_i of a test T_i and TAM width w_i w.r.t. test application time is investigated in Section 4. This relation is denoted by the function $\mathcal{F}_T$. Associated with each test is the time it takes to execute on a single bit wire, and the *speedup* $g(k)$ that can be obtained by executing it on a k-bit wide wire.

For convex $g(k)$ a $O(N_C)$ time optimum algorithm for minimizing the makespan is obtained. For a SOC with W such that the function $g(k)$ is convex for all cores, it is shown that a multiplexed TAM design is optimal for core-internal test. For concave $g(k)$ a reconfigurable wrapper design is proposed to derive efficient schedules based on the recent work of Błażewicz et al. [3] in the area of malleable task scheduling. We prove the complexity of this problem and show that it is harder than $\mathcal{NP}$-hard. For the constrained reconfiguration case when only one core in the SOC can have reconfigurability, a $O(N_C(N_C + W)\lg W)$ algorithm is presented which can compute near optimal schedules with zero idle time on TAM wires. To limit the overhead of reconfiguration another constrained schedule is also considered, in which the degree of reconfigurability of every core in the system is at most two, and no core is allowed to be reconfigured more than two times. A schedule for this problem can be computed in $O({N_C}^3)$. We show that although the theoretical complexity of scheduling with reconfigurable core wrappers is even worse than $\mathcal{NP}$-hard, approximate schedules to reconfigurable core wrapper SOC scheduling are better than static wrapper schedules.

The rest of the paper is organized as follows. In the next Section we survey the existing work in the area of core-based SOC test scheduling. In Section 3 we give an overview of the existing literature in the field of TAM and wrapper design. We also present the design of our *reconfigurable wrapper*. In Section 3.2 we explain an automatic procedure for creating the necessary DfT hardware for core access reconfiguration. The process of adding reconfiguration capabilities to existing P1500 style core wrappers is addressed in Section 3.3. A classification of existing scheduling formulations of ECTSP and their rectangle duals is presented in Section 4. We describe our scheduling algorithm for SOCs consisting of cores having reconfigurable wrappers in Section 5 and the algorithm for constrained reconfiguration scheduling in Section 6. We present details of the implementation and comparison with other algorithms on benchmark SOCs in Section 7. The experiments demonstrate that the proposed method outperforms state-of-are ILP formulations. We conclude in Section 8.

2. Previous Art on Test Scheduling and Wrapper Design

Recent work in the area of test scheduling can be found in [5, 6, 8, 10, 14, 17, 18, 19, 22, 24, 31, 33, 35, 37]. See Table 1. We present a brief overview of the

Table 1. Scheduling formulations for ECTSP (Notation: $Output_{Input}$).

Problem	Ref.	Comment
$\Sigma_{\pi BW}$	[4]	Number of partitions B and the partition π are given. Equivalent to $R\|\|C_{\max}$
$\pi\, \Sigma'_{BW}$	[22]	B is given, π is calculated for minimizing average completion time ($R\|\|\Sigma C_j$)
$\Sigma_{\pi w_j}$	[6]	TAMs are allowed to fork and merge
$\pi\, \Sigma_{BW}$	[18]	The number of partitions B is given, partition π is calculated
$\pi\, B \Sigma_W$	[18]	W is given, and the optimal B and π is calculated
$\Sigma^{Power}_{\pi BW}$	[24]	Scheduling with power constraints
$\Sigma^{Conflicts}_{\pi BW}$	[31]	Scheduling with test conflicts
$\Sigma^{Precedence}_{\pi BW}$	[17]	Scheduling with precedence constraints
$\Sigma_{\pi BW}$	[23]	B and π are given. Solved using network-flow algorithm
Σ_W	[19]	Rectangle packing method, non-partition
$\Sigma_{W\text{recon}}$	-Proposed-	Non-partition, requires reconfigurable wrappers

integer linear programming (ILP) approaches as these have received the most attention in recent years. In [5] Chakrabarty has analyzed the test scheduling problem by reducing the test scheduling problem to open-shop scheduling with m-processors. In his model a processor is a test resource like a test bus or BIST. Each test is treated as a job having two operations corresponding to externally applied tests (like scan tests) and BIST components, respectively. The *finish time* of the schedule is the latest completion time of the individual processor schedule. The objective is to minimize the finish time. The length of the job is taken as the amount of time for the test to execute. Only the case for uniform processors has been analyzed and solutions using Mixed Integer Linear Programming (MILP) techniques have been proposed. Recent improvements to this formulations have been presented by Iyengar et al. in [18]. The authors have proposed ILP based methods to solve the TAM and wrapper co-optimization problem (the formulation is still sequential, with the partition and scheduling steps combined in a single ILP). The authors have given a ILP formulation of the ECTSP for a fixed partition ($\Sigma_{\pi BW}$). This problem is same as the well known multiprocessor scheduling problem $R\|C_{\max}$ [25]. Then the approach is extended for variable width partitions ($\pi\, \Sigma_{BW}$) using ILP. This problem is computationally hard even for small instance size and in [18] the ILP solver was halted after 180 minutes. To speed up the computation, all partitions of W for small values of B are explicitly enumerated and evaluated by solving the resulting $\Sigma_{\pi BW}$ problem. This is clearly an exponential enumeration and hence is not practical for large B. To solve the general problem of $\pi B \Sigma_W$ (which is same as partition based ECTSP), the authors propose to enumerate all values of $1 \leq B \leq B_{\max}$, where $B_{\max}$ is experimentally determined. Their work is closest to the current work being presented and comparative analysis of the schedules is done in Section 7.

Precedence and power constrained scheduling has been analyzed in [8, 10, 17, 24, 31, 33, 37]. In this paper we do not discuss precedence constrained or power constrained scheduling, but these are directions of future research for us. Algorithmic approaches to ECTSP include combinatorial optimization based methods of Sugihara et al. [35], bipartite matching based methods of the author [22] and network flow based formulation of the author and Choudhary in [23]. All the above approaches have focused on *partition* based schedules; i.e., first the top level test pin width is divided into a number of candidate TAMs, and the core tests are scheduled on these partitioned TAMs. In the sequel of this paper we shall also discuss another solution class which is not based on partitions. A very recent development in test scheduling which does not use partitioning of top level TAM, has been using geometric packing dual, as presented by Huang et al. in [14]. An implementation of using rectangle packing for SOC test scheduling with encouraging results is presented by Iyengar et al. in [19]. These papers use methods similar to the ones explained in Section 4. To make the paper self contained we shall next discuss some of the TAM architectures that have been proposed in the literature, and explain the TAM model used in this paper. In particular we explain the term *reconfigurable wrapper*.

3. Design of Reconfigurable Core Wrappers

Recent years has seen a lot of activity in the area of wrapper and test access architecture design (cf. [4, 7, 11, 16, 26, 32, 36, 39]). A thorough analysis of the wrapper design problem in the context of scan chain partitioning is presented in [27]. As can be seen in the above references, solutions to the TAM design problem are built around two styles; (i) test-bus based architectures [7, 11, 16, 26, 36, 39], and (ii) transparent path based methods [11, 32]. In this paper a TESTBUS based TAM architecture is assumed, as its relatively higher DFT cost is justified for large SOCs due to its ease of in-

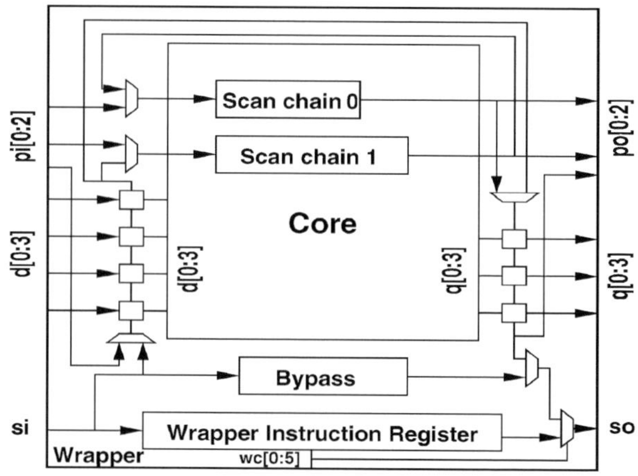

Fig. 1. Parallel internal test mode for P1500 style wrapper.

tegration and support for testmodes like core-internal test, interconnect logic test etc. An overview of the TAM models used for embedded core test scheduling is given in [18]. For the general ECTSP problem it is assumed in this paper that every core has a wrapper.

All previous approaches have designed test wrappers around cores assuming a static width of TAM. The proposed IEEE P1500 standard proposes a core wrapper design that is designed with the width of the external TAM fixed at a prior known value. An example is shown in Fig. 1. This wrapper design allows parallel loading of test patterns using the w_i TAM bits ($w_i = 3$ in Fig. 1), or the wrapper can be used in a serial manner. This is clearly sub-optimal, as a core can have a variety of tests. Each of tests will have varying bitwidth requirement. As an example a BIST test typically requires application only few parallel patterns from the TAM, and hence has a low bitwidth requirement. Compare this to scan test which has a high bitwidth requirement. Assigning a core a wider than required TAM can increase the SOC test time, as some other core could have benefitted from a wider TAM. It is more efficient that the wrapper can adjust to the bitwidth requirement dynamically; thus different tests of the same core can be executed on different width TAMs. When a core wrapper is designed in such a way that it admits the formation of balanced internal and surround scan chains for more than one TAM width, then the wrapper is said to be *reconfigurable*. In the next Section we describe our proposed design of reconfigurable wrappers.

The number of different TAM widths which can be admitted is denoted the *degree* of reconfigurability. Any core wrapper can be trivially connected to any TAM width, but the scan chains are not necessarily balanced.

Core A is said to have reconfiguration on TAM width w_1 and w_2 *iff* the following equations are satisfied:

$$L^{w_1}_{\max}(A) = \min\ L^{x}_{\max}(A) : \forall x \leq w_1$$

and

$$L^{w_2}_{\max}(A) = \min\ L^{y}_{\max}(A) : \forall y \leq w_2$$

where $L^{w_i}_{\max}(A)$ denotes the length of the longest scan chain of A when tested with a TAM of width w_i. In the above case core A would have a degree of reconfigurability equals two. Dynamic reconfiguration is the process by which a core-wrapper is reconfigured by special DFT hardware to admit a different TAM width, while still maintaining balanced scan chains.

3.1. *Graph Representation of Core Wrappers*

We propose to design reconfigurable core test wrappers using multiplexers at the input and output of each reconfigurable scan chain. The set of reconfigurable scan chains can be calculated using graph theoretic methods as explained below. We first describe the proposed method with the help of an example. Consider a core A which has a total of 6 scan chains of length {12, 5, 17, 9, 14, 4} respectively. We calculated partitions of these scan chains for some TAM widths. These partitions are shown in Table 2. The objective function used to compute these partitions is to minimize the length of the longest scan chain, i.e., min $L^{w_i}_{\max}(A)$.

The problem of calculating optimal scan chain partitions (PSC) was shown to be $\mathcal{NP}$-complete in [27] by a reduction from multiprocessor scheduling with uniform processors ($P\|C_{\max}$) [12]. In [27] a Best Fit Decreasing [9] heuristic was used to compute good approximate partitions. The reduction from $P\|C_{\max}$ to PSC, although theoretically correct, makes the scan chain problem seem unduly hard. This is because for the theoretical analysis of the multiprocessor scheduling problem($P\|C_{\max}$) the input set (which is the length of the job) is allowed to be an arbitrary real number between 0 and 1. When translating scan chain partitioning to multiprocessor scheduling, the length of the scan chain is treated as the length of the job. For scan chain partitioning we know that the minimum value is 1 and the maximum is bounded from above because of engineering reasons. Hence, $1 \leq l(S_i) \leq l_{\max}$, where $l_{\max}$ is typically a small number, e.g., 200. Secondly, the distribution of lengths is not random but follows a pattern. Consider Core 6, the largest core in the SOC p93791 referred in [18]; Core 6 has a total of 46 scan chains distributed as 7 scan chains of length 500, 30 scan chains of 520, and 9 chains of length 521. These classes of instances in which the number of scan chain lengths is bounded (independently of the number of scan chains) can be theoretically solved optimally in polynomial time [13]. This observation also explains the reported success of heuristics (like Best Fit Decreasing) for the scan chain partitioning problem.

Table 2. Scan chain partitions of core A.

No. of TAM bits	Partition of scan chain A	Max Len A $L^{w_i}_{\max}(A)$
1	[12, 5, 17, 9, 14, 4]	$L^{1}_{\max}(A) = 61$
2	[(12, 5, 14), (17, 9, 4)]	$L^{2}_{\max}(A) = 31$
3	[(12, 9), (5, 14), (17, 4)]	$L^{3}_{\max}(A) = 21$
4	[(17), (12, 4), (5, 9), (14)]	$L^{4}_{\max}(A) = 17$
5	[(17), (14), (9, 4), (12), (5)]	$L^{5}_{\max}(A) = 17$

We can also represent each of these partitions by means of directed graphs. We define $G = (V \cup I, E)$ to be a directed graph. For each scan chain S_i we denote a vertex i in G with label $l(S_i)$, where $l(S_i)$ is the length of the scan chain. I denotes the input/output TAM for the core and is assumed to have infinite length. Hence, $l(I) = \infty$. Two vertices are said to be *comparable iff* they are in the same partition. The special vertex I is assumed to be present in every partition. Then, the labels of the vertices which are comparable induce a partial order on $V \cup I$. The vertex set of G forms a partially ordered set (poset). A pairwise comparable subset of a poset is called a *chain*. A path from I to any other vertex forms a chain. We add an edge from a vertex $u \rightarrow v$ in G such that $\forall u, v \in V : l(u) \geq l(v)$ and $P(u) = P(v)$, i.e., u precedes v in the chain, where $P(i)$ denotes the partition number of vertex i. Informally, we add an edge between two vertices *iff* they are in the same partition and the direction of the edge is from the vertex representing the longer scan chain to the next shorter one in the poset. For each TAM width instance we create such a graph. Let the graph for TAM width i be G_i. Example graphs for core A with TAM width 1, 2, 3, and 4 are shown in Fig. 2. The shaded vertices are the scan chains that are also to be connected to the output TAM. The control signals for the output

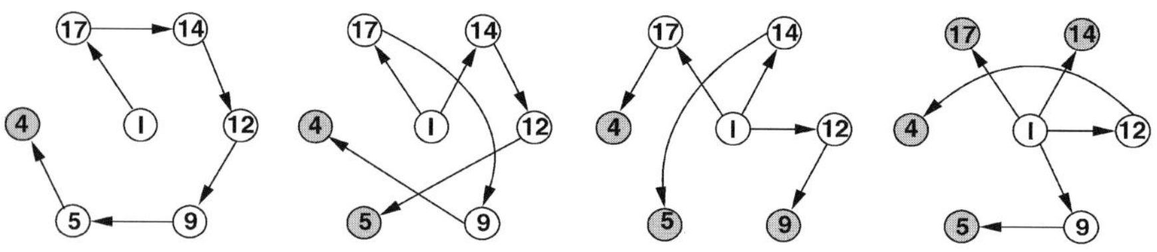

(a) G_1 (b) G_2 (c) G_3 (d) G_4

Fig. 2. Directed graph representation of G_1, G_2, G_3 and G_4.

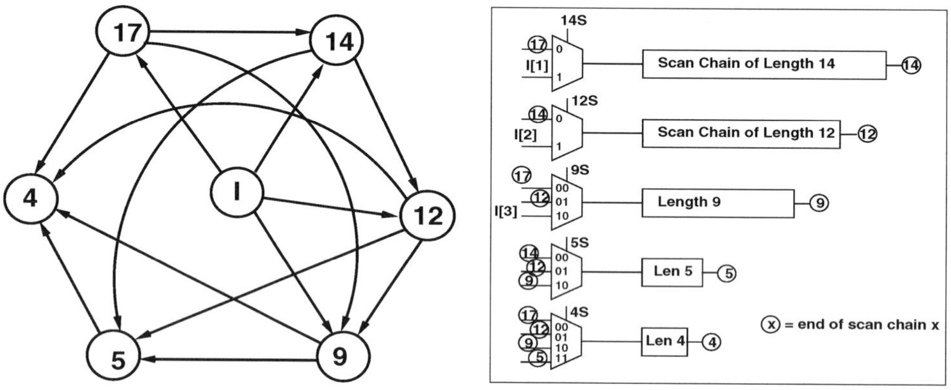

(a) G_A^* (b) Mux. implementation

Fig. 3. Combined directed graph and its multiplex implementation.

multiplexers are not calculated here, but it is easily seen that the shortest scan chain does not need to be multiplexed.

We form the combined directed graph representation for a core by taking the union of all directed graphs of the core. Formally, $G^* = G_1 \cup G_2 \ldots G_{\phi_i}$. For core A $G_A^* = G_1 \cup G_2 \cup G_3 \cup G_4$. The graph is shown in Fig. 3(a).

The combined directed graph representation captures the partitioning requirement per core. The indegree of any vertex i in G^* represents the number of signals that must be multiplexed into that scan chain. We can make some general observation about G^*. The longer a scan chain (or higher the label of the vertex) the less it needs to be multiplexed (muxed). Secondly, the longest chain of the core need not be muxed at all. This statement is easily provable as we can see that in any graph G_k for $1 \leq k \leq \phi_i$, there always exists one chain (vertex) which is connected only to I. This is the longest scan chain. In the next section we describe a method of automatically creating reconfigurable TAM partitions.

3.2. *Automatic Derivation of Reconfigurable TAM Structures*

Given the combined directed graph of a core G^* we can derive the scan chain access structure as follows. We design our architecture using multiplexers to switch the chain configuration. We observe that the number of select signals for each scan chain i is $\lceil \lg \text{indegree}(i) \rceil$. We take core A as a didactic example to explain our method. Referring to Fig. 3(a) we see that the number of mux select signals for each of the 6 scan chains is $\{17, 0\}, \{14, 1\}, \{12, 1\}, \{9, 2\}, \{5, 2\}, \{4, 2\}$, where the data should be read as {length of scan chain, no. of signals}. The muxing strategy for core A is shown in Fig. 3(b). Each scan chain vertex i which has indegree more than one is multiplexed. The multiplexer is controlled by select signals and the driving signals are precisely the neighbors of i. In the worst case the indegree of a vertex can be ϕ_i (the number of TAM configurations), but this is rare as for this to happen the vertex must be assigned to a different set of chains for each of the ϕ_i configurations. Also using the proposed

Table 3. Select signals for core A.

TAM width (x_2x_1)	14S	12S	9S	5S	4S
1(00)	0	0	01	10	11
2(01)	1	0	00	01	10
3(10)	1	1	01	00	00
4(11)	1	1	10	10	01

scheduling algorithm ECTSPSOL, ϕ_i is always equal to 2 on the average.

The control logic for the select signals can be derived from the directed graph representation G_1, G_2, G_3 and G_4. In Table 3 we show the control signals for core A. To read the table, note that when the core is assigned 2 bit wide TAM, the optimal scan chain configuration is having 2 scan chains of $\{12, 5, 14\}$, and $\{17, 9, 4\}$. This corresponds to setting the select signals of the multiplexers as shown below.

A simple synthesis procedure yields the control signals as a function of w_i. By implementing this control logic, we can dynamically restructure the scan chains for any w_i optimally. As an example if the scheduler at SOC calls for a 3 bit wide TAM for core A, we can set $x_1 = 0, x_2 = 1$ and the corresponding optimum scan chain structure is formed. The scan chain of length 17 (the longest scan chain) is not multiplexed at all and is always connected to bit 0 of the TAM. The control bits can be programmed per *test-mode* by the scheduler algorithm. The hardware implementation of the programmed control bits can be done using Test Control Mechanisms (TCM). TCMs provide a flexible and robust way of providing control for test to embedded cores. TCMs are described in detail in [26].

Note: It is to be remembered that we do not need to regenerate the patterns for all these configurations as the test assembler tool which optimizes scan patterns, can calculate which configuration is active for a particular TAM width. Test pattern reordering with reconfiguration is an interesting problem which has been solved before using the concept of *test protocols*, which are separate from *test patterns*. As the Core Test Language also seems to be moving in this direction, more tools will support this automatically. One such test plan generator is described in [29] and [30]. The tool uses the concept of *macro test* and was used successfully to expand the core level test patterns to top level with reconfigurable core wrappers.

3.3. Adding Reconfigurable Wrapper Cells to a P1500 Style Wrapper

The scan chain reconfiguration method is general and can be used with any scan chain partitioning and balancing method. Scan partitions created using clock domain analysis tools can also be reconfigured using the proposed directed graph approach. Reconfiguration capability can be added to existing P1500 style wrapper cells that are described in [26] (and elsewhere) as shown in Fig. 4(a). In the figure a conventional P1500 input wrapper cell is augmented with a reconfigurable mux connected to the scan multiplexer. The notation ⓧ denotes the end of scan chain x, and these multiplexer connections are computed using the directed graph method explained above. As can be seen from

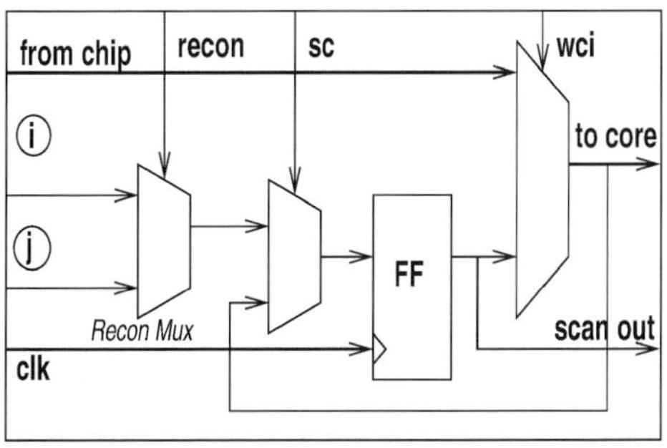

(a) P1500 Wrapper Input cell with Reconfiguration

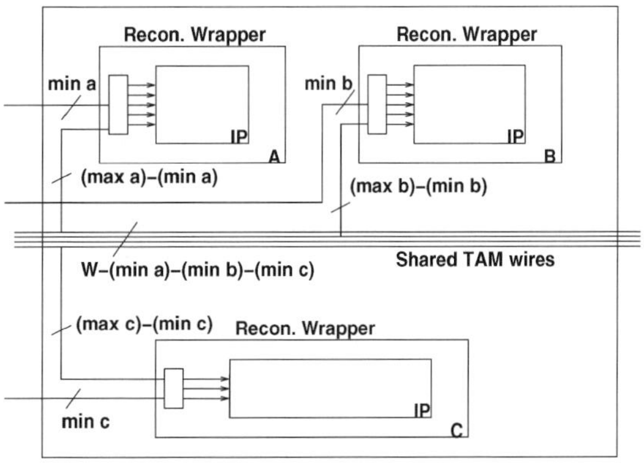

(b) SOC level TAM architecture

Fig. 4. Adding reconfiguration capabilities to existing P1500 style wrappers.

the figure the reconfigurable multiplexer is not on the functional path of the input signal and hence will not cause performance degradation. The sizes of the multiplexers depend only on the degree of reconfigurability of the core, and to limit this overhead, in Section 4 we present a constrained reconfiguration algorithm called ECTSPSOL which restricts the average degree of reconfiguration to be 2, hence every reconfiguration multiplexer will only have 2 data inputs and 1 select inputs, which will simplify the control logic needed for decoding the select signals. Also, all reconfigurable multiplexers will have the same size and hence not cause any timing problems. Since, the reconfiguration is done once at the start of a cluster of test pattern, timing issues with the select signals will not arise.

The TAM architecture assumed in this paper is shown in Fig. 4(b) (for brevity only TAM connections are drawn). The top-level TAM width is W. Let the set of TAM widths admitted by the cores A, B and C, be a, b and c, resp. Using ECTSPSOL we have $|a| = |b| = |c| = 2$. Let $\max\{a\}$ ($\min\{a\}$) denote the width of the maximum (minimum) TAM configuration. Each core is connected to chip level test pins by individual TAMs. In case the core wrapper is reconfigurable, the width of the individual physical TAM wire (the wire as seen in the layout) is the minimum of all TAM widths that are admitted by the reconfigurability ($\min\{a\}$). The remainder of the W TAM bits ($W - \min\{a\} - \min\{b\} - \min\{c\}$) are *shared TAM resources*. The shared TAM resources are the only wires that are dynamically controlled during reconfiguration. The width of the shared TAM to a core is simply the difference between the maximum and the minimum, $\max\{a\} - \min\{a\}$. These details are shown in Fig. 4(b). Hence, using reconfigurable wrappers gives us the ability to distinguish between physical TAM wires, and the use thereof in a test schedule. Previous approaches to solving the ECTSP have not exploited this degree of freedom in the scheduling formulation. In the sequel of this paper in Section 4 we propose our scheduling algorithm ECTSPSOL which exploits this dynamic property of TAM widths to arrive at better schedules. Since ECTSPSOL restricts the degree of reconfiguration to 2, only a single bit of control per core is needed at top level.

In this section we have seen how addition of simple DfT hardware allows for dynamic change in the number of TAM bits assigned to a core between tests. This property introduces an additional degree of freedom in the scheduling algorithm. In the next section we shall introduce our proposed scheduling algorithm for reconfigurable core test and analyze it. The algorithm (ECTSPSOL) is based upon some recent ideas developed in the field of parallel malleable task scheduling. We shall also define the ECTSP problem, its mathematical formulation, and the proposed algorithms in detail.

4. Algorithms for ECTSP

Above we have seen how addition of simple DfT hardware allows us to assign different TAM widths to the same core at different instants in the schedule. Using this property, ECTSP, which was hereto thought of as a static job scheduling problem can be modeled as a *parallel malleable task scheduling* problem, where it is allowed to change the resource set of a job dynamically. In this section we describe the scheduling algorithm which is inspired by the recent work done in the area of malleable task scheduling (MPTS) by Błażewicz et al. [3]. Although the notation and method of analysis is based on [3], there are important differences between MPTS and ECTSP ($\Sigma_{W\text{recon}}$) as explained in Section 5.2.

4.1. Notation and Preliminaries

Let us denote the test time calculator function as $\mathcal{F}_T(T_i, w_i)$, where i is a test and w_i is the width of the TAM assigned to execute this test. For scan tests $\mathcal{F}_T$ depends on the number of scan patterns and the length of the longest scan chain in the core. From [1] we can derive $\mathcal{F}_T(T_i, w_i) = L^{w_i}_{\max}(i)(p_i + 1) + p_i$; where $L^{w_i}_{\max}(i)$ denotes the length of the longest scan chain in core i when a TAM of width w_i is used to execute the test, and p_i denotes the number of patterns. $L^{w_i}_{\max}(i)$ is a monotonic non-increasing function w.r.t. w_i. An example of $L^{w_i}_{\max}(A)$ for core A was discussed above and is shown in Table 2.

With each core i we associate a parameter termed *bitwidth requirement* ϕ_i defined as follows:

$$\phi_i = k \text{ when } \mathcal{F}_T(T_i, k) \leq \mathcal{F}_T(T_i, k + x) \quad \forall x > 0 \quad (1)$$

Informally, ϕ_i is the TAM width value over which there is no decrease in the test time for test i. The test time calculator function can then be written as $\mathcal{F}_T(p_i, \phi_i, w_i)$, to represent the test time for a test i when executed on a TAM of width w_i such that $w_i \in [1 : \phi_i]$. We can make some observation about $\mathcal{F}_T$ as follows; due to the

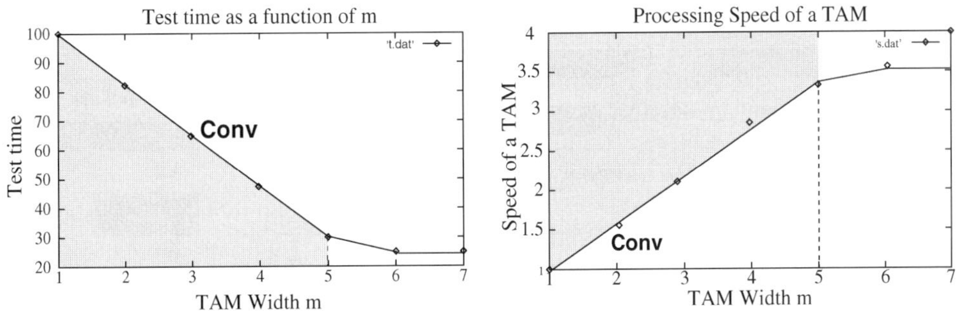

(a) Testing Time Vs TAM width

(b) Speedup Vs TAM width

Fig. 5. Testing time $\mathcal{F}_T(T_i, w_j)$ and speedup $g(k)$ functions.

inherent structure of the problem $\mathcal{F}_T$ is defined only at points where $p_i, \phi_i, w_i \in \mathbb{Z}^+$. This implies that the plot of test time for a core versus TAM width is a piecewise linear function. An example is shown in Fig. 5(a). The x-axis represents the TAM width assigned to the core. The y-axis represents the test time for a test. As can be seen the test time is a monotonic non-increasing function of the TAM width. When the TAM width becomes the bitwidth for that core, subsequent additions to the TAM width do not contribute to a reduction in test time. In the figure the bitwidth of the core is a TAM width of 6 bits. Adding an extra bit to make the width 7 does not reduce the test time. We also plot the processing speed function (or speedup) for a TAM in Fig. 5(b). Let us denote this speed function for a test i with a TAM width m by $g_i(m) \geq 0$. If m bits are assigned to execute test i then the test executes in time $\frac{\mathcal{F}_T(T_i,1)}{g_i(m)}$, where $g_i(m)$ is a non-decreasing *processing speed function* defined for $m \in 0, 1, \ldots, \phi_i, g_i(0) = 0$. The function is convex upto $m = 5$, the region is marked *Conv* in the figure. Informally, a region is convex *iff* the line joining any two points in the region is completely contained within the region. The convex region is a special case of the ECTSP, as discussed below.

For each test, a schedule specifies the time intervals within which this test is executed and the width of the TAM assigned to it. The objective is to find a schedule which satisfies the constraints that each TAM can service one core at one time, the total width of TAM is bounded by a constant W at top level, and the schedule has minimum makespan $C_{\max}$. Denote the minimum $C_{\max}$ value by $C^*_{\max}$. Our problem formulation is different from the well studied scheduling problem for unrelated parallel machine multiprocessor scheduling $(R\|C_{\max})$ [25]. In $R\|C_{\max}$ the processor set is well defined, and the problem is one of assignment. In the SOC test scheduling problem, there is a bound on the TAM width at top level, but neither the number, nor the processing speed of the TAM is fixed. In this sense SOC test scheduling can be thought of as a *generalized multiprocessor scheduling*.

In [3] the authors have investigated the problem for continuous resource allocation (P-CNTN) and discrete resource allocation (P-DSCR) with convex and non-convex processing speed functions for minimizing the makespan. They provide optimum schedules for P-CNTN and P-DSCR problems under convex processing speed functions in $O(n)$ time, where n is the number of tasks, and m is the number of resources. For P-CNTN problems with non-convex speed functions they have given a $O(n \max\{m, n \lg^2 m\})$ time algorithm. For P-DSCR with non-convex speed functions they analyze the case for $n = 2, 3$ and provide optimum schedule algorithms. In this paper the idea of discrete resource allocation is extended to test scheduling. The model in [3] is one of constant work, i.e., the work done on a job is constant. For the test scheduling formulation the number of patterns p_i for a test is constant and hence the model can be used. But we have to keep in mind that for ECTSP the cost of reconfiguration is high as most scan pattern generators optimize scan patterns for pipelining; which makes the reconfiguration operation consume non-zero time.

4.2. *Classification of Scheduling Formulations*

As an pictorial representation of the tests, each test i can be represented by a rectangle of width w and length $\mathcal{F}_T(p_i, \phi_i, w)$. By varying w in $[1 : \phi_i]$ a set of rectangles for each test is obtained. This is shown in Fig. 6(b). The width of the rectangle(w) is the number

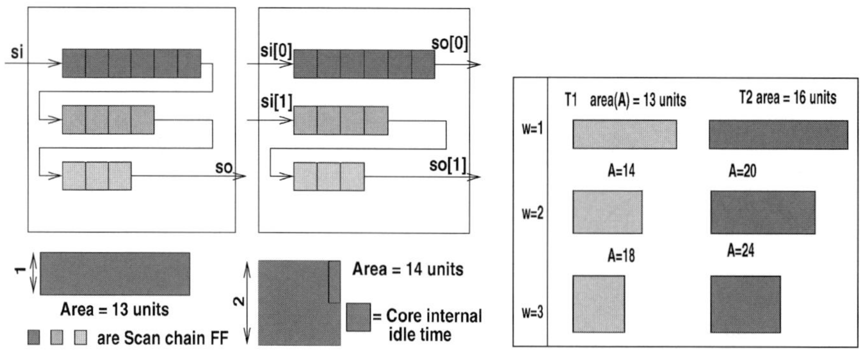

(a) Core idle time (b) Rectangle dual

Fig. 6. Representation of tests by (almost) constant area rectangles.

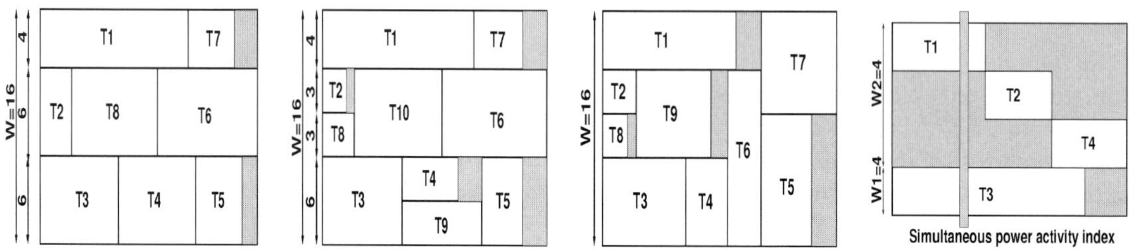

(a) $\Sigma_{\pi W}$ (b) $\Sigma_{\pi w_j}$ (c) Σ_W (d) $\Sigma_{\pi BW}^{Power}$

Fig. 7. Classification of ECTSP formulations using rectangle packing.

of TAM bits assigned to the test, the length of the rectangle is the time taken for the test to execute on a TAM of width w, and the area of the rectangle as a function of the work done by the test, which is a function of p_i. When $\mathcal{F}_T$ is linear, then the area of each rectangle (of the same test) is constant. As shown in 6(a) as w is increased the area of the rectangle can increase if the scan chains are not perfectly balanced. This implies that the rectangle with $w = 1$ has minimum area. This is useful in computing a lower bound for the SOC schedule. Let the set of rectangles for a test i be denoted $\mathbb{R}_i$. The SOC test scheduling problem ECTSP then has a natural geometrical formulation, which can be stated as follows; pack N_C rectangles $r_1, r_2, \ldots, r_{N_C}$ where $r_i \in \mathbb{R}_i$ into a rectangle of width W and length $C^*_{\max}$. Various formulations of the ECTSP can also be depicted using the rectangle dual. Some examples are shown in Fig. 7. Fig. 7(a) shows partition based ECTSP, the top-level test width of W is first partitioned into TAMs; core tests are then scheduled on these TAMs. Sub-figure (b) depicts the case considered by Chakrabarty in [6], where TAMs are allowed to *fork* and *join*. Recently a new non-partition based formulation has been considered by Huang et al. [14] and Iyengar et al. [19]. An example of such a non-partitioned schedule is shown in Fig. 7(c). Power constrained schedule is shown in sub-figure (d). In contrast an example of a schedule with reconfigurable wrapper is shown in Fig. 8.

Problems $\Sigma_{\pi W}$, $\Sigma_{\pi w_j}$ and Σ_W belong to the same class of problems as they do not require any change to the core wrapper design. We also see that w.r.t. solution space: $sol(\Sigma_{\pi W}) \subset sol(\Sigma_{\pi w_j}) \subset sol(\Sigma_W) \subset sol(\Sigma_{W\text{recon}})$.

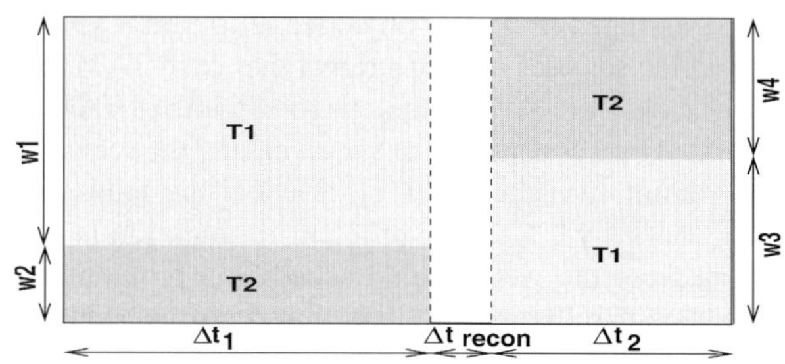

Fig. 8. Interpretation of schedule tuple returned from ECTSPSOL.

In the remainder of this paper we focus on providing efficient solutions to $\Sigma_{W\text{recon}}$. In the next sections we analyze the complexity of test scheduling problems and subsequently analyze the cases for convex and concave processing speed functions, respectively.

4.3. *Analysis of Complexity of Test Scheduling Problem*

It is well known that test scheduling for fixed partitions ($\Sigma_{\pi W}$) is a $\mathcal{NP}$-hard problem [18]. Since $\Sigma_{\pi w_j}$ is a particular case of ($\Sigma_{\pi W}$) by a method of *restriction* it has been shown in [6] that $\Sigma_{\pi w_j}$ is also $\mathcal{NP}$-hard. The problems Σ_W and $\Sigma_{W\text{recon}}$ are introduced here for the first time and we analyze their complexity. We state the decision variant of the embedded core test scheduling problem (ECTSP) below. Problem Σ_W is equivalent to the problem of packing r_{ij}, $i = 1, \ldots, \mathrm{N}_c$ rectangles into a rectangle of $W \times C_{\max}$, where $r_{ij} \in \mathbb{R}_i$. We prove that Σ_W is strongly $\mathcal{NP}$-complete (note that the problem itself is stated as a decision problem).

Problem 1 (ECTSP (Σ_W)). Given a chip with N_C cores each having a test T_i; where T_i takes time t_{ij} to execute on a TAM of width j, and a constraint W on the number of top-level test pins; does there exist a TAM assignment π and a schedule Σ for each test, such that the completion time of the full chip test is less than $C^*_{\max}$?

Theorem 1. *ECTSP* (Σ_W) *is strongly* $\mathcal{NP}$*-complete.*

Proof: Each test T_i is represented by a set of rectangles $\mathcal{R}_i$, where $|\mathcal{R}_i| \leq \phi_i$. Obviously, ECTSP $\in \mathcal{NP}$, as the depth of the decision tree is bounded by a polynomial in ϕ_i. Since $\phi_i \leq W$, and W is an input to ECTSP, ECTSP $\in \mathcal{NP}$. Any *YES* instance can be verified in polynomial time by examining the rightmost coordinates of each rectangle, and comparing them with $C^*_{\max}$. To prove that ECTSP is $\mathcal{NP}$-complete we use a restriction of $|\mathcal{R}_i| = 1$; denote this problem as ECTSP-1. The problem of 2-d rectangle packing can then be reduced to ECTSP-1, as for any instance of the 2-d rectangle packing an equivalent instance of ECTSP-1 can be made. Since 2-d rectangle packing is known to be $\mathcal{NP}$-complete, ECTSP-1 is also $\mathcal{NP}$-complete. Thus, ECTSP is strongly $\mathcal{NP}$-complete. □

Problem 2 (ECTSP-2Recon ($\Sigma_{W2\text{recon}}$)). Given a chip with $N_C > 3$ cores each having a test T_i; where T_i takes time t_{ij} to execute on a TAM of width j, one core wrapper has a degree of reconfigurability equals two, and a constraint W on the number of top-level test pins; does there exist a TAM assignment π and a schedule Σ for each test, such that the completion time of the full chip test is less than $C^*_{\max}$?

Theorem 2. *ECTSP-2Recon* ($\Sigma_{W2\text{recon}}$) $\notin \mathcal{NP}$.

Proof: We will show that even when a single core has degree of reconfigurability 2, ECTSP-2Recon is no longer solvable by even a non-deterministic algorithm. Consider the rectangle for the core with the reconfigurable wrapper; denote it $\mathcal{R}_A$. W.l.o.g. let the two widths admitted by the wrapper be w_1 and w_2. Let the length of $\mathcal{R}_A$ for w_1 and w_2 be l_1 and l_2, resp. Consider any position on $\mathcal{R}_A$; this position can be of width w_1 or w_2. Since, there can exist $\min\{l_1, l_2\}$ positions, the possible shapes of $\mathcal{R}_A$ are $2^{\min\{l_1,l_2\}}$. The cardinality of this set is not bounded by any polynomial in w_1 or w_2, hence the depth of the decision tree for ECTSP-2Recon is not bounded by any polynomial in its input, thus ECTSP-2Recon $\notin \mathcal{NP}$. □

Theorem 3. *ECTSP-Recon* ($\Sigma_{W\text{recon}}$) $\notin \mathcal{NP}$.

Proof: Follows immediately from Theorem 2. □

Intuitively, an $\mathcal{NP}$-hard problem can be solved by a "computer" which can make non-deterministic choices. Although, such a computer cannot be constructed, by enumerating all the choices (of which there are an exponential 2^n number of) we can find the optimum solution even on an ordinary machine. For small enough n, this might even be practical. By the result of Theorem 3, which proves that ECTSP $\notin \mathcal{NP}$, we have shown that ECTSP-Recon cannot be solved by even such a non-deterministic computer, and the potential search space on an ordinary machine is super-exponential (2^{n^n}), and hence, even for small n, enumeration would not be practical. Although the above results prove that generating optimum schedules for SOCs consisting of cores with reconfigurable task is very difficult, approximate schedules of $\Sigma_{W\text{recon}}$ are easily obtainable (as we show in the sequel of this paper) and can significantly improve upon the makespan of static core wrapper schedules. An advantage of the geometrical dual is exploited in Section 6 where Δt for the reconfiguration is implicitly calculated from the rectangle representation.

4.4. Scheduling with Convex g(m)

Denote the test scheduling problems with convex processing speed functions as P_{conv}. In Fig. 5(b) the processing speed function was plotted. Upto $m = 5$ the function is convex. Let W be such that the processing speed functions of *all* cores are convex.[2] For convex processing functions it is proved in [3] that the optimal schedule is one in which there is one time interval per task, and all the available resources are assigned to the task. We observe that the optimal schedule calls for assigning one processing interval of length Δ_j to test j, and within this interval allocating all W TAM bits to this test, $j = 1, \ldots, n$. The obtained schedule is also referred to as the *gang schedule*. This schedule can be obtained in $O(n)$ time. The resulting TAM architecture is the multiplexed architecture described in [1]. Hence problem P_{conv} can be solved optimally in time $O(n)$. In the context of core based testing the test time function can either be linear or sub-linear. Normally, it is not possible for the test time function to result in an super-linear speedup function. Hence, the above discussion holds only when all the cores have linear test time in the region $[1 : W]$, and in this case the method given above is optimal.

5. Scheduling with Concave $g(m)$ and Reconfigurable Wrappers

For most practical cases some or all of the processing speed functions will be concave for the W given in the problem. In this section we analyze the case for concave processing speed function. We first analyze the problem theoretically. In Section 5.3 we apply the method to the problem of test scheduling with reconfigurable wrappers.

5.1. Theoretical Analysis of P-CNTN with Concave Processing Speed Functions

In this section we theoretically analyze the problem of scheduling tests which do not have convex processing speed function in the domain $[1 : W]$. Let us call this problem P_{gen}. In order to define the function $g_j(m)$ at non-integer points the processing speed function $g_j(m)$ has to be extended to form a piecewise linear function such that $f_j(r) = \alpha_r g_j(\lfloor r \rfloor) + (1 - \alpha_r) g_j(\lceil r \rceil)$, where $\alpha_r = \lceil r \rceil - r, 0 \le r \le m$. Note that $f_j(s) = g_j(s)$ for $s \in [0 : m]$ and $j = 1, \ldots, n$. Thus, function f_j can be represented as:

$$f_j(r) = \beta_{j,\lceil r \rceil} r + \gamma_{j,\lceil r \rceil} \quad \text{for } 0 \le r \le m, \quad j = 1, \ldots, n \tag{2}$$

Since $f_j(r) = g_j(r)$ for $r = 0, 1, \ldots, m$ test scheduling with functions g_j is equivalent to test scheduling with functions f_j. Consider the *relaxed problem* P_{cntn} with functions f_j, $j = 1, \ldots, n$, in which the TAM width assigned to a core need not be integral. In problem P_{cntn} TAM widths are continuously divisible, but are bounded from above by W. Let $C^0_{\max}$ denote the minimum makespan for this problem. Clearly, $C^*_{\max} \ge C^0_{\max}$, where $C^*_{\max}$ is the optimum makespan for P_{gen}.

Let us define $\vec{r}$ and $\vec{u}$ as follows:

$$\vec{r} = (r_1, \ldots, r_n) \mid r_j \ge 0, \sum_{j=1}^{n} r_j \le m \tag{3}$$

$$\vec{u} = (u_1, \ldots, u_n) \mid u_j = f_j(r_j), \quad j = 1, \ldots, n, \ r_j \in \vec{r} \tag{4}$$

The vector $\vec{r}$ denotes the set of *feasible resource allocations* and $\vec{u}$ denotes the set of *feasible transformed resource allocations*. Let us denote $\vec{p} = (p_1, \ldots, p_n)$. We write conv$U$ to denote the convex hull of set $\vec{u}$. For an excellent introduction to the subject of linear programming, polytopes and their applications, see [34]. From the result in [3] we know that the minimum makespan value $C^0_{\max}$ for P_{cntn} is determined by the intersection point of the straight line $\vec{u} = \vec{p}/C, C > 0$, and conv$U$ in the n-dimensional space of transformed resource allocations. Let us denote such an intersection point by $\vec{u}^0$. We have $C^0_{\max} = p_j/u^0_j, j = 1, \ldots, n$.

Consider points $\vec{v}^{(j)}$ such that $v^{(j)}_j = f_j(m)$ and $v^{(j)}_l = 0, l \ne j, j = 1, \ldots, n, l = 1, \ldots, n$. We note that $\vec{v}^{(1)}, \ldots, \vec{v}^{(n)}$ represent all vertices of a convex polytope conv$U$, and the intersection point $\vec{u}^0$ belongs to this facet.

The intersection point $\vec{u}^0$ can therefore be represented as a linear combination of $\vec{v}^{(1)}, \ldots, \vec{v}^{(n)} : \vec{v}^0 = \sum_{i=1}^{n} \lambda_i \vec{v}^{(i)}$ where $\lambda_i \ge 0$, $i = 1, \ldots, n$, $\sum_{i=1}^{n} \lambda_i = 1$. The process interval for each test j, denoted $\Delta_j = \lambda_j C^0_{\max}$.

5.2. Calculation of processing intervals

In the context of test scheduling, reconfiguration is not a single clock cycle operation as assumed in the theoretical analysis. In order to reconfigure a core, we not only need to shift in the reconfiguration bits, but the test pattern must be aligned such that the complete pattern has been executed. Consider two cores A and B such that both A and B are reconfigurable. Assume that the optimal schedule calls for a reconfiguration of core A from w_1 to $w_1 - 1$ bit wide TAM, and similarly for core B from w_2 to $w_2 + 1$ bit wide TAM, at time instant t. At this time instant either (or both) cores might be executing a test pattern. Hence, the time instance would shift by Δt, where Δt depends on the length of the longest scan chain in A and B. Also, the next pattern stimuli cannot be pipelined with the current pattern's response. This observations lead us to believe that the number of reconfigurations per core should be limited, so as limit the overhead of reconfiguration. Hence, we constrain the search for facet vertices such that each core is not reconfigured more than twice (on the average). Since for test scheduling we are looking for an efficient algorithm (in addition to the optimal schedule) we use the following method (keeping in mind that the bound is W):

$$\mathbf{r}^{(1)} = \left(\lceil r_1^0 \rceil, \lfloor r_2^0 \rfloor, \ldots, \lfloor r_n^0 \rfloor\right)$$

$$\vdots$$

$$\mathbf{r}^{(\mathbf{n})} = \left(\lfloor r_1^0 \rfloor, \lfloor r_2^0 \rfloor, \ldots, \lceil r_n^0 \rceil\right)$$

Then we solve the following system of linear equations:

$$\mathbf{r}_\mathbf{1}^\mathbf{0} = r_1^1\lambda_1 + r_1^2\lambda_2 + \cdots + r_1^n\lambda_n$$

$$\vdots$$

$$\mathbf{r}_\mathbf{n}^\mathbf{0} = r_n^1\lambda_1 + r_n^2\lambda_2 + \cdots + r_n^n\lambda_n$$

to find $\lambda_1, \lambda_2, \ldots, \lambda_n$. Then we can calculate the processing interval for each test $\Delta_1 = \lambda_1 C_{\max}^0$, $\Delta_2 = \lambda_2 C_{\max}^0, \ldots, \Delta_n = \lambda_n C_{\max}^0$. Each test j is assigned $\mathbf{r}_\mathbf{j}^\mathbf{t}$ TAM bits in time interval Δ_t.

5.3. *Application to Test Scheduling*

Let us denote $f(\vec{r}) = \vec{u}$ and $f^{-1}(\vec{u}) = \vec{r}$ for $\vec{r}$ and $\vec{u}$ such that $f_j(r_j) = u_j,\ j = 1, \ldots, n$. The intersection point is $\vec{u}^{(0)}$ and the corresponding resource allocation $\vec{r}^{(0)} = f^{-1}(\vec{u}^{(0)})$. The processing time functions with reconfigurable wrappers can be written as $t_j(r) = L_{\max}^r(p_j+1) + p_j$. The length of the scan chain in the case of reconfigurable wrappers is the same as $L_{\max}$ for the cores without, the only advantage is that the width of the TAM can be changed dynamically. The processing speed function can be written as:

$$g_j(r) = \frac{p_j}{l_j(r)(p_j + 1) + p_j} \quad (5)$$

where the function $l_j(r)$ is the function responsible for calculating the longest scan chain involved in test j formed with a TAM width of r. We assume that the inverse function $l^{-1}(L_{\max}) = r$ is also available. Typically these will be implemented using lookup tables. Since for all practical cores we have $p_j \gg 1$ we can approximate Eq. (5) as $g_j(r) = \frac{1}{l_j(r)+1}$, which is a non-decreasing function in r. We know that the piecewise linear function can be written as $f_j(r) = \beta_{j,\lceil r \rceil} r + \gamma_{j,\lceil r \rceil}$. In order to calculate the resource allocation $\bar{\mathbf{r}}$ we have to calculate the intersection point of the line $\vec{u} = \vec{p}/C$ with the convex polytope formed by the convex hull of the points $f_j(r),\ j = 1, \ldots, n,\ r = 1, \ldots, m$. The parametric line equation can also be written as:

$$f_1(p_1, \bar{r}_1) = f_2(p_2, \bar{r}_2) = \cdots = f_n(p_n, \bar{r}_n) \quad (6)$$

$$\bar{r}_1 + \bar{r}_2 + \bar{r}_3 \leq W \quad (7)$$

We can calculate the intersection point of this line with the n-dimensional polytope. We approximate the vertices of the polytope facet, and calculate the processing intervals as described above in Section 5.2. Thus we can calculate the resource mapping and processing intervals for each test in the SOC.

5.4. *Interpretation of Output of ECTSPSOL*

If the schedule Σ returned by ECTSPSOL is:

$$\Sigma = \{(T_1, w_1, 0 \leftrightarrow \Delta t_1), (T_2, w_2, 0 \leftrightarrow \Delta t_1), (T_1, w_3, \Delta t_1 + \Delta t_{\text{recon}} \leftrightarrow \Delta t_2), (T_2, w_4, \Delta t_1 + \Delta t_{\text{recon}} \leftrightarrow \Delta t_2)\}$$

we can draw the corresponding schedule as shown in Fig. 8.

5.5. *Analysis of ECTSPSOL*

The input to the algorithm is information about the tests (p_i, $l_{\max}(T_i, m)$, $\beta(T_i, s)$, $\gamma(T_i, s)$) and the top-level constraint W. The algorithm returns the process intervals Δt_i and the resource allocations. The algorithm is separated into two phases; (i) the calculation of the intersection point of the parametric straight line $\vec{u} = \frac{\vec{p}}{C}$ with the polytope, and (ii) calculation of Δt_i. The intersection point can be calculated in time $O(N_C \max\{W, N_C \lg^2 W\})$ by the method given in [3]. In phase (ii) the intersection point is approximated by integer facet vertices. It is to be remembered that no efficient method is known to find the facet vertices, or to even bound their number. In the test scheduling case, to limit the overhead due to pattern alignment for reconfiguration we approximate the facet vertices by using the *floor* and *ceiling* functions. This introduces an approximation factor into our algorithm. Theoretical analysis of this approximation factor and to find better methods of determining the facet vertices are future directions of research. Then this resource matrix is inverted to calculate the processing intervals. The matrix inversion can be done in $O({N_C}^3)$. Hence, procedure ECTSPSol takes $O({N_C}^3)$ time.

6. Constrained Reconfiguration Scheduling

Designing reconfigurable wrappers for small cores might be relatively expensive in terms of area and routing. A question that can be asked is; is it possible to solve ECTSP when only some of the cores have reconfigurable wrappers? In Section 4 we have seen that the general problem of ECTSP-Recon is harder than $\mathcal{NP}$-hard problems, hence a answer to the above question is unknown. But in this section we analyze the case when only one of the cores (to be more precise; one of the larger cores) has a reconfigurable wrapper. We will show that under a reasonable assumption on the size of the core, close to optimal schedules can be found.

Assume one of the tests is *malleable*, i.e., it has a reconfigurable wrapper and can be dynamically reassigned TAM wires. Given a set $\mathcal{Z}$ of a set of rectangles $\mathcal{Z} = \mathcal{R}_1 \cup \cdots \cup \mathcal{R}_{N_C}$ we can then write $\mathcal{Z} = \mathcal{Z}_s \cup \mathcal{Z}_m$, where the subscripts s and m denote static and malleable, resp. Let us consider the case when $|\mathcal{Z}_m| = 1$. Let $\Sigma(s)$ denote the makespan of the schedule for the static tests. The area A_s occupied by the static schedule is then $W \cdot \Sigma(s)$. The idle time of the schedule is $A_s - \sum_{i \in \mathcal{Z}_s} \mathcal{A}_i$. When rectangle $\mathcal{R}_m$ is added we have:

$$\begin{aligned} C_{\max}(\mathcal{Z}) &= C_{\max}(\mathcal{Z}_s + \mathcal{R}_m) \\ &= \Sigma(s) + \frac{\mathcal{A}_m - \left(A_s - \sum_{i \in \mathcal{Z}_s} \mathcal{A}_i\right)}{W} \end{aligned} \quad (8)$$

From the above relation it is easy to see that if $\mathcal{A}_m \geq W \cdot \Sigma(s) - \sum_{i \in \mathcal{Z}_s} \mathcal{A}_i$, then $C_{\max}(\mathcal{Z}) = \frac{\sum_{i \in \mathcal{Z}} \mathcal{A}_i}{W}$, or that there is no idle time on the TAM wires. See Fig. 9. Using this result we can construct an algorithm which tries each malleable test (in decreasing order of area) as a candidate for the last addition. For any malleable test if the area condition is met, then we would have found a schedule with zero idle time on the TAM wires. The $N_C - 1$ cores have to be packed in a *LB-optimal* fashion (borrowing a term from the floorplanning literature; an LB-optimal packing is a packing of modules such that no module can be moved to the left or bottom without moving other modules). For ECTSP, such an algorithm exists based on the *partitionable independent task scheduling* problem discussed in [2]. The $N_C - 1$ cores are arranged using this method which we call `Belkhale` in the algorithm below. Barring the increase in the area of a test rectangle due to sub-linear speedup, this schedule is optimal. This result is not in conflict with the complexity result we proved above, as the non-inclusion of **ECTSP-Recon** in $\mathcal{NP}$ is assuming that all cores can be reconfigured at any instance

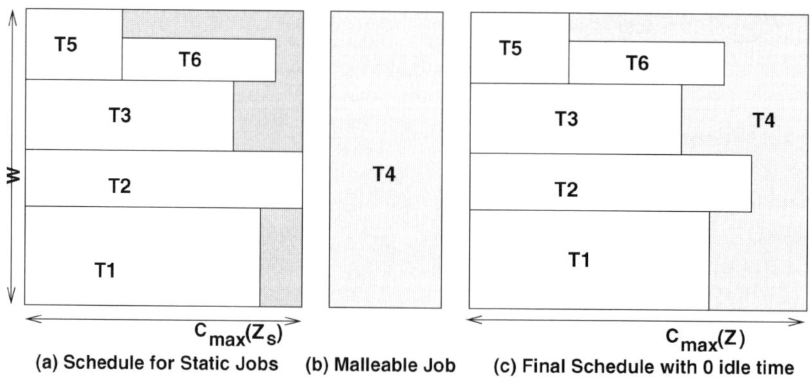

Fig. 9. Near optimal schedule obtained using a single malleable test.

```
procedure ECTSP1SOL;
input int N_C; // the number of tests
input int W; // the number of top-level test pins
input array Z_s; // static wrapper cores
input array Z_m; // reconfigurable wrapper cores
input array A[N_C]; // area of each test
begin
   dotimes i ∈ Z_m
      if (A[i] > Belkhale(|Z_s|, W, Z_s) ·W − Σ_{i∈Z_s} A[i]);
         return Σ = Belkhale(|Z_s|, W, Z_s) + i;
   end dotimes
end procedure ECTSP1SOL;
```

Fig. 10. Scheduling algorithm for SOCs with 1 reconf. wrapper core.

in time. In the figure above the Δt and $\vec{\mathbf{r}}$ for the malleable core are implicitly known from the schedule of the static tests. The algorithm is shown in Fig. 10.

6.1. *Analysis of ECTSP1SOL*

In the above algorithm, the call to `Belkhale` can be made $O(N_C)$ times in the worst case. From [2] we know that each call to `Belkhale` takes $O((N_C + W) \lg W)$ time, therefore the complexity of ECTSP1SOL is $O(N_C((N_C + W) \lg W))$. In the next section we explain the complete method with the help of a small example. Later in the section we compare our method to integer linear programming formulations on some benchmark SOCs.

7. Implementation Details and Comparative Studies

Before presenting the experimental results we explain the scheduling algorithm with an example. To limit the example we have used the processing speed function $g_j(r) = \frac{r}{\phi_j}$. The design considered is mentioned as SOC $S2$ in [6]. Details of this SOC and another benchmark SOC $p93791$ appear in [18] and are given in Tables 4 and 5 respectively. The top level TAM width W is analyzed for $W = 24$. We then get the following equations:

$$\frac{\phi_1 p_1}{\bar{r}_1} = \frac{\phi_2 p_2}{\bar{r}_2} = \cdots = \frac{\phi_{10} p_{10}}{\bar{r}_{10}} \tag{9}$$

$$\sum_{i=1}^{n} \bar{r}_i = W \tag{10}$$

Table 4. Test data for the cores in $d695$ [18].

	Number of			Scan chain lengths	
Circuit (core)	Test patterns	Primary IOs	Scan chains	Min	Max
c6288	12	64	–	–	–
c7552	73	315	–	–	–
s838	75	35	1	32	32
s9234	105	75	4	52	54
s38584	110	344	32	44	45
s13207	234	214	16	39	41
s15850	95	227	16	33	34
s5378	97	84	4	44	46
s35932	12	355	32	54	54
s38417	68	134	32	51	55

Solving for $W = 24$ we compute $\bar{\mathbf{r}}$. Since $\beta_{j,s} = 0$ we get $\mathbf{r}^0 = \bar{\mathbf{r}} = \{0.0017, 0.4095, \ldots, 2.2788\}$.

Let us write the facet vertices for the 10-dimensional polytope for the case of $W = 24$. The vertices are shown in Table 6. Since some of the rows in the Table 6 have sum less than 24, we can make the resource allocation of some test higher. As an example $\mathbf{r}_5$ and $\mathbf{r}_6$ should be assigned 8 or more TAM bits when possible. After we have calculated the vertices of the polytope we can formulate the system of linear equations from Table 6. We get the following system of linear equations for the case of $W = 24$ for SOC $S2$. *Note*: some of the entries in the system had to be modified to reduce the number of negative λ_i.

$$\begin{aligned}
0.0017 &= \lambda_1 \\
0.4095 &= \lambda_2 + \lambda_3 \\
0.4076 &= \lambda_3 \\
1.1115 &= \lambda_1 + \lambda_2 + \lambda_3 + 2\lambda_4 + \lambda_5 + \lambda_6 + \lambda_7 \\
&\quad + \lambda_9 + \lambda_{10} \\
&\ \ \vdots \\
2.2788 &= 2\lambda_1 + 2\lambda_2 + 2\lambda_3 + 2\lambda_4 + 2\lambda_5 + 2\lambda_6 \\
&\quad + 2\lambda_7 + 2\lambda_8 + 2\lambda_9 + 3\lambda_{10}
\end{aligned} \tag{11}$$

The solution to the above system may have $\lambda_i < 0$ for some vertices. When constructing the schedule these values have to be dropped.

Solving for λ_i we get $\vec{\lambda} = (0.0017, 0.0019, 0.4076, 0.0009, 0.1779, 0.0812, 0.3118, -0.0163,$

Table 5. Test data for the cores in $p93791$ [18].

Circuit (core)	Number of Test patterns	Number of Primary IOs	Number of Scan chains	Scan chain lengths Min	Scan chain lengths Max
Core 1	409	213	46	1	168
Core 2	192	74	–	–	–
Core 3	648	69	–	–	–
Core 4	11	117	23	4	5
Core 5	6127	248	–	–	–
Core 6	218	813	46	500	521
Core 7	177	41	–	–	–
Core 8	177	41	–	–	–
Core 9	192	77	–	–	–
Core 10	1164	395	–	–	–
Core 11	187	286	11	17	82
Core 12	391	369	46	92	93
Core 13	194	214	46	173	219
Core 14	194	214	46	173	219
Core 15	288	78	–	–	–
Core 16	396	201	–	–	–
Core 17	216	283	43	145	150
Core 18	42	113	–	–	–
Core 19	210	903	44	97	100
Core 20	416	220	44	132	181
Core 21	42	113	–	–	–
Core 22	42	76	–	–	–
Core 23	234	205	46	1	175
Core 24	3072	21	–	–	–
Core 25	2688	45	–	–	–
Core 26	96	76	–	–	–
Core 27	916	109	46	50	68
Core 28	396	159	–	–	–
Core 29	172	231	35	185	189
Core 30	192	77	–	–	–
Core 31	204	218	–	–	–
Core 32	3084	396	–	–	–

0.0400, 0.0884). We remove the term λ_8, to get $\sum \lambda_i \mid \lambda_i > 0 = 1.1115$. We know that $C^0_{\max}(W = 24) = 5313504/24 = 221396$. We calculate the processing intervals $\Delta_j = \lambda_j C^0_{\max}$. As an example $\Delta_1 = 0.0017 * 221396 = 377$, $\Delta_2 = 0.0019 * 221396 = 421$, and so on. The resource mapping vector for interval t is $\mathbf{r}^{(\mathbf{t})}$ as given in Table 6. As an example in interval Δ_1 the resource mapping is (1, 0, 0, 1, 7, 7, 2, 1, 1, 2). Which can be read as test T1 gets 1 bit, test T4 gets 1 bit, test T5 gets 7 bits and so on. In this schedule if we see the reconfiguration requirement, we see that cores `s38584`, `s13207`, `s15850`, `s5378`, `s35932` and `s38417` need 2 different widths of TAM somewhere in the schedule. Hence, the generation of reconfigurable wrappers for these cores can be constrained with $\phi_i = 2$ for all cores. Since the number of different configurations is only 2, we would need only one TCM control signal (each) for these cores.

Table 6. Facet vertices for the convex polytope in 10 dimensions for core $S2$ when $W = 24$.

$\mathbf{r}$	$\mathbf{r}_1$	$\mathbf{r}_2$	$\mathbf{r}_3$	$\mathbf{r}_4$	$\mathbf{r}_5$	$\mathbf{r}_6$	$\mathbf{r}_7$	$\mathbf{r}_8$	$\mathbf{r}_9$	$\mathbf{r}_{10}$
$\mathbf{r}^{(1)}$	1	0	0	1	7	7	2	1	1	2
$\mathbf{r}^{(2)}$	0	1	0	1	7	7	2	1	1	2
$\mathbf{r}^{(3)}$	0	1	1	1	7	7	2	1	1	2
$\mathbf{r}^{(4)}$	0	0	0	2	7	7	2	1	1	2
$\mathbf{r}^{(5)}$	0	0	0	1	8	7	2	1	1	2
$\mathbf{r}^{(6)}$	0	0	0	1	7	8	2	1	1	2
$\mathbf{r}^{(7)}$	0	0	0	1	7	7	3	1	1	2
$\mathbf{r}^{(8)}$	0	0	0	1	7	7	2	2	1	2
$\mathbf{r}^{(9)}$	0	0	0	1	7	7	2	1	2	2
$\mathbf{r}^{(10)}$	0	0	0	1	7	7	2	1	1	3

7.1. *Comparison of Schedules on Benchmark SOCs*

For the benchmark comparison we substitute the processing speed function by the more realistic $g_j(r) = \frac{1}{l_j(r)+1}$. The algorithms presented in this paper were implemented with Common Lisp running on a Pentium III machine. The comparison is shown in Table 7 for SOC $S2$ (also referred to as $d695$ in [18]) and Table 8 for a Philips SOC $p93791$. The details for these benchmark SOCs are given in Tables 4 and 5 respectively. All runtimes marked with † are cases when the program was halted, and the best result is shown. All the cores were assumed to be individually tested blocks, and no hierarchy was used. Functional inputs, outputs and inouts were assumed to be equivalent to scan chain of unit length. For comparison with enumeration based methods, we have compared our results with those presented in [18]. We also implemented our own ILP formulation using CPLEX (available from `www.ilog.com`). In some cases our ILP formulation provides better results than the enumeration algorithm in [18]. This is because the enumeration in [18] is done only for small values of

Table 7. Comparative study for SOC *d*695, $|\Sigma|$ denotes the makespan.

	Results of [18]				Our experimental results					
	lpsolve		Enumerative		CPLEX		ECTSPSOL		ECTSP1SOL	
W	$\|\Sigma\|$	Time (min)	$\|\Sigma\|$	Time (sec)	$\|\Sigma\|$	Time (min)	$\|\Sigma\|$	Time (sec)	$\|\Sigma\|$	Time (sec)
16	41969	4.5	41949	4	42293	1	41299	2	41258	2
20	33458	16.0	33458	6	33917	1	33110	2	32937	2
24	28327	32.5	28327	10	28305	1	28288	2	28714	2
28	24021	52.3	24021	12	24021	3	23100	2	26986	2
32	21423	64.7	21423	16	21474	3	20858	2	24870	2
36	19573	85.0	19573	21	19034	5	18381	2	22753	2
40	17210	104.6	17210	24	17612	15	16799	2	21505	2
44	16975	180†	16975	30	15811	180†	15467	2	20096	2
48	16975	180†	16403	40	14317	180†	14331	2	18528	2
52	16928	180†	14028	44	13482	180†	13202	2	17396	2
56	15694	180†	13023	53	12466	180†	12148	2	16536	2
60	15694	180†	12544	63	11432	360†	11090	2	14878	2
64	15694	180†	12327	84	9869	720†	10462	2	13111	2

Table 8. Comparative study for SOC *p*93791 $|\Sigma|$ denotes the makespan.

	Results of [18]				Our experimental results					
	lpsolve		Enumerative		CPLEX		ECTSPSOL		ECTSP1SOL	
W	$\|\Sigma\|$	Time (min)	$\|\Sigma\|$	Time (sec)	$\|\Sigma\|$	Time (min)	$\|\Sigma\|$	Time (sec)	$\|\Sigma\|$	Time (sec)
16	1883150	180†	1775090	5	1818466	1	1755886	2	1807200	2
20	1520840	180†	1435650	8	1479472	1	1445577	2	1557938	2
24	1288380	180†	1192980	10	1350462	3	1164023	2	1228766	2
28	1163100	180†	1030920	13	1038325	10	1014358	2	1075825	2
32	944881	180†	899807	17	922679	180†	919354	2	967274	2
36	924909	180†	797940	23	812318	180†	783924	2	890474	2
40	929848	180†	705164	26	790434	180†	707812	2	890768	2
44	873276	180†	659856	33	772109	180†	645540	2	885318	2
48	835526	180†	602613	42	659281	180†	585771	2	631115	2
52	807909	180†	567306	46	581454	180†	541699	2	594433	2
56	537891	180†	521806	54	540268	180†	517707	2	562376	2
60	545154	180†	495518	65	498770	180†	482476	2	526748	2
64	551111	180†	463707	88	466813	180†	453868	2	498763	2

B. Since the version of CPLEX used was a demonstration version (able to support only 100 variables) for the Philips SOC not all values of B could be experimented with. The maximum value of B used was $B = 3$. In the case of ECTSP1SOL core `s13207` and `c6` were chosen to have reconfigurable wrappers for SOC *d*695 and *p*93791, resp. The ECTSPSOL algorithm computes schedules that are almost 18% better than the enumerated schedule computed by [18] for some values of W for the SOC *d*695. For the SOC *p*93791 the schedules almost 5% better in some instances of W. To test the scalability of the algorithms we generated a random

Table 9. Summary of ECTSP algorithms.

Algorithm	Time complexity	Applicability
ECTSPCONVEX	$O(N_C)$	All cores have convex $g(m)$, $m \in [1 : W]$
ECTSPSOL	$O(N_C{}^3)$	All cores have reconf. wrappers
ECTSP1SOL	$O(N_C((N_C + W) \lg W))$	At least 1 core has reconf. wrapper

SOC of 80 cores, and the ECTSP* algorithms were able to produce a schedule within 5 seconds of runtime. An important factor contributing to ECTSPSOL performance is the independence of the runtime from the top level TAM width, W. In contrast, the enumerative approach of [18] does not scale well with W, the number of partitions that have to be analyzed are exponential (for $B = 3$ in [18] the authors calculate the number of partitions $p_3(W) = \sum_{i=0}^{\lfloor \frac{W}{3} \rfloor} \lfloor \frac{W-(3i+1)}{2} \rfloor$). The results obtained from the experiments with ECTSP1SOL were not as good as expected. Further analysis convinced us that the quality of the schedule depends more on the choice of the core which is made reconfigurable. More experiments are underway to determine the exact nature of this dependency and to build efficient algorithms around it.

8. Conclusion and Future Work

In this paper we have presented a formulation of the embedded core test scheduling problem (ECTSP). We classified existing approaches to the ECTSP by formulating ECTSP as a geometrical problem for generalized rectangle packing. We have described the design of reconfigurable test wrappers for cores; using these wrappers the width of a TAM connected to a core may be allowed to change dynamically. A procedure for the automatic derivation of the necessary DfT hardware needed for reconfiguration using a graph theoretic representation of core wrappers was also presented. The area cost of the proposed design is of the order of few logic gates. We have proposed new non-partition based scheduling formulations, and analyzed their theoretical complexity. We show that determining the optimal test schedule with reconfigurable wrappers is harder than $\mathcal{NP}$-hard. For cores having convex processing speedup functions in the domain $[1 : W]$, where W is the top-level TAM width bound at SOC level we present optimal schedules which can be computed in $O(N_C)$ time. For concave processing speed functions we propose to use reconfigurable wrapper design alongwith a $O(N_C{}^3)$ scheduling algorithm to arrive at good schedules. The ECTSP algorithms presented in this paper were implemented. A summary of the algorithms presented above is given in Table 9. For an integrator, the choice of the scheduling algorithm depends on a number of factors. If testing time is of prime importance, then all the core wrappers should be modified to be reconfigurable. For SOCs employing a combination of static and reconfigurable wrappers, an easy to implement scheme ECTSP1SOL is shown which is a natural derivation from the geometrical representation. In our opinion this method has the most promise as we feel that large cores which have a high number of scan chains and patterns are ideal candidates for addition of reconfiguration capability.

Experimental results containing the comparison with state-of-art ILP formulations indicate that reconfigurable wrappers can alleviate the test scheduling problem. Future work will concentrate on benchmarking this method against static wrapper test methods on the axes of test time, DfT area and routing congestion. On the theoretical side, an analysis of the approximation factor introduced by the relaxation technique used in this paper to approximate the vertices of the polytope facet is required. We have not considered the effect of power constraints, or test conflicts on reconfiguration based scheduling. Extension of this scheme to incorporate test conflicts, power constraints and precedence relations are also topics of future research.

Acknowledgments

I would first especially like to thank Maciej Machowiak for providing a copy of his paper [3] and other references, which form the kernel of the scheduling algorithm in this paper. I would like to take this opportunity to thank my colleagues at Philips Research, Erik Jan Marinissen, Sandeep Kumar Goel and Maurice Lousberg for valuable discussions on the subject of embedded core-based SOC test. Vikram Iyengar and

Krishnendu Chakrabarty of Duke University reviewed a preliminary version of this paper and provided details about the Duke SOC mentioned in this paper. I would also like to thank Mario Konijnenburg and Om Prakash Gangwal for constructively reviewing an earlier version of this paper. I also express my gratitude to the anonymous reviewers who helped improve the presentation of this paper.

Notes

1. Bitwidth for a core is defined as the minimum TAM width over which there is no decrease in the test time for a core.
2. This situation although unrealistic, is not impossible. Imagine a SOC having 4 cores of bitwidth 64, 64, 32 and 32 respectively. If $W = 6$, then all the cores will probably have convex processing speed functions in the domain $[1 : W]$, and hence the schedule for convex functions can be applied. *Note*: for a SOC which has W such that all processing speed functions are convex, the *multiplexed* architecture for TAMs [1] is provably the optimum TAM design for core internal test application.

References

1. J. Aerts and E.J. Marinissen, "Scan Chain Design for Test Time Reduction in Core-Based ICs," in *Proc. IEEE Intl. Test Conf. (ITC)*, Oct. 1998, pp. 448–457.
2. K.P. Belkhale and P. Banerjee, "Approximate Algorithms for the Partitionable Independent Task Scheduling Problem," in *Proceedings of the International Conference on Parallel Processing*, 1990, pp. 72–75.
3. J. Blazewicz, M.Y. Kovalyov, M. Machowiak, D. Trystram, and J. Weglarz, "Scheduling Malleable Tasks on Parallel Processors to Minimize the Makespan," in *Proc. of ECCO XIV*, May 2001.
4. K. Chakrabarty, "Design of System-on-a-Chip Test Access Architectures Under Place-and-Route and Power Constraints," in *Proc. ACM/IEEE Design Automation Conf. (DAC)*, June 2000, pp. 432–437.
5. K. Chakrabarty, "Test Scheduling for Core-Based Systems Using Mixed-Integer Linear Programming," *IEEE Trans. on CAD*, vol. 19, no. 10, pp. 1163–1174, Oct. 2000.
6. K. Chakrabarty, "Optimal Test Access Architectures for System-on-a-Chip," *ACM Trans. Design Automation of Electronic Systems*, Jan. 2001.
7. R. Chandramouli and S. Pateras, "Testing Systems on a Chip," *IEEE Spectrum*, pp. 42–47, Nov. 1996.
8. R.M. Chou, K.K. Saluja, and V.D. Agrawal, "Scheduling Tests for VLSI Systems Under Power Constraints," *IEEE Trans. VLSI Systems*, vol. 5, no. 2, pp. 175–185, 1997.
9. M.R. Garey and D.S. Johnson, *Computers and Intractability*, New York: W. H. Freeman, 1979.
10. M. Garg, A. Basu, T.C. Wilson, D.K. Banerji, and J.C. Majithia, "A New Test Scheduling Algorithm for VLSI Systems," in *Proc. of the Symp. on VLSI Design*, Nov. 1999, pp. 148–153.
11. I. Ghosh, N.K. Jha, and S. Dey, "Low Overhead Design for Testability and Test Generation Technique for Core-Based Systems-on-a-Chip," *IEEE Trans. on CAD*, vol. 18, no. 11, p. 1661, Nov. 1999.
12. R. Graham, "Bounds on Multiprocessor Timing Anomalies," *SIAM Journal on Appl. Math*, vol. 17, pp. 263–269, 1969.
13. D. Hochbaum, *Approximation Algorithms for NP-Hard Problems*, PWS Publishing, 1996.
14. Y. Huang, W.T. Cheng, C.C. Tsai, N. Mukherjee, O. Samman, Y. Zaidan, and S.M. Reddy, "Resource Allocation and Test Scheduling for Concurrent Test of Core-Based SOC Design," in *Proc. Asian Test Symposium (ATS)*, Nov. 2001, p. 5A.3.
15. IEEE P1500 Web Site, http://grouper.ieee.org/groups/1500/.
16. V. Immaneni and S. Raman, "Direct Access Test Scheme—Design of Block and Core Cells for Embedded ASICs," in *Proc. IEEE Intl. Test Conf. (ITC)*, Sept. 1990, pp. 488–492.
17. V. Iyengar and K. Chakrabarty, "Precedence-Based, Preemptive, and Power-Constrained Test Scheduling for System-on-a-Chip," in *Proceedings IEEE VLSI Test Symposium (VTS)*, Marina del Rey, CA, IEEE Computer Society Press, May 2001, pp. 368–374.
18. V. Iyengar, K. Chakrabarty, and E.J. Marinissen, "Test Wrapper and Test Access Mechanism Co-Optimization for System-on-a-Chip," in *Proc. IEEE Intl. Test Conf. (ITC)*, Oct. 2001, pp. 1023–1032.
19. V. Iyengar, K. Chakrabarty, and E.J. Marinissen, "On Using Rectangle Packing for SOC Wrapper/TAM Co-Optimization," in *Proc. IEEE VLSI Test Symposium (VTS)*, 2002.
20. R. Kapur, M. Lousberg, T. Taylor, B. Keller, P. Reuter, and D. Kay, "CTL, the Language for Describing Core-based Test," in *Proc. IEEE Intl. Test Conf. (ITC)*, Oct. 2001, pp. 131–139.
21. M. Keating and P. Bricaud, *Reuse Methodology Manual For System-on-Chip Designs*, Kluwer Academic Publishers, 1998.
22. S. Koranne, "On Test Scheduling for Core-Based SOCs," in *Proc. of VLSI Design/ASP-DAC 2002*, Jan. 2002, pp. 505–510.
23. S. Koranne and V.S. Choudhary, "Formulating SOC Test Scheduling as a Network Transportation Problem," in *Proc. Design, Automation, and Test in Europe (DATE)*, March 2002.
24. E. Larsson and Z. Peng, "An Integrated System-on-Chip Test Framework," in *Proc. Design, Automation, and Test in Europe (DATE)*, March 2001, pp. 138–144.
25. J.K. Lenstra, D.B. Shmoys, and E. Tardos, "Approximation Algorithms for Scheduling Unrelated Parallel Machines," in *Proc. 28th Annual IEEE Symposium on Foundations of Computer Science (FOCS)*, 1987, pp. 217–224.
26. E.J. Marinissen, R. Arendsen, G. Bos, H. Dingemanse, M. Lousberg, and C. Wouters, "A Structured and Scalable Mechanism for Test Access to Embedded Reusable Cores," in *Proc. IEEE Intl. Test Conf. (ITC)*, Oct. 1998, pp. 284–293.
27. E.J. Marinissen, S.K. Goel, and M. Lousberg, "Wrapper Design for Embedded Core Test," in *Proc. IEEE Intl. Test Conf. (ITC)*, Oct. 2000, pp. 911–920.
28. E.J. Marinissen, R. Kapur, and Y. Zorian, "On Using IEEE P1500 SECT for Test Plug-n-Play," in *Proc. IEEE Intl. Test Conf. (ITC)*, Oct. 2000, pp. 770–777.
29. E.J. Marinissen, K. Kuiper, and C. Wouters, "Test Protocol Expansion in Hierarchical Macro Testing," in *Proc. IEEE European Test Conf. (ETC)*, April 1993, pp. 28–36.
30. E.J. Marinissen and M. Lousberg, "The Role of Test Protocols in Testing Embedded-Core-Based System ICs," in *Proc. IEEE*

European Test Workshop (ETW), May 1999, pp. 70–75.
31. V. Muresan et al., "A Comparison of Classical Scheduling Approaches in Power-Constrained Block-Test Scheduling," in *Proc. IEEE Intl. Test Conf. (ITC)*, Oct. 2000, pp. 882–891.
32. S. Ravi, G. Lakshminarayana, and N.K. Jha, "Testing of Core-Based Systems-on-a-Chip," *IEEE Trans. on CAD*, vol. 20, no. 3, pp. 426–439, March 2001.
33. P. Rosinger, B. Al-Hashimi, and N. Nicolici, "Power Constrained Test Scheduling Using Power Profile Manipulation," in *Proc. Intl. Symposium on Circuits and Systems (ISCS)*, May 2001, pp. 251–254.
34. A. Schrijver, *Theory of Linear and Integer Programming*, John Wiley & Sons, 1986.
35. M. Sugihara, H. Date, and H. Yasuura, "A Novel Test Methodology for Core-Based System LSIs and a Testing Time Minimization Problem," in *Proc. IEEE Intl. Test Conf. (ITC)*, Oct. 1998, pp. 465–472.
36. P. Varma and S. Bhatia, "A Structured Test Re-Use Methodology for Core-Based System Chips," in *Proc. IEEE Intl. Test Conf. (ITC)*, Oct. 1998, pp. 294–302.
37. Y. Zorian, "A Distributed BIST Control Scheme for Complex VLSI Devices," in *Proc. IEEE VLSI Test Symposium (VTS)*, April 1993, pp. 6–11.
38. Y. Zorian, "Test Requirements for Embedded Core-Based Systems and IEEE P1500," in *Proc. IEEE Intl. Test Conf. (ITC)*, Nov. 1997, pp. 191–199.
39. Y. Zorian, E.J. Marinissen, and S. Dey, "Testing Embedded-Core Based System Chips", in *Proc. IEEE Intl. Test Conf. (ITC)*, Oct. 1998, pp. 130–143.

Sandeep Koranne was born in New Delhi, India on 8th January 1976. He received his Bachelor of Engineering degree in Computer Science and Engineering from Maulana Azad College of Technology (R.E.C.), Bhopal, India in 1997 (where he received the university Gold Medal) and his Masters of Technology degree in VLSI Design, Tools and Technology from Indian Institute of Technology Delhi, India in 1998. From July 1999 till September 2001 he was working with ED&T/Test, Philips Research Labs, Eindhoven, where he was responsible for development of CAD algorithms and tools for enabling core-based testing of system-on-a-chip ICs.

Presently, he is a Senior Scientist with Tanner Research, Inc., Pasadena, CA. His research interests include combinatorial optimization, graph theory, discrete algorithms, SoC test and physical design automation. He is a member of the IEEE P1500 task force on mergable cores and has published many papers in the area of graph theory, parallel and distributed computing and layout analysis. He is a member of the IEEE.

JOURNAL OF ELECTRONIC TESTING: Theory and Applications 18, 435–454, 2002

The Role of Test Protocols in Automated Test Generation for Embedded-Core-Based System ICs

ERIK JAN MARINISSEN
Philips Research Laboratories, IC Design—Digital Design & Test, Prof. Holstlaan 4, M/S WAY-41, 5656 AA Eindhoven, The Netherlands
Erik.Jan.Marinissen@philips.com

Received September 24, 2001; Revised January 15, 2002

Editor: Krishnendu Chakrabarty

Abstract. Modular testing is an attractive approach to testing large system ICs, especially if they are built from pre-designed reusable embedded cores. This paper describes an automated modular test development approach. The basis of this approach is that a core or module test is dissected into a test protocol and a test pattern list. A test protocol describes in detail how to apply one test pattern to the core, while abstracting from the specific test pattern stimulus and response values. Subsequent automation tasks, such as the expansion from core-level tests to system-chip-level tests and test scheduling, all work on test protocols, thereby greatly reducing the amount of compute time and data involved. Finally, an SOC-level test is assembled from the expanded and scheduled test protocols and the (so far untouched) test patterns. This paper describes and formalizes the notion of test protocols and the algorithms for test protocol expansion and scheduling. A running example is featured throughout the paper. We also elaborate on the industrial usage of the concepts described.

Keywords: system chips, embedded cores, test generation, test protocol, expansion, test protocol scheduling

1. Introduction

Modern semiconductor process technologies enable the manufacturing of a complete system on one single die, the so-called *system chip* or SOC. Such system chips typically are very large ICs, consisting of millions of transistors, and containing a variety of hardware modules. In order to design these large and complex system chips in a timely manner and to leverage from external design expertise, increasingly reusable cores are utilized [14]. *Cores* are pre-designed and pre-verified design modules, meant to be reused in multiple SOC designs [20]. Examples of cores are CPUs, DSPs, media co-processors, communication modules, memories, and mixed-signal modules. Core-based design divides the IC design community into *core providers* and *core users*. Core-based design takes place within companies, as well as between companies, i.e., one company acts as core provider, while another company is core user. Especially in the latter scenario, intellectual property rights and liability issues need to be addressed.

Due to imperfections in their manufacturing process, all electronic systems need to be tested for manufacturing defects. System chips are no exception to that rule. In traditional system *boards*, the components (ICs) used are tested by their provider (the IC manufacturer), and hence the system integrator only needs to test the rest of the board (often mainly interconnects). For system *chips*, this situation is quite different. The components (cores) cannot be tested for manufacturing defects by their providers, as what is transferred from provider to user is merely a description, and hence not yet manufactured [42]. Manufacturing only takes place once the entire system chip design is assembled.

Therefore, a system chip integrator is responsible for testing not only the user-defined circuitry surrounding the embedded cores, but also the embedded cores themselves. Even though the core provider might have tested its product for design errors and other users of the same core might have tested it for manufacturing defects in their SOCs, that does not take away the obligation of the core user to test his/her system chip for manufacturing defects.

Testing of core-based system chips is best done in a core-based fashion: the core providers prepare their products with the right design-for-testability hardware and create test patterns for the cores, while the system chip integrator adds SOC-level design-for-testability and creates a system chip test using the core-level tests as building blocks [42, 43]. In this core-based way of SOC test development, the system chip integrator does not need to be bothered with the content of the cores he/she is using in order to develop a test of sufficient quality. In many cases, core providers also do not want to share the implementation details of their cores in order to protect their intellectual property. Furthermore, this scenario allows for test reuse. A test developed once for a given core, can be reused for many instantiations of that core in the same or different system chips, hence optimizing the usage of scarce engineering resources. And, if proper procedures and tools are in place, this way of working should also save precious development time on the side of the system chip integrator, hence contributing to shorter time-to-market.

Core-based testing is relatively new, and therefore bears with it several new challenges [32]. These include (1) the transfer of information about the core's test(s) from core provider to core user, (2) test access to the, often deeply embedded, cores, and (3) chip-level optimization of test-related costs such as test time, silicon area, power dissipation, etc.

This paper elaborates on the role that *test protocols* can have in core-based SOC testing. A test protocol describes how to apply test stimuli to and capture test responses from a core. Together with the test patterns it refers to, a test protocol describes a complete test for a core. A test protocol can be described at any level of design hierarchy surrounding the core-under-test, starting with the core itself and ending with the SOC pins. Depending on a specific SOC design and the embedding of a core in that SOC design, the test protocol for that core might be different at the various hierarchy levels, while the test patterns remain unchanged. Test protocols play a role in all three of the abovementioned challenges, and this paper provides the details of that.

1. A core's test should be described by the core provider such that the test protocol and the test patterns are separated. The core provider does not necessarily have knowledge about the embedding of the core and can describe the test protocol at the core's terminals. This paper describes generic test protocol and test pattern formats and formalizes the information contained herein.
2. The core user is responsible for providing sufficient test access to the embedded core, e.g., by means of a dedicated test access infrastructure consisting of *wrappers* and *test access mechanisms* (TAMs) [27, 41, 42, 43]. Also, the core user should translate a core's test(s), described by the core provider at the core's terminals, to the SOC pins. In this paper, we show that this can be done by manipulation of the test protocol only. We describe an automated procedure, called *test protocol expansion*, and outline its advantages compared to an approach in which a test is not separated into protocol and patterns.
3. One of the chip-level optimization issues is the scheduling of the various core tests, such that test application time and power consumption are minimized. In this paper, we outline a procedure called *test protocol scheduling* that uses the test protocol as its basic schedule unit and demonstrate its advantages over other approaches.
4. Finally, the various SOC-level tests need to be written out for simulation or actual application on an IC tester. Hereto, the expanded and possibly scheduled test protocols are combined with their original test patterns in a process that we call *test assembly*.

The sequel of this paper is organized as follows. Section 2 provides a short overview of the prior work done on test protocols, test expansion, and test scheduling. Section 3 gives the formal definition of cores and their hierarchical netlists that we use throughout the paper. Section 4 introduces and formalizes the concept of test protocols. Sections 5–7 describe subsequently the details of test protocol expansion, test protocol scheduling, and test assembly. Section 8 puts the industrial usage of the methods described into perspective, and Section 9 concludes this paper.

2. Prior Work

The field of automation for modular, core-based test development is relatively new, and hence not much literature is available on methods or tools that use core-level tests as building blocks for SOC-level tests. Many still perform test pattern generation at SOC-level, with only dedicated tests for embedded modules with different circuit structures, such as embedded memories or analog modules [33]. It is obvious that black-boxed IP cores break this traditional method. However, very large SOC designs also become increasingly intractable for test generation tools, even if all implementation details are known. A modular test development approach breaks the overall problem into better digestable chunks, and puts the responsibility of DfT insertion and test generation there where most knowledge is available to execute these tasks. Also, modular, core-based testing enables reuse of tests over a family of SOC designs, where subsequent derivates only differ in a few modules.

The concepts of test protocols and test assembly as described in this paper have, to the best of our knowledge, not been published by others before. The concept of separation of test protocol and test patterns has recently also been adopted by IEEE P1500 (SECT) [26], the industry-wide standard-under-development for embedded core test, and its Core Test Language (CTL), now standardized as IEEE P1450.6 [19]. The main difference between the approach described in this paper and the approach of CTL, is that in our case, a test protocol refers to test patterns, whereas the reverse is the case in CTL. Also, it is not clear yet how automation for CTL will work out.

The translation of core-level tests into SOC-level tests has received some attention. Monzel and Orosz [33] describe very briefly how entire tests are 'migrated' from core-level to SOC-level. In [11], Ghosh et al. describe another method, based on 'design-for-transparency.' Test access to embedded cores is obtained by propagation of stimuli and responses through neighboring cores. For this, the core designs are, if necessary, augmented with design-for-transparency, such that every core terminal has transparent paths through the core. The transparency is limited to scan chains and multiplexers only, and does not include combinational logic. The requirement for transparency on all core terminals will be overkill in many practical applications. In a subsequent paper [10], the same authors extend this concept by describing a trade-off between the silicon area consumed by the design-for-transparency and its latency. They suggest that the core provider offers a catalogue of area/latency versions per functional core design, from which the core user can select one in order to meet the test access needs of the neighboring cores. Although interesting, this method seems rather impractical: a large effort is required on the part of the core provider to maintain the catalogue of core versions, and, more importantly, if test pipelining is allowed, latency has actually no impact on test throughput. Mentor Graphics has developed an extension to its ATPG tool FASTSCAN, called FASTSCAN MACROTEST, for test expansion of small non-scannable embedded modules [13]. It uses the ATPG engine to translate the tests of the embedded modules to the scan chains of the surrounding logic and then up to the SOC pins. Successful usage of this tool on 22 small embedded memories of AMD's Athlon processor has been reported in [36]. FASTSCAN MACROTEST is not based on test protocols, and therefore has to expand all bits of all test patterns to SOC pins. In this paper, we demonstrate that our approach to test expansion is (1) more generic, because we can handle basically any type of transparency, including transparency through both combinational and sequential circuitry, and (2) more efficient, because it is executed on test protocols only.

From all topics addressed in this paper, test scheduling has received the most attention in the literature so far. Sugihara et al. [37], Chakrabarty [7, 8], Larsson and Peng [21, 22], and Rosinger et al. [35] have all published papers that are based on the Test Bus Architecture [12]. This model assumes that only one core at a time can be connected to a TAM, and hence, cores connected to the same TAM cannot execute externally-applied tests in parallel. Some of these papers consider a combination of BIST and externally-applied tests for cores, and then BIST engines which are shared between cores can cause another type of resource conflict. Several papers assume that each core test consumes a certain amount of power, and in those cases, the scheduling takes into account that a global power budget is never exceeded. On top of this, Iyengar and Chakrabarty [17] also take pre-emption and precedence constraints into account. In this paper we show that our approach to test scheduling is efficient, as it is executed on expanded test protocols only, and hence we need to check for resource conflicts only at the SOC pins, while we can abstract from the actual stimulus and response values. Our scheduling approach is more generic than other approaches, as it is not limited to any TAM or BIST model.

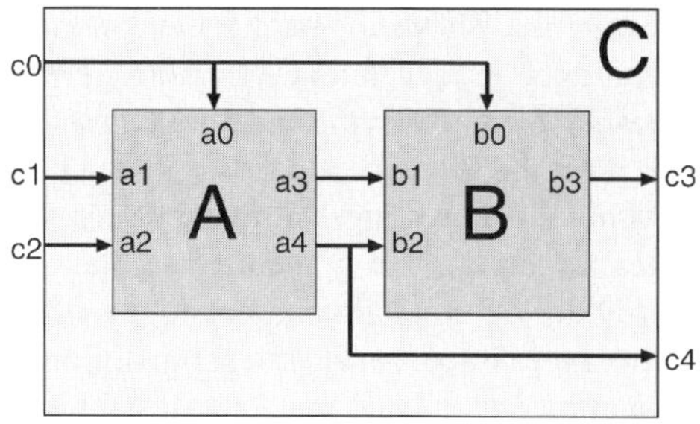

(a)

- $A = (\{a0, a1, a2\}, \{a3, a4\}, \emptyset, \nu_{\text{id}})$, where ν_{id} is the identity function.
- $B = (\{b0, b1, b2\}, \{b3\}, \emptyset, \nu_{\text{id}})$.
- $C = (\{c0, c1, c2\}, \{c3, c4\}, \{A, B\}, \nu_C)$, with
  $\nu_C(c0) = \nu_C(a0) = \nu_C(b0) = \{c0, a0, b0\}$,
  $\nu_C(c1) = \nu_C(a1) = \{c1, a1\}$,
  $\nu_C(c2) = \nu_C(a2) = \{c2, a2\}$
  $\nu_C(a3) = \nu_C(b1) = \{a3, b1\}$,
  $\nu_C(a4) = \nu_C(b2) = \nu_C(c4) = \{a4, b2, c4\}$,
  $\nu_C(b3) = \nu_C(c3) = \{b3, c3\}$.

(b)

Fig. 1. Simple example circuit (a) and its formal notation (b).

We do take into account pre-emption and precedence constraints, and additionally also user steering with respect to which tests are candidates for parallel execution. Our current approach does *not* take into account the power consumption during testing.

3. Cores and Hierarchy

This section defines the hierarchical model of cores, SOCs, and their interconnections, that we will use throughout the paper. Our model of netlists is based on the following definition of a core.

Definition 1 [core]. Core C is four tuple (PI, PO, I, ν), where

- $PI \subseteq TERM$[1] is a finite set of input terminals, called the *primary inputs*,
- $PO \subseteq TERM$ is a finite set of output terminals, called the *primary outputs*,
- I, the *instance set*, is a finite set of cores, and
- Let $TERM_C = PI \cup PO \cup I.PI \cup I.PO$, i.e., all terminals of Core C and its instances.
  $\nu: TERM_C \to \mathcal{P}(TERM_C)$ is the *net function*, where, for $p \in TERM_C$,
  $p \in \nu(p) \wedge \forall_{q \in \nu(p)} \{\nu(q) = \nu(p)\}$. □

In Definition 1, $I.PI$ refers to the union of all sets of primary inputs of all cores in set I, i.e., $I.PI = \bigcup_{c \in I}\{c.PI\}$. Likewise, $I.PO = \bigcup_{c \in I}\{c.PO\}$.

A Core $C = (PI, PO, I, \nu)$ has primary input and output terminals, possibly lower-level cores, called instances, and nets that connect the terminals of Core C and the terminals of its instances. A net is defined as a set of electrically-connected terminals; every terminal is associated with the net of which it forms part by means of net function ν. For reasons of simplicity, we exclude bidirectional terminals from the discussion in this paper. Hence, $PI \cap PO = \emptyset$. Note that this definition of a core is a recursive one; the instance set of a core consists of other cores. The definition defines a hierarchical tree of cores. The leafs of the tree are cores with empty instance sets. A system chip itself is the core in the root of such a hierarchy tree.

Fig. 1 illustrates the usage of Definition 1 by means of a simple example with three cores in a hierarchical setting. Fig. 1(a) shows Core C, which contains two 'atomic' Cores A and B. Fig. 1(b) contains the formal notation for this circuit.

Based on the previous definition, we can formally define the notion of a core being *included* in another core.

Definition 2 [inclusion]. Let X, Y be cores.
The *inclusion* of X in Y, denoted by $X \sqsubseteq^* Y$, is defined by

$$X \sqsubseteq^* Y \equiv (X = Y) \vee \exists_{m \in Y.I}\{X \sqsubseteq^* m\}. \qquad \square$$

In Definition 2, $Y.I$ refers to the instance set of Core Y. Note that the base of this recursive definition is formed by the leaf cores in the design hierarchy, for which the instance set is empty and hence the second half of the definition resolves to *false*.

In the example of Fig. 1, Cores A and B are both included in Core C, i.e., $A \sqsubseteq^* C$ and $B \sqsubseteq^* C$.

4. Test Protocols

In our approach, we distinguish between test patterns and test protocols. A *test pattern* is an array with stimulus and response values. For example, in case of a scan-testable core, a test pattern consists of all bits involved in a full load, apply/capture, and unload of the scan chains. A *test pattern list* is the list of all patterns needed to test a core. The order of the test patterns in the list corresponds to the order in which the various test patterns should be applied. A *test protocol* specifies formally how a single test pattern should be applied. A *test* for a certain core consists of the repeated execution of its test protocol, with every time another test pattern from the corresponding test pattern list filled in.

A test protocol of a core describes in detail how to apply the stimuli and capture the responses of a single test pattern of that core. It contains information about at which terminals and relative clock cycles to apply or capture test data. A test protocol is by definition pattern-value independent. It contains no information about the number of test patterns nor their content. It does contain pattern indices, which refer to the stimulus or response values in a corresponding test pattern, and it contains test control data.

4.1. Test Protocol Example

Fig. 2 illustrates the test protocol concept by means of a simple scan test example. Fig. 2(a) shows a Core A with two scan chains, consisting of three and six flip flops respectively. The terminal *sc* controls the operation of both scan chains; a logic '1' on this input makes the scan chains shift, while a logic '0' sets the scan flip flops in their functional (non-scan) mode. Apart from two scan inputs and two scan outputs, Core A also has primary inputs, $a[0:4]$, and primary outputs, $z[0:2]$.

Fig. 2(b) shows a table with four of the possibly very many test patterns for Core A. Every test pattern is an array of stimulus values (by convention denoted by '0' and '1') and response values (by convention denoted by 'L' and 'H'). A test pattern for Core A contains fourteen stimulus values; three to be shifted into scan chain 0, six to be shifted into scan chain 1, and five to be applied on input terminals $a0 \ldots a4$. Likewise, a test pattern for Core A contains twelve response values.

Fig. 2(d) lists the test protocol for Core A. It consists of five parts, of which the semantics are informally defined below.

1. *Patternfile*: Defines the test pattern file for Core A, to which is referenced by the test pattern indices in this protocol.

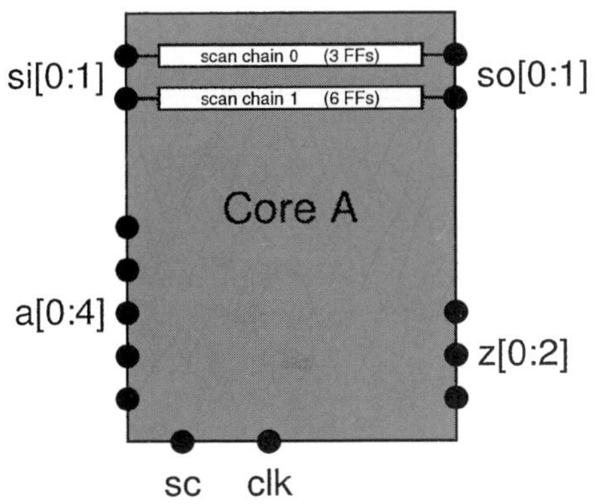

(a)

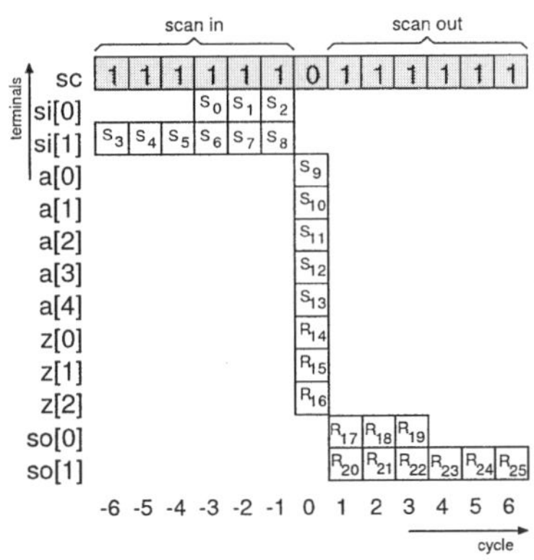

(c)

Pat.	Stimulus/Response Values																									
	0	1	2	3	4	5	6	7	8	9	10	11	12	13	14	15	16	17	18	19	20	21	22	23	24	25
1	0	0	0	1	1	0	1	0	0	1	1	0	0	1	H	H	L	L	H	L	L	H	L	H	L	L
2	1	0	0	1	1	0	0	1	1	0	1	1	0	0	H	H	H	H	L	L	H	L	H	H	L	H
3	1	1	1	1	0	0	0	1	0	1	0	0	1	0	L	L	H	H	L	H	H	L	H	L	L	L
4	0	0	0	1	1	1	1	1	1	1	0	0	0	1	H	L	H	L	L	H	H	L	L	H	L	H

(b)

```
Cell A {
 Protocol at A {
  Patternfile = "scan.pat";
  Initialization { }
  Condition {
   sc<-6:-1> = [1]*;
   sc<0> = [0];
   sc<1:6> = [1]*;
  }
  Stimulus {
   si[0]<-3:-1> = S[0:2];
   si[1]<-6:-1> = S[3:8];
   a[0]<0> = S[9];
   a[1]<0> = S[10];
   a[2]<0> = S[11];
   a[3]<0> = S[12];
   a[4]<0> = S[13];
  }
  Response {
   z[0]<0> = R[14];
   z[1]<0> = R[15];
   z[2]<0> = R[16];
   so[0]<1:3> = R[17:19];
   so[1]<1:6> = R[20:25];
} } }
```

(d)

Fig. 2. Example of a simple scan test protocol: (a) core-under-test A, (b) test patterns `scan.pat` for A, (c) graphical representation of the test protocol for A, and (d) the textual representation of the test protocol for A.

2. *Initialization*: Defines an initialization sequence which is to be executed once before the repetitive part of the test starts. In this example, the initialization sequence is empty.
3. *Condition*: Defines the control values required during the test. The test protocol in our example defines that *sc* has to be set to '1' for six consecutive clock cycles, followed by a '0' for one clock cycle, and finally '1' for another six consecutive clock cycles. These control values correspond with the three phases of a conventional scan test: scan in, apply/capture, and scan out.
4. *Stimulus*: Defines which stimuli have to be supplied to which input terminals at which (relative) clock cycles. The actual test stimulus values are not defined; instead the test protocol refers to test pattern indices. For example, `S[9]` refers to the stimulus value at position 9 of a test pattern array in the test pattern file.
5. *Response*: Defines which responses have to be observed at which output terminals at which (relative) clock cycles. The reference mechanism to the pattern file works as described above for stimuli.

Fig. 2(c) depicts the test protocol for Core *A* graphically. At the horizontal axis of this graph the time (in clock cycles) is plotted. Vertically is set out which terminals of Core *A* are involved in the test. The test control values for terminal *sc*, represented by the gray-shaded squares, are specified by the test protocol; they are equal for every test pattern. The white squares represent the test-pattern-dependent stimuli and responses. The values S_x and R_y correspond to the entries `S[x]` and `R[y]` in the textual version of the test protocol in Fig. 2(d). These values are only references to the test pattern file; their actual values depend on the contents of the test patterns.

It is important to note that all timing values relate to an arbitrarily chosen reference point in time. In case of scan testing, by convention we choose the 'apply/capture' clock cycle as reference (null) point, such that scanning in takes place in the negative time, and scanning out in the positive time. However, this is just a convention and nothing prohibits us to choose our reference point differently. Note also that upon actual test execution, the initialization sequence is placed in front of the repeated execution of the rest of the protocol. Hence, the clock cycle references in the *initialization* part of the test protocol only determine the relative timing of the initialization sequence and are unrelated to the other clock cycle references in the test protocol.

A test protocol defines *necessary* and *sufficient* conditions to apply a test pattern. This means that terminal – clock cycle combinations which are unspecified in a test protocol (i.e., the unboxed white space in Fig. 2(c)) are *don't care* with respect to the application of test patterns.

4.2. Formalization of Test Protocols

Test protocols are described in a formal language, as shown in our example in Fig. 2(d). On the other hand, test protocols can also be formally defined in mathematical terms. We show this formalization below, as it helps to understand better the nature and expression power of test protocols, as well as the algorithms for test protocol expansion and test protocol scheduling in the sequel of this paper.

Definition 3 defines the values that occur in a test protocol. On the one hand we have conditions, represented by the logic values '0', '1', and 'Z'; the latter denotes a tri-state mode. On the other hand a test protocol might contain references to stimuli and responses specified in its corresponding test pattern list.

Definition 3 [set of values]. The *set of values*, denoted by *VALUE*, is defined by $VALUE = \{0, 1, Z\} \cup \bigcup_{i \in \mathbb{N}} \{S_i, \overline{S_i}, R_i, \overline{R_i}\}$. □

Value S_i is a reference to the ith index location in a corresponding test pattern; this should be a stimulus. Similarly, R_i is a reference to a response in a corresponding test pattern. We allow for inversions of reference values, which are denoted by $\overline{S_i}$ and $\overline{R_i}$. S_i may refer to the logic value '0' for one pattern, and a '1' or 'Z' for another. Similarly, R_i refers to either 'L', 'H', or 'U'. As a result we consider the values '0', '1', 'Z', S_i, S_j, R_i, and R_j (for $i \neq j$) to be pairwise different.

Definition 4 [set of test protocol elements]. The *set of test protocol elements*, denoted by *ELEM*, is defined by

$$ELEM = TERM \times \mathbb{Z} \times VALUE. \qquad \square$$

A test protocol element is a three tuple (t, z, v). It expresses that in order to test a certain core, a value v has to be applied (in case of a condition or stimulus) or observed (in case of a response) at terminal t at clock cycle z. In the graphical representation of a test protocol (cf. Fig. 2(c)), a test protocol element is denoted as a small square.

Two test protocol elements are conflicting if they require different values at a certain terminal at the same clock cycle. We want to prevent conflicts, and therefore we define the notion of conflict-free sets of test protocol elements.

Definition 5 [conflict free]. Let e and e' be two test protocol elements, i.e., $e, e' \in ELEM$, with $e = (t, z, v)$ and $e' = (t', z', v')$. We call e and e' *conflict free*, denoted by $cf(e, e')$, if and only if $(t \neq t' \vee z \neq z' \vee v = v')$.

A set of test protocol elements E is called *conflict free*, denoted by $cf(E)$, if and only if $\forall_{e,e' \in E}(cf(e, e'))$. □

Note that the above definition implies that $cf(E) \wedge E' \subseteq E \Rightarrow cf(E')$.

We consider a test protocol for Core A, expanded to Core C, as a conflict-free set of test protocol elements, of which all terminals are terminals of C and of which execution guarantees implicit execution of the original, unexpanded test protocol of A.

Definition 6 [test protocol]. Let A and C be two cores with $A \sqsubseteq^* C$.
A *test protocol* for Core A defined at Core C, denoted by $\mathcal{T}_{A,C}$, is defined as follows.

- $\mathcal{T}_{A,C} \in \mathcal{P}(ELEM)$, i.e., $\mathcal{T}_{A,C}$ is a set of test protocol elements.
- $cf(\mathcal{T}_{A,C})$, i.e., $\mathcal{T}_{A,C}$ is conflict free.
- $\forall_{(t,z,v) \in \mathcal{T}_{A,C}}\{t \in (C.PI \cup C.PO)\}$, i.e., all terminals of $\mathcal{T}_{A,C}$ are of Core C.
- Execution of $\mathcal{T}_{A,C}$ guarantees implicit execution of $\mathcal{T}_{A,A}$. □

5. Test Protocol Expansion

5.1. Test Protocol Expansion Concept

Test *expansion* is the process of translating a test defined at the terminals of an embedded core into a test defined at the terminals of the embedding entity. In case of a simple embedding of a core into an SOC, the translation is from the core terminals to the SOC pins. However, more complex cases with multiple layers of hierarchy also occur and hence should be handled. In these cases, test expansion repeatedly translates the test from one level of design hierarchy to a higher level of hierarchy, until the SOC pins are reached.

Test expansion involves searching for test access paths to and from the core-under-test along which stimuli and responses can be transported. The expansion needs to make sure that every stimulus value reaches the core-under-test at the right moment and every response value is observed at the right moment. Test expansion requires a test access path per test pattern bit. Depending on the actual value of this bit, the test access path needs to be capable of transporting either a '0' or a '1'. Test expansion searches for paths such that the combination of all test access paths for one test pattern is conflict-free. This process has to be repeated for every test pattern. A test for a core in a real-life, industrial SOC may easily consist of hundreds or thousands of test patterns, each consisting of tenthousands of stimulus and response values. This can make test expansion a very compute-intensive operation.

In order to prevent from an explosion in compute time, we perform expansion on test protocols, also referred to as *Test Protocol Expansion* (TPE). Instead of transforming the entire test from the core-under-test to the chip level, we only transform its test protocol. Test protocol expansion abstracts from the pattern values. Hence, per test protocol element, one path is required to be capable of transporting both '0' and '1'. Test protocol expansion searches for paths such that the combination of test access paths for all elements of the test protocol is conflict-free. The same set of paths can then be used for all possible combinations of test pattern values that might occur in the test pattern list. Hence, at the expensive of a stronger requirement with respect to the individual test access paths, we typically save two or three orders of magnitude in compute time, as the expansion process only needs to be done once for the test protocol, and does not need to be repeated for each test pattern. Another benefit is that all test patterns of the same test are routed over the same test access paths to the embedded core, thereby easing the processes of silicon debug and failure analysis.

Fig. 1 gives a simple example of a hierarchical circuit. Chip C consists of two cores, viz. Core A and Core B. For both A and B test protocols have been defined at their own respective terminals. These test protocols are named $\mathcal{T}_{A,A}$ (i.e., a test protocol which tests Core A and is defined at the terminals of Core A) and $\mathcal{T}_{B,B}$ respectively. Test protocol expansion is the transformation of $\mathcal{T}_{A,A}$ into $\mathcal{T}_{A,C}$, i.e., a test protocol which tests Core A and is defined at the pins of Chip C.

```
Cell And {
 Transfer in0_out {
  Condition {
   in1 = [1];
  }
  Effect {
   out = in0;
  }
 Transfer in1_out {
  Condition {
   in0 = [1];
  }
  Effect {
   out = in1;
 } } }
```

(a)

```
Cell RAM {
 Transfer Store5 {
  Condition {
   address<0> = [0101];
   address<5> = [0101];
   ce<0> = [1];
   ce<1:4> = [0]*;
   ce<5> = [1];
   rwn<0> = [0];
   rwn<5> = [1];
  }
  Effect {
   data_out<5>=data_in<0>;
 } } }
```

(b)

Fig. 3. Example transfers for (a) a two-input and gate and (b) a RAM.

Bottom line requirement for the expanded test protocol $\mathcal{T}_{A,C}$ is that, when executed, it implicitly executes test protocol $\mathcal{T}_{A,A}$. Likewise, test protocol $\mathcal{T}_{B,B}$ should be expanded to a test protocol $\mathcal{T}_{B,C}$.

During TPE no use is made of the logic values in the test patterns. In fact, TPE can take place even before the actual list of test patterns is generated. Only the stimulus and response references are used. All references that occurred in the original test protocol, will also occur in the expanded test protocol, so that both original and expanded test protocol match the same test pattern list.

5.2. *Transfer Through Other Circuitry*

Test access paths to the core-under-test inevitably use other parts of the SOC. Often, these test access paths do not simply consist of wires to SOC pins, but go to width-adapting core test wrappers, travel over TAMs, contain logic, or pass through other modules and cores. Our TPE algorithm can handle all of those by means of its flexible concept of *transfer*. Transfer is a way to describe the data transportation capability of a module at any level of circuit hierarchy. Gate-level transfer can be described for basic gates, such as AND- and OR-gates. For example, a two-input AND-gate has a transfer path from one input to the output, if the other input is held at logic '1'. Transfer can also invert its data; our TPE algorithm keeps track of these inversions. Furthermore, transfer at higher-level modules can be described. For instance, a multiplier has a transparent mode: multiplying by 1. Transfer can also be sequential in nature, meaning that there is a latency of one or more clock cycles between input and output. Even transfers through RAMs can be described by a variable latency.

Fig. 3 shows two examples of transfer information. The syntax of the transfer information is very similar to the one for test protocols, shown in Section 4. The transfer information in Fig. 3(a) is for a combinational two-input AND-gate, with inputs `in0` and `in1`, and output `out`. Two transfers are described. The logic value present at `in0` is transferred to `out` if a '1' is applied at `in1`. The second transfer is similar, with `in0` and `in1` swapped. The fact that no (relative) clock cycles are specified means that condition and effect both happen in the same clock cycle. The transfer information in Fig. 3(b) is for a RAM. In this example, relative clock cycles are specified. This example describes a transfer through the RAM with a latency of five clock cycles. The logic value present at RAM input `data_in` at clock cycle 0 can be read out at output `data_out` at clock cycle 5, provided we use address 0101 for writing and reading, the RAM is enabled during writing and reading by asserting `ce`, and the RAM is disabled during clock cycles 1 through 4 by de-asserting `ce`.

Transfer can be formally defined as follows.

Definition 7 [transfer]. A transfer set, denoted by *TF*, is defined by

$$TF = \mathcal{P}(TERM \times (TERM \times \mathbb{N}) \times \mathcal{P}(TERM \times \mathbb{Z} \times \{0, 1, Z\})).2 \qquad \square$$

A single transfer is a three-tuple $(i, (o, x), c)$. It expresses that a data value, be it either a stimulus, response, or condition value, can be transferred from terminal i to terminal o with x clock cycles delay, under condition set c. i and o are both terminals, x is a non-negative integer indicating the transfer delay in clock cycles, and c is a set of test protocol elements, i.e., three-tuples (t, z, v), of which the values v are restricted to control values. During TPE, a transfer can be used in two directions. A stimulus or condition value can be back-traced from o to i under condition set c. Likewise, a response value can be propagated forward from i to o under condition set c. The transfer example in Fig. 3(a) can be formally expressed as follows: $TF = \{(in0, (out, 0), \{(in1, 0, 1)\}), (in1, (out, 0), \{(in0, 0, 1)\})\}$.

The transfer set for a single module defines the transfer options through that particular module. In case of a TPE run on an entire SOC, all transfer sets that should be taken into account can be united into one combined transfer set for the whole SOC.

By means of the flexible notion of transfer, basically all possible test access paths to the core-under-test can indeed be utilized as such by the TPE tool. This ranges from dedicated TAMs such as the TESTRAIL [2, 25] to cores for which no dedicated TAM was implemented and test access is only possible by propagating signals through neighboring modules. In the latter case, TPE is in principle as difficult and as compute-time-consuming as Sequential ATPG.

Transfer information describes which circuit elements we allow to be part of the access paths for TPE. The more we limit the transfer information, the more we confine the search space, and consequently, the more we confine the compute time needed for TPE. However, we then also restrict the designer in his/her design freedom to make a circuit testable. For example, allowing only scan flip flops in the access paths eases TPE's computation, but restricts the designer who would like to use functional access paths in order to avoid extra DfT hardware. Transfer information provides a means to the IC designer and/or test engineer to optimize between DfT freedom and CPU time needed for TPE.

5.3. *Test Protocol Expansion Algorithm*

The TPE algorithm [28] is outlined in Algorithm 1. Three variables, S, R, and T, all representing sets of test protocol elements, play a key role in the algorithm. S represents the test protocol elements related to stimuli and conditions that still need to be expanded. R represents the test protocol elements related to responses that still need to be expanded. During initialization,

Algorithm 1 [TPE]

/* Initialization */
$S := \{(t, z, v) \mid (t, z, v) \in \mathcal{T}_{A,A} \wedge v \notin \{R_i, \overline{R_i}\}\}$;
$R := \{(t, z, v) \mid (t, z, v) \in \mathcal{T}_{A,A} \wedge v \in \{R_i, \overline{R_i}\}\}$;
$T := \emptyset$;
/* Loop 1: forward propagation of responses */
while $R \neq \emptyset$ **do** {
  let $r \in R$;
  $RP_{\text{feas}} := \{(t, z, v) \cup c \mid$
    $((t \in \nu(r) \backslash \{r\} \wedge z = r.z \wedge v = r.v \wedge c = \emptyset) \vee$
    $((r.t, (t, x), c) \in TF \wedge z = r.z + x \wedge v = r.v)) \wedge$
    $cf(S \cup R \backslash \{r\} \cup T \cup \{(t, z, v)\} \cup c)\}$;
  if $RP_{\text{feas}} \neq \emptyset$ {
    let $rp \in RP_{\text{feas}}$;
    $T := T \cup \{(t, z, v) \mid (t, z, v) \in rp \wedge t \in (C.PI \cup C.PO)\}$;
    $S := S \cup \{(t, z, v) \mid (t, z, v) \in rp \wedge t \notin (C.PI \cup C.PO) \wedge v \notin \{R_i, \overline{R_i}\}\}$;
    $R := R \backslash \{r\} \cup \{(t, z, v) \mid (t, z, v) \in rp \wedge t \notin (C.PI \cup C.PO) \wedge v \in \{R_i, \overline{R_i}\}\}$;
  }
  else { backtrack }
}
/* Loop 2: backward tracing of stimuli and conditions */
while $S \neq \emptyset$ **do** {
  let $s \in S$;
  $RP_{\text{feas}} := \{(t, z, v) \cup c \mid$
    $((t \in \nu(s) \backslash \{s\} \wedge z = s.z \wedge v = s.v \wedge c = \emptyset) \vee$
    $((t, (s.t, x), c) \in TF \wedge z = s.z - x \wedge v = s.v)) \wedge$
    $cf(S \backslash \{s\} \cup T \cup \{(t, z, v)\} \cup c)\}$;
  if $RP_{\text{feas}} \neq \emptyset$ {
    let $rp \in RP_{\text{feas}}$;
    $T := T \cup \{(t, z, v) \mid (t, z, v) \in rp \wedge t \in (C.PI \cup C.PO)\}$;
    $S := S \backslash \{s\} \cup \{(t, z, v) \mid (t, z, v) \in rp \wedge t \notin (C.PI \cup C.PO)\}$;
  }
  else { backtrack }
}

the test protocol elements of $\mathcal{T}_{A,A}$ are divided over S and R. T represents the test protocol elements of which expansion has been completed, because their terminals are at top-level C; initially, T is empty. After initialization, the algorithm has two main iterative loops. In the first loop, test responses are propagated forward to the terminals of C. In the second loop, test stimuli and conditions are traced backward to the terminals of C. These two loops show large resemblance.

During the first iterative loop, every time an element r of R is selected, for which a set of feasible replacements RP_{feas} is computed. A *replacement* is a set of test protocol elements that, if executed, guarantee the execution of the selected test protocol element r, that is to be replaced. A replacement (Line 8) is made up of a test protocol element (t, z, v) that will replace r, and a (possibly empty) condition set c of test protocol elements. The replacements are based on the connectivity as described in the net element function ν (Line 9, cf. Definition 3) or the transfer set TF (Line 10, cf. Definition 5.2). A *feasible* replacement is a replacement that maintains the conflict-freeness of the (partially expanded) test protocol, i.e., that maintains the invariant $cf(S \cup R \cup T)$. If RP_{feas} is not empty, one of its feasible replacements is selected and applied. Element r is deleted from R. The replacing test protocol elements are added to either S, R, or T. Elements with their terminal at C are added to T, others need further expansion and are added to either S (in the case of conditions) or R (in the case of responses). If RP_{feas} is empty, there are no possible replacements for the currently selected elements r; backtracking on previous decisions is then needed.

The second iterative loop is similar to the first, but here stimuli and conditions are handled. Hence, every time an element s of S is selected, for which a set of feasible replacements is computed and one of them applied. The propagation of a response in the first loop typically yields both a new element of R (the response propagation itself), as well as zero or more new elements of S (the conditions that need to be set to open up the response propagation path). However, the backtracing of a stimulus or condition value in the second loop only yields new elements of S. The responses are all handled before the stimuli and conditions are dealt with, and therefore, variable R does not play a role any more in the second loop.

Per loop, there are two points in the algorithm where selections are made.

1. **let** $r \in R$ (Line 7) and **let** $s \in S$ (Line 22): Selection of the next test protocol element to be replaced.
2. **let** $rp \in RP_{\text{feas}}$ (Lines 13 and 28): Selection of one of several feasible replacements for r or s.

These selections often provide a choice, as in many cases R, S, and RP_{feas} are not singletons. During the course of the execution, the two types of selections are made in alternating order. These selections form the nodes in the *execution tree* of the algorithm; backtracking is done with respect to this tree. The backtracking mechanism is not explained here; it works as one would intuitively expect.

Note that during the iteration cycles, the following invariants are maintained.

- INV_1: $T \in \mathcal{P}(ELEM)$, i.e., T is a set of test protocol elements.
- INV_2: $cf(T)$, i.e., T is conflict free. This is true because the algorithm specifically tests for $cf(S \cup R \cup T)$ and $cf(S \cup R \cup T) \Rightarrow cf(T)$.
- INV_3: $\forall_{(t,z,v) \in T} \{t \in (C \cdot PI \cup C \cdot PO)\}$, i.e., all terminals in T are at Core C.

The algorithm terminates if one of the following two conditions holds.

1. $R = \emptyset \wedge S = \emptyset$. The empty set R and S imply that all elements of $\mathcal{T}_{A,A}$ have been successfully expanded. Execution of T guarantees implicit execution of $\mathcal{T}_{A,A}$. In that case, $T = \mathcal{T}_{A,C}$.
2. The entire search tree has been traversed. Traversal of TPE's entire search space has taken place without R or S becoming empty. Hence, expansion is not feasible, given the netlist, transfer information, and original test protocol $\mathcal{T}_{A,A}$.

TPE can be compute-intensive. Boosten and Jacobs [5] proved that the Test Protocol Expansion Problem TPEP is $\mathcal{NP}$-hard [9] by a reduction from Satisfiability (SAT). While selecting the next element to be replaced and while selecting a feasible replacement, often multiple options are encountered. The algorithm uses heuristics to determine which branch seems most attractive to explore first [5, 39]. Dedicated TAMs, such as our Philips-internal TestRail [25], if recognized, enjoy the preference of the TPE algorithm for test data transfer. Additional heuristics providing intelligent selection criteria for r, s, and rp reduce the need for backtracking [5, 39]. Boosten and Jacobs were able to achieve impressive speed-ups on several known-hard test cases by implementing the following heuristics [5].

- Reduction of RP_{feas} by removal of redundant transfers (‘*possibility reduction*’).
- Reduction of RP_{feas} by administration of side effects of the expansion (‘*propagation of effects*’).
- Definition of an ordering of computations such that the algorithm can take advantage of previous results (‘*learning*’).

Furthermore, the tool is equipped with many handles to provide user control over the TPE process in terms of global test conditions, combinatorial trace depth, and compute time.

5.4. *Test Protocol Expansion Example*

Fig. 4(a) shows Core *A*, described in the previous section, now wrapped in a TestShell *SA* [25]. This TestShell allows for easy connection to a TestRail [25] that serves as our SOC-wide TAM. In our example, the TestRail is three bits wide; *tri*[0 : 2] and *tro*[0 : 2] are the TestRail inputs and outputs, respectively. The TestShell has four modes of operation: (1) normal (non-test) mode, in which the TestShell is transparent, (2) core-internal test mode, (3) core-external test mode, and (4) bypass mode. The test control signals for these modes come from a Test Control Block (TCB), which can be serially loaded via TestShell input *si*. Fig. 4(b) and (c) show the test protocol of Core *A*, expanded to the TestShell boundary. In the initialization part (only shown as pseudo-code in Fig. 4(b)), the TCB is put into its core-internal test mode by shifting in the control sequence '010'. Note that the expanded protocol contains the same references to test pattern values as the original protocol in Fig. 2, but that the terminals and clock cycles at which they are applied have changed.

Fig. 5(a) shows Core *A* plus its TestShell, together with another Core *B* and its TestShell, as part of a larger SOC *C*. The test control lines of the two TCBs are connected to an IEEE 1149.1 TAP controller ('JTAG') [16]. After reset, the TAP controller can be brought into a state in which it transfers its role of test controller to the core-level TCBs. Fig. 5(b) and (c) show the test protocol of Core *A*, expanded to the chip pins. The initialization part of the test protocol (only shown as pseudo-code in Fig. 4(b)) shows that in order to test Core *A*, first the TAP controller needs to be configured such the core-level TCBs are accessible. After this, Core *A* is brought into its core-internal test mode, while Core *B* is brought into its bypass mode by shifting in the control sequence '000' into the TCB of *B*. The bypass mode of Core *B* uses a bypass register, which results in one additional clock cycle latency before the responses of Core *A* are visible at the chip output pins. This is reflected in the expanded test protocol.

6. Test Protocol Scheduling

6.1. *Test Protocol Scheduling Concept*

Two important and related factors in the recurring costs of testing SOCs are the size of the external test vector set and the test application time [31]. The external test vector set size dictates the required size of drive/compare memories of the test equipment, which is one of the key price parameters of such equipment. The test application time per SOC determines the total cost of test and handling equipment necessary to obtain a certain throughput of tested products. Reduction of both cost factors directly reduces the costs of testing.

Both the test vector set and the test application time can be reduced by simply not executing all core test patterns; however, assuming that these test patterns were necessary to obtain fault coverage, this will

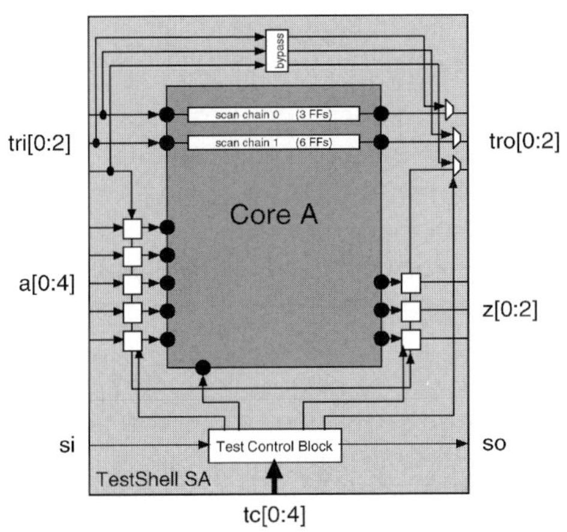

(a)

```
Cell A {
  Protocol at SA {
    Patternfile = "scan.pat";
    Initialization {
      TCB to 'mode receive';
      si<0:2> = [010];
      TCB to 'mode apply';
    }
    Condition {
      tc[2]<-6:-1> = [1]*;
      tc[2]<0> = [0];
      tc[2]<1:6> = [1]*;
    }
    Stimulus {
      tri[0]<-3:-1> = s[0:2];
      tri[1]<-6:-1> = s[3:8];
      tri[2]<-5:-1> = s[13:9];
    }
    Response {
      tro[0]<1:3> = r[17:19];
      tro[1]<1:6> = r[20:25];
      tro[2]<1:3> = r[14:16];
    }
  }
}
```

(b)

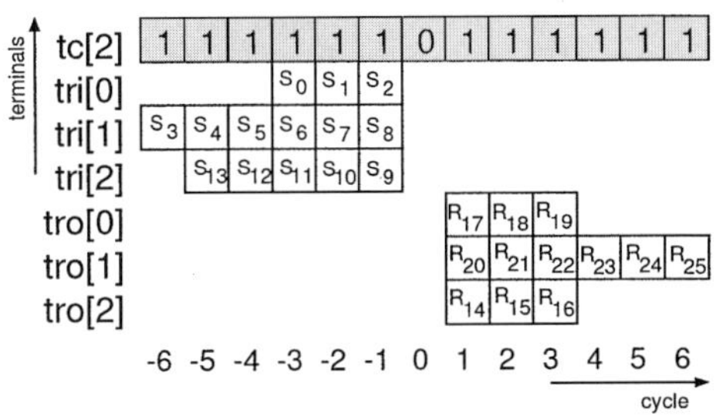

(c)

Fig. 4. (a) Example Core *A* with TestShell *SA*, (b) test protocol $\mathcal{T}_{A,SA}$, and (c) its graphical representation.

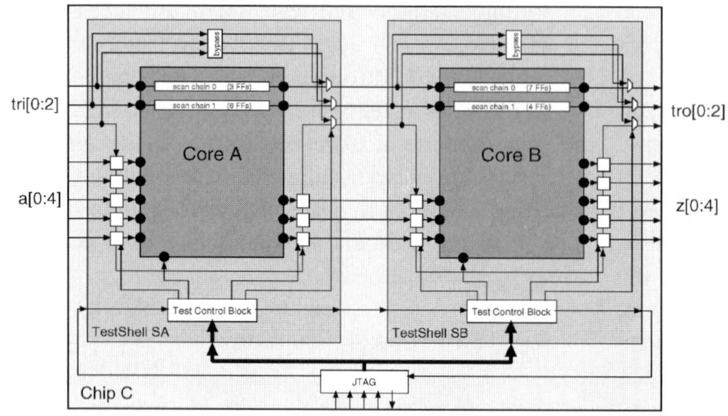

(a)

```
Cell A {
 Protocol at C {
  Patternfile = "scan.pat";
  Initialization {
   trstn<-1> = [0];
   JTAG to 'TCB mode';
   TCBs to 'mode receive';
   tdi<0:2> = [000];
   tdi<3:5> = [010];
   TCBs to 'mode apply';
  }
  Condition {
   tdi<-6:-1> = [1]*;
   tdi<0> = [0];
   tdi<1:7> = [1]*;
  }
  Stimulus {
   tri[0]<-3:-1> = s[0:2];
   tri[1]<-6:-1> = s[3:8];
   tri[2]<-5:-1> = s[13:9];
  }
  Response {
   tro[0]<2:4> = r[17:19];
   tro[1]<2:7> = r[20:25];
   tro[2]<2:4> = r[14:16];
  }
 }
}
```

(b)

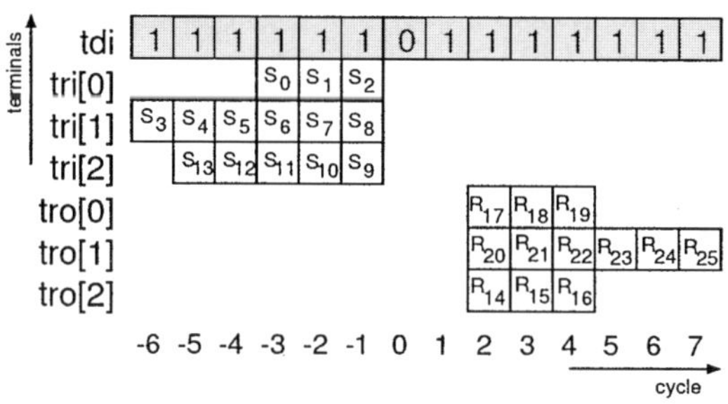

(c)

Fig. 5. (a) Example Core A with TestShell SA embedded in SOC C, (b) test protocol $\mathcal{T}_{A,C}$, and (c) its graphical representation.

compromise the test quality and hence we reject this solution. Built-In Self Test (BIST) is an attractive way to reduce the external test vector set of a core; the stimulus generator and/or response evaluator might be built into the core or reside on-chip outside the core. Unfortunately, BIST is not always applicable. Not all cores come with BIST capability inside, and also not all cores are 'BIST Ready', i.e., specifically prepared for BIST. These unprepared cores might generate values which corrupt the function of the BIST response evaluator or might not obtain sufficient fault coverage with pseudo-random test stimuli. Other cores might not contain any design-for-testability hardware and can only be tested through functional test patterns, which are difficult to generate by on-chip resources. Furthermore, whereas the external test vector set is reduced, many BIST methods increase the internal test vector set and hence the test application time.

The external test vector set and the test application time can also be reduced by careful test scheduling. For a core-based design, with a given set of expanded test protocols and corresponding lists of test patterns, *test scheduling* is the determination of the start times of all core tests, such that the completion time of the last core test is minimized, and hence the total vector set of the SOC is minimized. The test vector set reduction is achieved by executing (parts of) tests simultaneously. An important constraint for test scheduling is that the tests of which the executions overlap do not have *resource conflicts*, neither at the SOC pins, nor at SOC-internal wires or logic. A resource conflict occurs when conflicting logical values need to be transported over the same wire at the same moment in time. Test scheduling does not affect the quality level of the tests, and is applicable to all kinds of cores and their tests, including structural tests applied through scan or built-in self test, as well as functional tests.

TAM design and test scheduling are strongly interrelated. The TAM architecture, i.e., the TAM type, the number of TAMs, their widths, and the assignment of cores to TAMs all have a major impact on the total resulting test time [18]. In our discussion on test scheduling, we assume that the test access infrastructure, whether or not based on dedicated TAMs, is given, and even that a particular expansion for the core tests to SOC level is given. Test scheduling then tries to minimize the overall completion time of the tests by overlapping as much as possible the execution of the various tests.

Test scheduling can be done at various granularity levels. Types of test scheduling vary in (1) the size of the schedule unit, and (2) whether or not the schedule objects contain the actual values of the stimuli and responses, or only references to those. We use *Test Protocol Scheduling* (TPS), where the schedule units are test protocols. Compared to using entire tests as schedule units [7, 8, 17, 21, 22], the fine grain of test protocols allows to better exploit parallelism opportunities. Nevertheless, TPS is not very computationally demanding, as we abstract from the actual values of the test patterns.

TPS uses expanded test protocols as its basic schedule unit. TPS has to determine the start times of the executions of the expanded test protocols for all test patterns of all cores, such that (1) no resource conflicts occur with respect to the test access infrastructure from SOC pins to core terminals and vice versa, and (2) the overall test completion time is minimized.

TPE guarantees that execution of an expanded test protocol implies execution of the corresponding original, unexpanded test protocol. TPE preserves the properties 'necessity' and 'sufficiency' of test protocols throughout its expansion process. This means that if two expanded test protocols of different cores can be executed without a conflict at the SOC pins, the expansion process of TPE guarantees that both original test protocols are executed at the terminals of their respective cores. Hence, the resource conflict checking for test protocols expanded by TPE can be limited to checking that no conflicts occur at the SOC pins, and does not require knowledge of the full test access infrastructure to the respective cores. A conflict at an SOC pin between two expanded test protocols is defined in line with Definition 5; an SOC pin cannot have two different values, required by two different test protocols, at the same clock cycle.

6.2. *Formal Problem Definition*

Consider a set of tests defined at SOC level, consisting of a set of expanded test protocols T and corresponding lists of test patterns. For every test protocol $\mathcal{T} \in T$, the corresponding test pattern list contains $p_{\mathcal{T}}$ test patterns, $p_{\mathcal{T}} \in \mathbb{N}$. Adequate testing of the entire SOC requires execution of all test protocols in T; each test protocol $\mathcal{T} \in T$ should be executed $p_{\mathcal{T}}$ times, with each time another test pattern from the corresponding test pattern list filled in.

The time in test protocol elements is relative to an arbitrarily chosen reference point. Hence, without loss of generality, we can assume that every test protocol starts at relative time zero, i.e.,

$$\forall_{\mathcal{T} \in T} \left\{ \min_{(t,z,v) \in \mathcal{T}} (z) = 0 \right\}.$$

The test time necessary for one execution of test protocol $\mathcal{T}$ is then given by

$$tt_{\mathcal{T}} = \max_{(t,z,v) \in \mathcal{T}} t[z] + 1.$$

In TPS, one execution of one test protocol is considered as schedule unit. We denote for $\mathcal{T} \in T$ the start time of execution i by $s_{\mathcal{T}}(i)$, for $i = \{1, \ldots, p_{\mathcal{T}}\}$. All start times are non-negative integers, so $s_{\mathcal{T}}(i) \in \mathbb{N}$. The objective of TPS is to minimize the total test time, which is given by

$$\max_{\mathcal{T} \in T,\ i \in \{1,\ldots,p_{\mathcal{T}}\}} \{s_{\mathcal{T}}(i) + tt_{\mathcal{T}}\}.$$

To obtain a feasible solution, the start times should be chosen such that they do not result in conflicts between test protocol elements of any two test protocol executions. This means that

$$\forall_{\mathcal{T},\mathcal{T}' \in T} \forall_{(t,z,v) \in \mathcal{T},(t',z',v') \in \mathcal{T}'} \forall_{i \in \{1,\ldots,p_{\mathcal{T}}\}, j \in \{1,\ldots,p_{\mathcal{T}'}\}}$$
$$\{t \neq t' \ \vee \ s_{\mathcal{T}}(i) + z \neq s_{\mathcal{T}'}(j) + z' \ \vee \ v = v'\}.$$

The TPS problem can be formally defined as follows.

Problem 1 [Test Protocol Scheduling]. Given a set of test protocols and corresponding number of test patterns. Assign a starting time $s_{\mathcal{T}}(i) \in \mathbb{N}$ to each execution $i \in \{1, \ldots, p_{\mathcal{T}}\}$ of all test protocols $\mathcal{T} \in T$ such that the completion time $\max_{\mathcal{T} \in T, i \in \{1,\ldots,p_{\mathcal{T}}\}}\{s_{\mathcal{T}}(i) + tt_{\mathcal{T}}\}$ is minimized, and

$$\forall_{\mathcal{T},\mathcal{T}' \in T} \forall_{(t,z,v) \in \mathcal{T},(t',z',v') \in \mathcal{T}'} \forall_{i \in \{1,\ldots,p_{\mathcal{T}}\}, j \in \{1,\ldots,p_{\mathcal{T}'}\}}$$
$$\{t \neq t' \vee s_{\mathcal{T}}(i) + z \neq s_{\mathcal{T}'}(j) + z' \vee v = v'\}. \quad \square$$

6.3. *Relation* TPS *with Job Shop Scheduling*

Scheduling in general can be defined as the allocation of limited resources to tasks over time [34]. For TPS we consider the SOC pins as the resources and the tests as the tasks. In this section we describe the relation between the well-known Job Shop Scheduling (JSS) problem [9, 34] and TPS . We start with a problem definition of JSS.

Problem 2 [Job Shop Scheduling]. Given are a number of machines and a number of jobs. Each job consists of a number of operations, which have to be processed in a pre-defined order. Furthermore each operation has to be processed on a specified machine for a given amount of time. The machines can only handle one operation at a time and the execution of an operation cannot be interrupted. The problem is to assign a non-negative starting time to each operation, such that the completion time of the last operation is minimized. $\square$

To translate a TPS instance into a JSS instance, we consider the SOC pins as the machines and the tests as the jobs. Furthermore we consider an operation to be a subset with maximal cardinality of elements of a test protocol, for which the terminal is the same for all elements and the relative time points of the elements are consecutive. The above correspondence is shown in Fig. 6. The processing time of an operation is defined as the number of test protocol elements it corresponds to.

In contrast to JSS, where only the order in which the operations have to be executed is defined, in TPS the starting times of the operations are strictly related to each other, in such a way that the assignment of a starting time to one operation fixes the starting times of all operations of the job. Scheduling literature refers to this as the *no-wait constraint* [15]. For JSS this means that once the execution of the first operation of a job is started, all operations of this job have to be executed without interruption, as described in the following problem definition.

Problem 3 [No-Wait Job Shop Scheduling]. Given is an instance of job shop scheduling. Find an assignment of non-negative starting times that minimizes the completion time of the last operation under the additional constraint that the starting time of an operation equals the completion time of its predecessor, except for the first operation of a job. $\square$

The above shows a large resemblance between TPS and no-wait JSS, but we can distinguish two differences.

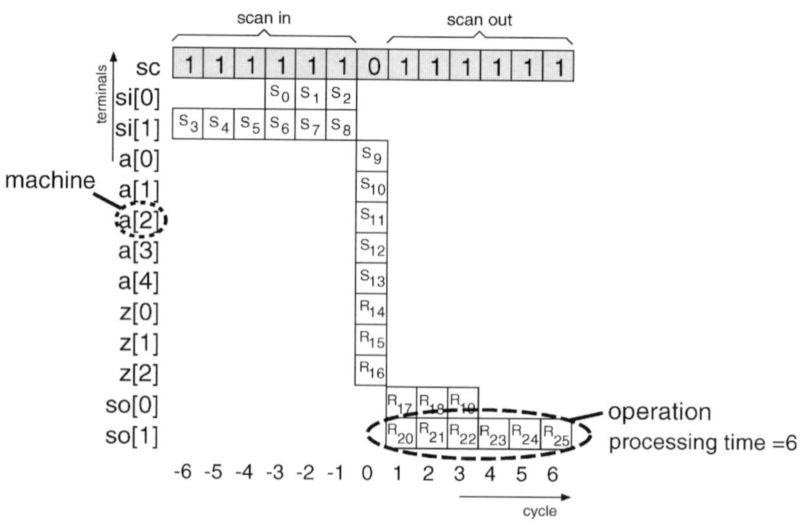

Fig. 6. Relation between TPS and JSS.

The operations of one job in no-wait JSS are executed strictly after each other, where in TPS idle time can occur between operations of the same job and also operations of the same job can overlap. Furthermore, in JSS each machine can only handle one operation at a time, but in TPS it is allowed to use a pin for more than one control element at the same clock cycle if the same value has to be applied.

We used the correspondence between the two problems to prove that the TPS problem is $\mathcal{NP}$-hard in the strong sense [9] by a reduction from No-Wait Job Shop Scheduling [1, 24]. A survey of scheduling problems with no-wait constraints is given by Hall and Sriskandarajah [15]. However, they note in their conclusions that No-Wait Job Shop Scheduling has received surprisingly little attention. Consequently, the heuristic that is discussed in the next section is not based on general scheduling algorithms from literature, but specifically developed for our application.

6.4. *Test Protocol Scheduling Algorithm*

This section describes our TPS algorithm. We first outline the overall behavior of our algorithm, after which we provide more details on its various elements.

In overview, TPS works as follows. We compute for all pairs of expanded test protocols the potential gain in test application time of parallel execution. These potential gains are based on the overlap with which two protocols can be executed without conflicts and the minimum of the sizes of the two corresponding test pattern lists. The results of this computation are stored in a half matrix, the so-called *gains matrix*. A simple greedy heuristic iteratively selects the two test protocols with the highest potential gain in test time, implements the parallelization of the two test protocols involved, and updates the gains matrix. This process continues until no further test time reduction can be obtained. In this, we take into account whether or not pre-emption of the tests is allowed for the cores in question.

The user can determine which test protocols are candidates for parallel execution by specifying so-called *groups* of expanded test protocols. The tool only tries to overlap the execution of test protocols within one group and never parallelizes test protocols from different groups. Also, the user can specify *precedence constraints*, by specifying an execution order for groups of protocols. These user inputs can best be illustrated by means of an example. Consider a set of ten expanded test protocols, named $\mathcal{T}_1$, $\mathcal{T}_2, \ldots, \mathcal{T}_{10}$. A user could specify a set of scheduling and precedence constraints as follows: $\{\mathcal{T}_1, \mathcal{T}_2, \mathcal{T}_3, \mathcal{T}_4\}$; $\{\mathcal{T}_{10}\}$; $\{\mathcal{T}_5, \mathcal{T}_6, \mathcal{T}_7\}$; $\{\mathcal{T}_8, \mathcal{T}_9\}$. In this example, four groups are specified. Group 1 contains test protocols $\mathcal{T}_1$, $\mathcal{T}_2$, $\mathcal{T}_3$, and $\mathcal{T}_4$. Group 2 contains $\mathcal{T}_5$, $\mathcal{T}_6$, and $\mathcal{T}_7$. Group 3 contains $\mathcal{T}_8$ and $\mathcal{T}_9$. Group 4 contains only $\mathcal{T}_{10}$. The specified execution order amongst these groups is: Group 1, followed by Group 4, followed by Group 2, followed by Group 3. Within each group, TPS gets a free hand to obtain any sequential or parallel execution order, as long as the test application time is minimized. For Group 4, this is very straightforward; as the group consists of one protocol only, there is no additional scheduling work to be done. This is different however for the other three groups. One extreme of the above specification mechanism is that every protocol forms its own singleton group; in this case, the user allows no parallelism and specifies that all test protocols need to be executed sequentially according to the specified order. The other extreme is that all protocols are in one group, in which the user gives no constraints to the scheduler. The latter case is the default behavior.

Consider the case in which we want to schedule two tests, say A and B, with test protocols $\mathcal{T}_A$ and $\mathcal{T}_B$ and p_A and p_B test patterns respectively. We need to find a new composite test protocol $\mathcal{T}_{AB}$, of which execution guarantees the execution of both $\mathcal{T}_A$ and $\mathcal{T}_B$. The new test protocol $\mathcal{T}_{AB}$ is the conflict-free union of the sets of test protocol elements $\mathcal{T}_A$ and $\mathcal{T}_B$.

Let us assume $p_A < p_B$. In that case, it only makes sense to execute the composite test protocol for p_A patterns. The remaining $p_B - p_A$ test patterns for Test B can better be executed with test protocol $\mathcal{T}_B$ only, as that test protocol can only be equal or shorter in execution length and offers equal or better opportunities to be combined with yet other test protocols than the composite test protocol $\mathcal{T}_{AB}$. We refer to the fact that Test B will be divided into a part which will be executed in parallel with Test A and a remainder which will be executed on its own as *pattern list breaking*.

The scheduling of tests with broken test pattern lists depends on whether or not the test pattern list in question is *pre-emptable*. In case pre-emption is allowed, the execution of the test can be suspended for a certain amount of time. In case of pattern list breaking, if pre-emption is allowed, the execution of the remaining test patterns can be delayed in order to first execute one or more other tests. This degree of freedom might be beneficial for the overall test time of the SOC. Whether or not a test can be pre-empted is determined by the requirements for the integrity of the test in question and is specified by the user in the form of a *test pattern type* per core test. Most tests can be pre-empted. A DRAM test is an example of a test which often cannot be pre-empted, as interruption might lead to leakage and the loss of the memory contents.

In a test protocol, time is defined in a relative manner. When combining two test protocols, we need to fix the timing relation between the two. We utilize this in our algorithm for gain calculation to maximize the overlap between the two protocols, yet guaranteeing their conflict-freeness. This procedure is described in Algorithm 2. We left-align the two protocols in order to create maximum overlap o_{right} (Line 6). We iteratively shift one test protocol, say $\mathcal{T}_B$, towards the right until (i) the composition of the two test protocols is conflict free, or (ii) the two protocols do not overlap any longer (Line 9). The maximum overlap o_{right} is stored. Subsequently, we again left-align the two protocols (Line 18), and now determine the maximum overlap o_{left} when test protocol $\mathcal{T}_B$ is shifted to the left. The gain is the product of the maximum of o_{right} and o_{left} and the minimum of the pattern counts for A and B (Line 27).

Algorithm 2 [Gain Calculation for Tests A and B]

1 /* Initialization 1 */
2 $\mathcal{T}_{B,orig} := \mathcal{T}_B$;
3 $l_A := \min_{(t,z,v)\in\mathcal{T}_A}\{z\}$; $r_A := \max_{(t,z,v)\in\mathcal{T}_A}\{z\}$;
4 $l_B := \min_{(t,z,v)\in\mathcal{T}_B}\{z\}$; $r_B := \max_{(t,z,v)\in\mathcal{T}_B}\{z\}$;
5 /* Left-align protocols of A and B */
6 **for all** $(t, z, v) \in \mathcal{T}_B$ **do** $\{ z := z - (l_B - l_A) \}$;
7 /* Shift protocol of B to the right */
8 $o_{\text{right}} := \min(r_A, r_B) - \max(l_A, l_B)$;
9 **while** $\neg cf(\mathcal{T}_A \cup \mathcal{T}_B) \wedge o_{\text{right}} \neq 0$ **do** {
10   **for all** $(t, z, v) \in \mathcal{T}_B$ **do** $\{ z := z + 1 \}$;
11   $l_B := l_B + 1$; $r_B := r_B + 1$;
12   $o_{\text{right}} := \min(r_A, r_B) - \max(l_A, l_B)$
13 }
14 /* Initialization 2 */
15 $\mathcal{T}_B := \mathcal{T}_{B,orig}$;
16 $l_B := \min_{(t,z,v)\in\mathcal{T}_B}\{z\}$; $r_B := \max_{(t,z,v)\in\mathcal{T}_B}\{z\}$;
17 /* Left-align protocols of A and B */
18 **for all** $(t, z, v) \in \mathcal{T}_B$ **do** $\{ z := z - (l_B - l_A) \}$;
19 /* Shift protocol of B to the left */
20 $o_{\text{left}} := \min(r_A, r_B) - \max(l_A, l_B)$;
21 **while** $\neg cf(\mathcal{T}_A \cup \mathcal{T}_B) \wedge o_{\text{left}} > o_{\text{right}}$ **do** {
22   **for all** $(t, z, v) \in \mathcal{T}_B$ **do** $\{ z := z - 1 \}$;
23   $l_B := l_B - 1$; $r_B := r_B - 1$;
24   $o_{\text{left}} := \min(r_A, r_B) - \max(l_A, l_B)$
25 }
26 /* Calculate maximal gain */
27 gain $:= \max(o_{\text{right}}, o_{\text{left}}) \times \min(p_A, p_B)$

Algorithm 3 [Greedy Heuristic on Gains Matrix]

while $\exists$ { matrix entry > 0 } **do** {
  find highest matrix entry, say for Tests A and B;
  compose $\mathcal{T}_A$ and $\mathcal{T}_B$;
  delete gain matrix entries for Tests A and B;
  add gain matrix entries for Test AB (with $\mathcal{T}_{AB}$ and $p_{AB} = \min(p_A, p_B)$);
  if $p_A < p_B$ **do** {
    add gain matrix entries for Test B (with $\mathcal{T}_B$ and $p'_B = p_B - p_A$)
  }
  if $p_A > p_B$ **do** {
    add gain matrix entries for Test A (with $\mathcal{T}_A$ and $p'_A = p_A - p_B$)
  }
}

The potential gain of composing two test protocols is computed for all pairs of expanded test protocols and stored in the lower half of a matrix, known as the *gains matrix*. A simple greedy heuristic iteratively selects a pair of test protocols with the highest gain as candidates for parallel execution. This procedure is outline in Algorithm 3. The two tests that are to be merged, say Test A and B, are deleted from the gains matrix. In their place we add entries for the combined Test AB, and possibly for the remainder of either Test A or Test B. This procedure is repeated until the gains matrix does not contain any positive gains any more. This algorithm always terminates, as per iteration step, either the number of matrix entries decreases, or the number of matrix entries remains equal, but the gain values reported decrease.

7. Test Assembly

In our approach, at delivery of a core, its test is broken down into a test protocol and a test pattern list. All subsequent tasks, such as expansion and scheduling, are performed on test protocols only, while the test patterns remain untouched. Finally, the system chip test is put together by filling in the original test patterns into the expanded and scheduled test protocols. This task is called *test assembly*.

We illustrate the concept of test assembly by elaborating on the scan test example of the previous sections. For reasons of simplicity we suppose that we have a test pattern list for Core A which consists of only four test patterns; actual test pattern lists contain much more

Pat.	Stimulus/Response Values																									
	0	1	2	3	4	5	6	7	8	9	10	11	12	13	14	15	16	17	18	19	20	21	22	23	24	25
1	0	0	0	1	1	0	1	0	0	1	1	0	0	1	H	H	L	L	H	L	L	H	L	H	L	L
2	1	0	0	1	1	0	0	1	1	0	1	1	0	0	H	H	H	H	L	L	H	L	H	H	L	H
3	1	1	1	1	0	0	0	1	0	1	0	0	1	0	L	L	H	H	L	H	H	L	H	L	L	L
4	0	0	0	1	1	1	1	1	1	1	0	0	0	1	H	L	H	L	L	H	H	L	L	H	L	H

(a)

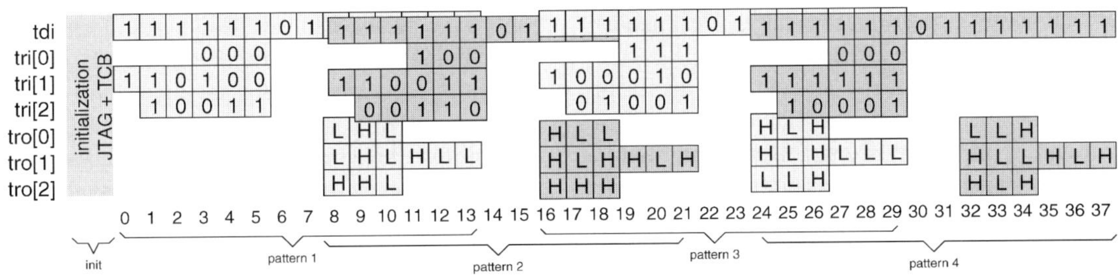

(b)

Fig. 7. Test patterns `scan.pat` for Core A (a) and the test assembly result for $\mathcal{T}_{A,C}$ (b).

patterns. Fig. 7(a) lists the stimulus and response values of all four test patterns.

Every test pattern is an array of stimulus values (denoted by '0' and '1') and response values (denoted by 'L' and 'H'). A test pattern for Core A contains fourteen stimulus values; three to be shifted into scan chain 0, six to be shifted into scan chain 1, and five to be applied on input terminals $i0 \ldots i4$. Likewise, a test pattern for Core A contains twelve response values. Which stimuli have to be applied at which inputs and which responses are expected at which outputs is described in the expanded test protocol $\mathcal{T}_{A,C}$ (see Fig. 5).

Fig. 7(b) is a graphical representation of the result of the test assembly process for Core A. In this example, only test protocol expansion and test assembly were applied; test protocol scheduling in order to test Cores A and B simultaneously was not taken into account.

The test for Core A defined at SOC C consists of a fourfold execution of test protocol $\mathcal{T}_{A,C}$. The four test patterns of `scan.pat` are filled in into the pattern-dependent part of this test protocol. Note that the details of the initializations of the JTAG TAP controller and TCBs are not shown in Fig. 7(b).

In order to reduce test time, consecutive executions of the same test protocol with subsequent test patterns are scheduled with as much overlap as possible without resource conflicts. For scan-testable Core A, this means that scanning out the responses of one test pattern is combined with scanning in the stimuli of the next pattern. Even in our small example, the total test time is reduced from 56 to 38 clock cycles, a reduction of 32%. For large sets of scan test patterns, this pipelining of scan out with scan in yields up to 50% reduction in test time. Our test assembly tool TASS handles this optimization automatically.

8. Industrial Usage of Test Protocols in SOC Testing

The concept of test protocols and their expansion and scheduling has been developed by Philips since the mid 1980s under the name 'Macro Test' [3, 4]. The original goal of Macro Test was to enable modular testing of ICs composed of 'macros' with different circuit structures, such as logic, memories, PLAs, register files, etc. [29]. Different circuit structures exhibit different defect behavior, and hence require dedicated test pattern generation and independent testing. Hence, the original focus of Macro Test was on improving test quality.

In recent years, semiconductor design and manufacturing technologies have enabled the creation of very large system chips. These SOCs have increased the need for a modular test strategy. Many SOCs use third-party cores as embedded design modules. For many of these cores, the implementation details are not known to their user, and hence he/she can only rely on the test patterns that come with the core and needs to test the embedded core as a separate entity. Next to the emphasis on test quality, reduction of test development time has become an important motivation for modular, core-based test development. The sheer size of many SOCs is impractical or intractable for DfT insertion and/or test generation tools. Also, many SOCs are designed as part of 'families', in which derivative SOC designs differ in only a few modules from their predecessors. This has boosted the need for a 'divide-and-conquer' test strategy, in contrast to having test generation tools choke on gigantic netlists.

In the early days of Macro Test usage, the focus of TPE was on utilizing existing functional transfer paths through neighboring macros. Both scan chains and functional transparent paths in neighboring macros were used. The reason for the emphasis on re-using existing functionality was that adding dedicated additional test access hardware was considered too expensive at that time. Over the years, area costs for design-for-testability hardware have decreased in importance, while reducing test development time has gained in importance. Hence, the focus in finding test access paths to embedded cores has shifted from re-used functional paths through neighboring modules (path tracing through "residential areas") to dedicated test access mechanisms (TAMs) around cores and other on-chip modules (path tracing via dedicated "highways") [23]. Dedicated TAMs have the benefit that TPE does not depend any longer on the amount of transfer information available for neighboring modules, which for black-boxed third-party cores is often not available at all. At the expense of a few percent silicon area, we reduce the test development time, as the TPE process becomes efficient and guaranteed successful.

The attractiveness of our test-protocol-based solution for testing system chips with embedded cores, as described in this paper, is that (1) it works very efficiently with dedicated TAMs such as our TestShell and TestRail [25], and (2) can still handle cases where such a dedicated TAM infrastructure is not or only partially available and we need to resort to exploiting functional transfer paths through neighboring modules [30].

Test protocols and their expansion and scheduling have been successfully used for numerous industrial ICs. Some of these cases have been published: the CarDSP [6], the Digital Still Camera IC [2], the Video Co-Processor Array [38], and the Nexperia™ Digital Video Platform Processor PNX8525 [40]. Many other applications remain unpublished.

9. Conclusion

In this paper we presented an automated modular test generation approach that uses pre-defined module tests. This test approach is meant for modular testing of large system chips, especially, but not exclusively, if the SOC has reusable pre-designed embedded cores. The methods divides a module or core test into a test protocol and a list of test patterns. The tasks of expansion and scheduling of the various core tests both take place on test protocols only, thereby saving orders of magnitude in compute time and memory usage. Only at the end of the test development trajectory, the expanded and scheduled test protocols and the original, untouched test pattern lists are brought together for test assembly. In this paper we have described the concepts, formalization, and algorithms of both test protocol expansion and test protocol scheduling. Tools based on the approach described have been and continue to be used for successful and efficient test generation of the tests of numerous industrial SOCs and their successor derivatives in various application domains. The approach described in this paper provides an impression of a new generation of hierarchical test development tools that is expected to come into usage throughout the industry once IEEE P1500 SECT [26] is approved and used.

Acknowledgments

I would like to acknowledge the cooperation and comments of my colleagues Emile Aarts, Krijn Kuiper, Paul Merkus, and Clemens Wouters. M.Sc. students Joep Aerts, Marcel Boosten, Harro Jacobs, Micle Moerenhout, and Michael van Wijngaarden and MTD student Florin Zavoianu (all with the Department of Mathematics and Computing Science of Eindhoven University of Technology) have all contributed to this work through their respective graduation projects, carried out at Philips Research, and for which I thank them. Thanks are also due to Bart Vermeulen and Harald Vranken for their useful comments on a draft version of this paper. This work was supported by Philips Semiconductors, and partly funded by Esprit 2318 (Everest), Jessi AC6, and Medea A403-2.2.

Note

1. *TERM* denotes the set of all terminals. Terminals are considered to be primitive entities and are not defined further.

References

1. J. Aerts, "Test Time Reduction Algorithms for Core-Based ICs," Master's thesis, Eindhoven University of Technology, Eindhoven, The Netherlands, April 1998.
2. R. Arendsen and M. Lousberg, "Core Based Test for a System on Chip Architecture Framework," In *Digest of Papers of IEEE International Workshop on Testing Embedded Core-Based Systems (TECS)*, Washington, DC, Oct. 1998, Paper 5.1.
3. F. Beenker, B. Bennetts, and L. Thijssen, *Testability Concepts for Digital ICs—The Macro Test Approach*, vol. 3 of *Frontiers in Electronics Testing*, Boston, USA: Kluwer Academic Publishers, 1995.
4. F. Beenker, K. van Eerdewijk, R. Gerritsen, F. Peacock, and Max van der Star, "Macro Testing: Unifying IC and Board Test," *IEEE Design & Test of Computers*, vol. 3, no. 4, pp. 26–32, Dec. 1986.
5. M. Boosten and H. Jacobs, "Test Protocol Expansion: Memory Handling and Efficiency Improvements," Master's thesis, Eindhoven University of Technology, Eindhoven, The Netherlands, June 1994.
6. F. Bouwman, S. Oostdijk, R. Stans, B. Bennetts, and F. Beenker, "Macro Testability; The Results of Production Device Applications," in *Proceedings IEEE International Test Conference (ITC)*, Sept. 1992, pp. 232–241.
7. K. Chakrabarty, "Test Scheduling for Core-Based Systems," in *Proceedings International Conference on Computer-Aided Design (ICCAD)*, San Jose, CA, Nov. 1999, pp. 391–394,
8. K. Chakrabarty, "Test Scheduling for Core-Based Systems Using Mixed-Integer Linear Programming," *IEEE Transactions on Computer-Aided Design*, vol. 19, no. 10, pp. 1163–1174, Oct. 2000.
9. M.R. Garey and D.S. Johnson, *Computers and Intractability—A Guide to the Theory of NP-Completeness*, San Francisco: W.H. Freeman and Company, 1979.
10. I. Ghosh, S. Dey, and N.K. Jha, "A Fast and Low Cost Testing Technique for Core-Based System-on-Chip," in *Proceedings ACM/IEEE Design Automation Conference (DAC)*, San Francisco, CA, June 1998. Association for Computing Machinery, Inc., pp. 542–547.
11. I. Ghosh, N.K. Jha, and S. Dey, "A Low Overhead Design for Testability and Test Generation Technique for Core-Based Systems," in *Proceedings IEEE International Test Conference (ITC)*, Washington, DC, Nov. 1997, pp. 50–59.
12. S.K. Goel and E.J. Marinissen, "Cluster-Based Test Architecture Design for System-on-Chip," in *Proceedings IEEE VLSI Test Symposium (VTS)*, Monterey, CA, April 2002, pp. 259–264.
13. Mentor Graphics, "Solving the Challenges of Testing Small Embedded Cores and Memories Using FastScan MacroTest," White Paper, Jan. 2000.
14. R.K. Gupta and Y. Zorian, "Introducing Core-Based System Design," *IEEE Design & Test of Computers*, vol. 14, no. 4, pp. 15–25, Dec. 1997.
15. N.G. Hall and C. Sriskandarajah, "A Survey of Machine Scheduling Problems with Blocking and no-Wait in Process," *Operations Research*, vol. 44, pp. 510–525, 1996.
16. IEEE Computer Society, *IEEE Standard Test Access Port and Boundary-Scan Architecture—IEEE Std. 1149.1-1990*, New York: IEEE, June 1993.
17. V. Iyengar and K. Chakrabarty, "Precedence-Based, Preemptive, and Power-Constrained Test Scheduling for System-on-a-Chip," in *Proceedings IEEE VLSI Test Symposium (VTS)*, Marina del Rey, CA, May 2001, pp. 368–374.
18. V. Iyengar, K. Chakrabarty, and E.J. Marinissen, "Co-Optimization of Test Wrapper and Test Access Architecture for Embedded Cores," *Journal of Electronic Testing: Theory and Applications*, vol. 18, no. 2, pp. 213–230, April 2002.
19. R. Kapur et al., "CTL—The Language for Describing Core-Based Test," in *Proceedings IEEE International Test Conference (ITC)*, Baltimore, MD, Oct. 2001, pp. 131–139.
20. M. Keating and P. Bricaud, *Reuse Methodology Manual for System-on-a-Chip Designs*, Norwell, Massachusetts: Kluwer Academic Publishers, June 1999.
21. E. Larsson and Z. Peng, "An Integrated System-on-Chip Test Framework," in *Proceedings Design, Automation, and Test in Europe (DATE)*, Munich, Germany, March 2001, pp. 138–144.
22. E. Larsson and Z. Peng, "System-on-Chip Test Parallelization under Power Constraints," in *Digest of Papers of IEEE European Test Workshop (ETW)*, Saltsjobaden, Sweden, May 2001, pp. 281–283.
23. E.J. Marinissen, "Philips' Approach to Core-Based System Chip Testing," in *Proceedings IFIP International Conference on Very Large Scale Integration (VLSI-SOC)*, Montpellier, France, Dec. 2001, LIRMM, France, pp. 201–210.
24. E.J. Marinissen and J. Aerts, "Test Protocol Scheduling for Embedded-Core Based System ICs," in *Digest of Papers of IEEE International Workshop on Testing Embedded Core-Based Systems (TECS)*, Washington, DC, Oct. 1998, Paper 5.3.
25. E.J. Marinissen et al., "A Structured and Scalable Mechanism for Test Access to Embedded Reusable Cores," in *Proceedings IEEE International Test Conference (ITC)*, Washington, DC, Oct. 1998, pp. 284–293.
26. E.J. Marinissen et al. "On IEEE P1500's Standard for Embedded Core Test," *Journal of Electronic Testing: Theory and Applications*, vol. 18, no. 4, Aug. 2002, pp. 365–383.
27. E.J. Marinissen, R. Kapur, and Y. Zorian, "On Using IEEE P1500 SECT for Test Plug-n-Play," in *Proceedings IEEE International Test Conference (ITC)*, Atlantic City, NJ, Oct. 2000, pp. 770–777.
28. E.J. Marinissen, K. Kuiper, and C. Wouters, "Test Protocol Expansion in Hierarchical Macro Testing," in *Proceedings IEEE European Test Conference (ETC)*, Rotterdam, The Netherlands, April 1993, pp. 28–36.
29. E.J. Marinissen and M. Lousberg, "Macro Test: A Liberal Test Approach for Embedded Reusable Cores," in *Digest of Papers of IEEE International Workshop on Testing Embedded Core-Based Systems (TECS)*, Washington, DC, Nov. 1997, Paper 1.2.
30. E.J. Marinissen and M. Lousberg, "The Role of Test Protocols in Testing Embedded-Core-Based System ICs," in *Proceedings IEEE European Test Workshop (ETW)*, Konstanz, Germany, May 1999, pp. 70–75.
31. E.J. Marinissen and H. Vranken, "On the Role of DfT in IC–ATE Matching," in *Digest of IEEE International Workshop on Test Resource Partitioning*, Paper 1.2, Baltimore, MD, Nov. 2001.
32. E.J. Marinissen and Y. Zorian, "Challenges in Testing Core-Based System ICs," *IEEE Communications Magazine*, vol. 37, no. 6, pp. 104–109, June 1999.
33. J. Monzel and E. Orosz, "Testing 'Systems-on-a-Chip' in a Low Cost ASIC Test Environment," in *Digest of Papers of IEEE International Workshop on Testing Embedded Core-Based Systems (TECS)*, Washington, DC, November 1997, Paper 5.1.

34. M. Pinedo, *Scheduling—Theory, Algorithms, and Systems*, Englewood Cliffs, New Jersey: Prentice Hall, 1995.
35. P. Rosinger, B. Al-Hashimi, and N. Nicolici, "Power Constrained Test Scheduling Using Power Profile Manipulation," in *Proceedings International Symposium on Circuits and Systems (ISCAS)*, vol. V, May 2001, pp. V251–V254.
36. D.E. Ross, T. Wood, and G. Giles, "Conversion of Small Functional Test Sets of Nonscan Blocks to Scan Patterns," in *Proceedings IEEE International Test Conference (ITC)*, Atlantic City, NJ, Oct. 2000, pp. 691–700.
37. M. Sugihara, H. Date, and H. Yasuura, "A Novel Test Methodology for Core-Based System LSIs and a Testing Time Minimization Problem," in *Proceedings IEEE International Test Conference (ITC)*, Washington, DC, Oct. 1998, pp. 465–472.
38. J. van Beers and H. van Herten, "Test Features of a Core-Based Co-Processor Array for Video Applications," in *Proceedings IEEE International Test Conference (ITC)*, Atlantic City, NJ, Sept. 1999, pp. 638–647.
39. M. van Wijngaarden, "Test Protocol Expansion: Models and Solution Approaches," Master's thesis, Eindhoven University of Technology, Eindhoven, The Netherlands, Aug. 1993.
40. B. Vermeulen, S. Oostdijk, and F. Bouwman, "Test and Debug Strategy of the PNX8525 Nexperia^{TM} Digital Video Platform System Chip," in *Proceedings IEEE International Test Conference (ITC)*, Baltimore, MD, Oct. 2001, pp. 121–130.
41. Y. Zorian and E.J. Marinissen, "System Chip Test: How Will It Impact Your Design?" in *Proceedings ACM/IEEE Design Automation Conference (DAC)*, Los Angeles, June 2000. Association for Computing Machinery, Inc., pp. 136–141.
42. Y. Zorian, E.J. Marinissen, and S. Dey, "Testing Embedded-Core Based System Chips," in *Proceedings IEEE International Test Conference (ITC)*, Washington, DC, Oct. 1998, pp. 130–143.
43. Y. Zorian, E.J. Marinissen, and S. Dey, "Testing Embedded-Core-Based System Chips," *IEEE Computer*, vol. 32, no. 6, pp. 52–60, June 1999.

Erik Jan Marinissen is Principal Scientist at Philips Research Laboratories in Eindhoven, The Netherlands. He holds an M.Sc. degree in Computing Science (1990) and an MTD (Master of Technological Design) degree in Software Technology (1992), both from Eindhoven University of Technology. Marinissen's research interests include all topics in the domain of test and debug of digital VLSI circuits. He has published over 50 journal and conference papers, holds two US patents, and has several US and EP patents pending in the domain of core test and other digital test fields. He is the recipient of the Best Paper Award of the Chrysler-Delco-Ford Automotive Electronics Reliability Workshop 1995. Marinissen is a Senior Member of IEEE, and member of VIE, XOOTIC, and Philips CTAG. He serves as Editor-in-Chief of the IEEE P1500 Standard for Embedded Core Test and is a member of the organizing and program committees of DATE, DDECS, ETW, ITSW, LATW, SBCCI, and TECS. He has presented numerous tutorials on core-based testing within Philips and at international conferences. Marinissen serves as Member of the Editorial Board of JETTA.

JOURNAL OF ELECTRONIC TESTING: Theory and Applications 18, 455–473, 2002

CAS-BUS: A Test Access Mechanism and a Toolbox Environment for Core-Based System Chip Testing

MOUNIR BENABDENBI, WALID MAROUFI* AND MERYEM MARZOUKI
LIP6 Laboratory, Couloir 55-65, 4 Place Jussieu, 75252 Paris Cedex 05, France
Mounir.Benabdenbi@lip6.fr
Meryem.Marzouki@lip6.fr

Received September 1, 2001; Accepted January 14, 2002

Editor: Krishnendu Chakrabarty

Abstract. As System on a Chip (SoC) testing faces new challenges, some new test architectures must be developed. This paper describes a Test Access Mechanism (TAM) named CAS-BUS that solves some of the new problems the test industry has to deal with. This TAM is scalable, flexible and dynamically reconfigurable. The CAS-BUS architecture is compatible with the IEEE P1500 standard proposal in its current state of development, and is controlled by Boundary Scan features.

This basic CAS-BUS architecture has been extended with two independent variants. The first extension has been designed in order to manage SoC made up with both wrapped cores and non wrapped cores with Boundray Scan features. The second deals with a test pin expansion method in order to solve the I/O bandwidth problem. The proposed solution is based on a new compression/decompression mechanism which provides significant results in case of non correlated test patterns processing. This solution avoids TAM performance degradation.

These test architectures are based on the CAS-BUS TAM and allow trade-offs to optimize both test time and area overhead. A tool-box environment is provided, in order to automatically generate the needed component to build the chosen SoC test architecture.

Keywords: SoC test control, test access mechanism, TAPed cores, P1500 wrappers, I/O bandwidth

1. Introduction and State of the Art

With the ever growing integration density, current and future ICs include more and more capabilities. Complete systems on a single die, chips containing numerous IP cores are commonplace today. These cores can be DSPs, CPUs, memories... and can be provided to the system integrator by different companies. The reuse of these IP cores significantly reduces the time to market of such Systems on a Chip (SoCs) [29].

However, while geometry shrinking and design reuse allow impressive gains, SoC testing faces new set of problems and has become one of the bottlenecks of the IC industry progress. Cores coming from different companies with different functionalities, various Design for Testability (DFT) features and delivered at different levels of description (soft, firm or hard) harden the work of the system test integrator.

Solving these problems needs new types of test architectures, able to manage the test of up to 100 million transistors cores while allowing the high fault coverage required before signing-off a design to manufacturing. Moreover, highly standardized solutions are needed in such a context. The efforts of an IEEE working group [9] have resulted in the IEEE P1500 standard proposal of a SoC test architecture,

*Present address: Nortel Networks, Ottawa.

which main elements are, in its current development status:

- **Test Sources** to generate the core test stimuli and **Test Sinks** to compare the test responses to the expected ones.
- A **Test Access Mechanism (TAM)** in charge of transporting test data between sources, cores and sinks.
- A **Core Test Wrapper** [18] which is the interface between the embedded core and the TAM. Through different modes, it provides test functions at the core terminals.

The major effort of the working group focuses on the wrapper standardization problem, while source, sink and TAM design is left to the system designer. So new test architectures must be developed to access core I/Os from the SoC primary I/Os, some cores being deeply embedded in the circuit. TAM architectures can be based on the use of the system bus [7] or on a specific test bus [16, 23]. A detailed study of existing proposed architectures can be found in [8].

The SoC test integrator should be provided with different TAM architectures, in order to decide which one is best adapted to a given core-based SoC. Various criteria may be taken into account, such as:

- The level or core access and test preparation. The core may be wrapped or not, or include Boundary Scan features easing its accessibility (TAPed cores). It may also include different internal DFT capabilities.
- The complexity of the whole test control.
- The amount of test patterns to be applied to the circuit. With the integration density growth comes the increase of the number of patterns needed to test the whole SoC and test patterns at core level must be transferred at SoC level [17]. This obviously leads to a test time increase.
- The packaging of the SoC. The number of IC pins is sometimes a constraint since it defines the number of pins dedicated to test capabilities.
- The area overhead induced by test feature inclusion.
- The external test equipment that will be used to test the SoC. Depending on the Automated Test Equipment (ATE) capabilities (memory, bandwidth...), the SoC test integrator will select the best TAM architecture for the SoC design.

These constraints lead the test engineer to make multiple trade-offs regarding the choice of the appropriate TAM and its characteristics.

Among these constraints, testing time is a key issue. Methods and algorithms to optimize TAM width have been presented in [5] and in [8]. This optimization has been extended to include power and routing constraints [4]. To complete these approaches, a wrapper/TAM co-optimization has been developed and presented in [10]. These approaches reduce test application time but are not sufficient if not associated with test scheduling. A test framework including a set of design algorithms, dealing with test scheduling and test resource placement has been recently proposed in [13].

Some TAMs are designed to access cores with standard or proprietary wrappers [16, 23, 26]. We have developed a scalable and reconfigurable TAM architecture, named **CAS-BUS** [1], which falls into this category. This TAM is detailed in Section 2, as well as the developed tools to generate and use such an architecture.

The main benefit of this TAM over the other approaches is that it is dynamically reconfigurable. As others it allows test time optimizations during its design by optimizing the bus width. However what is different is that it provides test time optimization capability during the test pattern application step by changing the test data routing scheme, from one test session to the other.

Some others TAMs access TAPed cores, where the Boundary Scan features of the core are used instead of a wrapper to access the core I/Os [3, 6, 25].

But, to our knowledge, very few of the existing TAMs deals with both wrapped cores and TAPed cores integrated in the same SoC. However, a system integrator may need to integrate some TAPed cores together with P1500 wrapped cores, when these TAPed cores have primarily been designed as stand-alone chips. This need may arise in order to keep the SoC time-to-market as short as possible, thus avoiding the redesign of already TAPed cores to suppress their boundary scan features and replace them with P1500 wrappers. In addition, a TAPed core cannot simply be wrapped on top of IEEE 1149.1 features, because of area overhead as well as boundary scan rules regarding hierarchical global test control. In [27] the problem began to be adressed but the approach is limited to a TAM width equal to one (one TDI/TDO couple). To cope with these real-world needs, we have extended our basic CAS-BUS TAM in such a way that SoC populated with both wrapped cores and TAPed cores may be easily tested. This extension, which has been first presented in [2], is now detailed, together with the presentation of its design environment, in Section 3 of this paper.

Another real-world concern deals with available SoC I/O bandwidth, when the TAM width is larger than the number of SoC test pins. In this case, the expansion of one or more external I/O pair of serial test pins into parallel internal signals would be mandatory, in order to connect all test bus wires. Otherwise, this would obviously increase test time and degrade test performances of the SoC. To overcome this constraint we developed a test pin expansion mechanism [19]. It is based on a compression/decompression method, allowing to limit test time increase. This compression/decompression method is primarily dedicated to our architecture (CAS-BUS TAM), however it can be adapted to other architectures. The main benefit of this method is that the gain in test time is completely independent from any test data correlation. Many approaches based on statistical coding [14, 28] take advantage of the probabilistic distribution of test data. The approach we propose here is aimed to be used when test data samples have the same probability of occurence. This method and its implementation tools are detailed in Section 4 of this paper.

The control of the basic CAS-BUS TAM architecture and its two instances (the one considering TAPed cores inclusion and the one solving the bandwidth problem) has also been addressed [20], and is described in the corresponding sections of the paper. Some comments on the peculiarities of the control process of each instance are provided in Section 5. Experimental results for the implemented architectures are presented in Section 6.

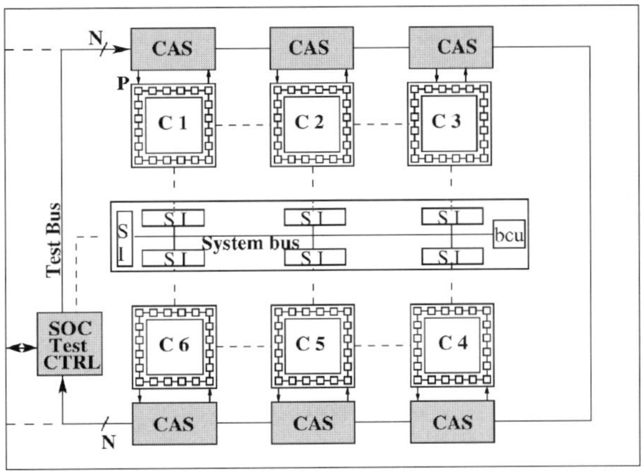

Fig. 1. Basic CAS-BUS TAM architecture.

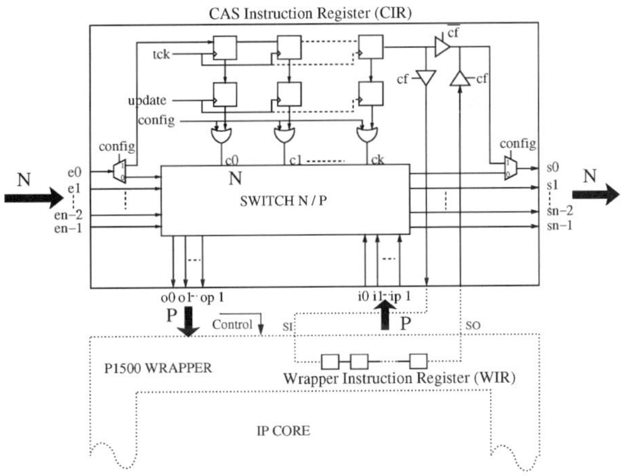

Fig. 2. CAS router architecture.

2. The Basic CAS-BUS TAM

The CAS-BUS (Fig. 1) is a TAM which main function is to provide test access to embedded cores whatever the wrapper is. As an example, the wrappers described in this paper are IEEE P1500 wrappers in the current state of development of the standard proposal, since the CAS-BUS architecture is compliant with this ongoing standard.

2.1. Architecture

This TAM is made up of two main elements:

- A **Core Access Switch (CAS)** (Fig. 2), which is a simple programmable router connected to each testable core through the wrapper at its test terminals,
- A **test bus** which is a set of wires transporting test data through the SoC and connecting CASes to each other.

Let N be the width of the test bus, and P the number of test pins for a given core. N is greater or equal to 1 and P is lower or equal to N. The CAS chooses among the N wires composing the test bus the P wires which will be connected to the test pins of the core. P depends on the core test method:

- For **scannable** cores, P is the number of integrated scan chains (Fig. 3(a)),
- For **BISTed** cores, P is generally equal to 1 (Fig. 3(b)),

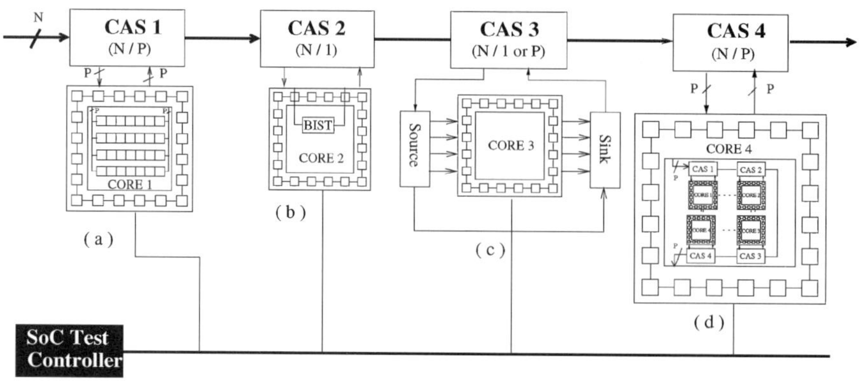

Fig. 3. Test types supported by the CAS-BUS.

- For cores tested using **external** sources and sinks, P depends on the nature of these source and sink, e.g. $P = 1$ when the source is a simple LFSR and the sink a simple MISR (Fig. 3(c)),
- For **hierarchical** cores, we consider that internal cores can be CASed, and in this configuration P is equal to the width of the internal test bus (Fig. 3(d)).

All test control signals, either for the CAS or for the testable cores, are connected to a central SoC test controller which is in charge of synchronizing test data and control. Each CAS selects from the N wires of the bus (e_i) the P ones that will be applied to the core wrapper inputs (o_i). It also connects the P outputs of the wrapper (i_i) to the CAS outputs (s_i). Unselected inputs (e_i) are bypassed to the outputs (s_i).

The CAS is composed by a **switch** and a **CAS Instruction Register (CIR)**. It is controlled by several signals. For example signal c_f can connect the CIR to the Wrapper Instruction Register (WIR) and signals c_i control the switch.

Controlled by the CIR, the switch is in charge of routing the test stimuli and the test responses to/from the core wrapper. Two kinds of switch implementation are possible:

(a) The switch can be made up of N decoders (one per input wire) and some logic gates (Fig. 4(a)).
(b) The switch can alternatively be made up of only one decoder and some logic gates (Fig. 4(b)).

The decoder controls the multiplexing of the N/P I/Os. By selecting P wires from the N ones existing at the switch inputs, the decoder must be able to address all the possible N/P combinations.

To avoid timing problems, a flip flop cell is inserted before each CAS output (s_i). This kind of cell is not present at (o_i) and (i_i) I/Os since they are connected to the internal wrapped core registers. The critical delay may appear during the updating of the routing scheme

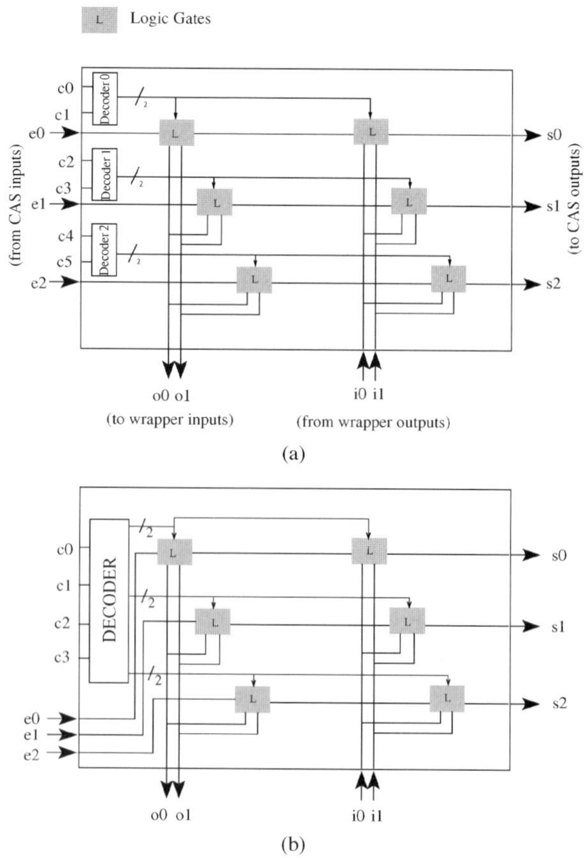

Fig. 4. Two types of switch architectures. (a) Switch with many decoders; (b) Switch with one decoder.

when using a switch with one decoder and with great values of N and P. In this case delay grows with the decoder area. However this kind of switch is not intended to be used for large values of N and P.

The number of combinations as well as the switch area obviously depend on N and P values and on the kind of implementation. Results obtained for different values of N and P can be found in Section 6.1.

2.2. *Control*

The control of the CAS-BUS architecture (Fig. 5) must be simple without degrading the TAM advantages. Two options have been considered: defining an ad hoc test control mechanism or reusing a standard control architecture. The decision to reuse the IEEE 1149.1 Test Access Port (TAP) naturally came to manage our TAM functionalities. This solution fits well with our SoC test architecture, since it allows TAM control, wrapper control and usual Boundary Scan functionalities at the same time. The test control must allow to manage three mandatory test steps:

- The configuration of the different core wrappers
- The configuration of all the CASes
- The application and the analysis of the test vectors.

The normal Boundary Scan architecture has been extended with a **CAS-Wrapper Coupling Register** (CWCR). The CWCR width is equal to the number

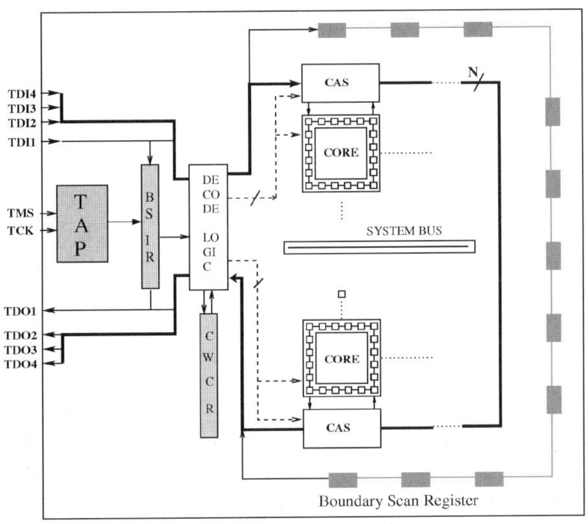

Fig. 5. CAS-BUS test control.

of CASes. Each bit of this register corresponds to the value of the configuration signal for each CAS.

Three new instructions are needed to control the three mandatory test steps. They have been defined as Boundary Scan optional instructions.

- The **CAS-Wrapper_Coupling** instruction (Fig. 6(a)). It defines which CIR/WIR pairs must be connected in order to configure at the same time a core wrapper and the corresponding CAS. TDI1 and TDO1 are connected to the CWCR. Configuration values are shifted in the CWCR to indicate which WIRs will be connected to the CIRs.
- The **CAS_CONFIG** instruction (Fig. 6(b)). It configures all the CASes, together with the wrappers selected with the instruction CAS-Wrapper_Coupling. The config signals provide access to the CIR of each CAS. TDI1 and TDO1 are respectively connected to first input of the first CAS and to first output of the last one. The CIR and the WIR are then loaded with the appropriate values, putting each CAS in the correct scheme and each wrapper in the chosen mode.
- The **CAS_TEST** instruction (Fig. 6(c)). When loaded, it allows the test vectors to be applied to the different cores and the test stimuli to be propagated to the SoC test outputs. All TDIs and TDOs are connected to the I/Os of the CAS-BUS. The set of test vectors can be concurrently shifted into the IP core inputs and the responses shifted out from the IP outputs. In our modified TAP architecture, TDI and TDO is a bus. This TDI/TDO bus width may vary from 1 (normal Boundary Scan architecture) to N (width of the CASBUS). This width can be chosen by the SoC integrator, depending on SoC pin availability.

Using the IEEE 1149.1 control part with scalable TDI/TDO lets the CAS-BUS TAM keep its main benefits and adds to it flexibility.

2.3. *Developed Tools*

Two independent software modules have been developed to ease the integration of this TAM in the SoC architecture.

2.3.1. CAS Generator. The SoC test designer first has to decide the width of the test bus (N) and the number of switched wires for each CAS (P).

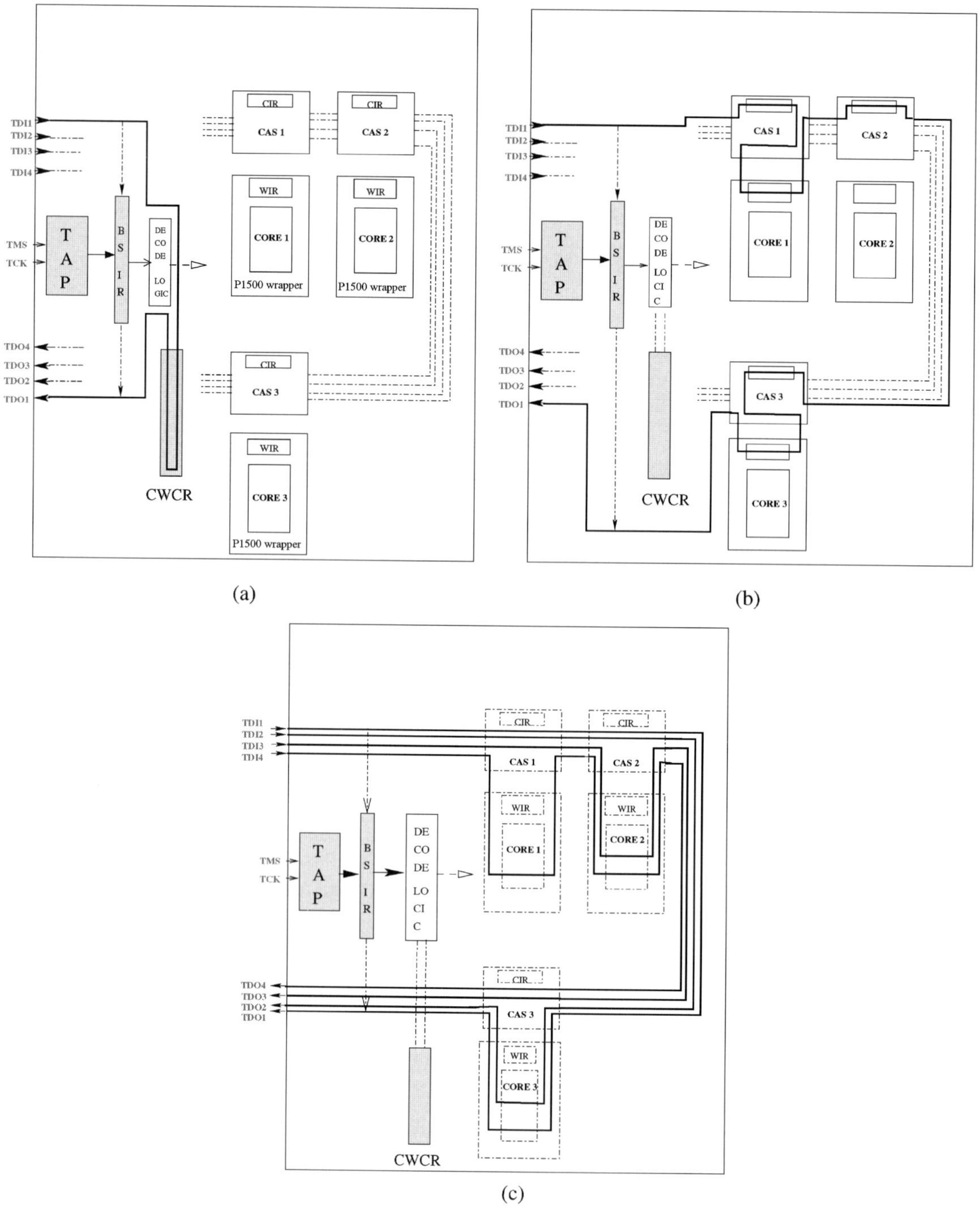

Fig. 6. The three new mandatory instructions. (a) CAS-Wrapper coupling; (b) CAS Config; (c) CAS test.

In order to reduce the number of combinations and thus the width of the instruction register, the following heuristic has been defined (Fig. 2): *When an input e_i is switched to an output o_j, the corresponding i_j CAS input is switched to the s_i output.*

The use of this heuristic obviously limits the number of combinations needed and reduces the control complexity. Then we can completely construct a test bus wire path from an e input to an s output with the same control word $(c_0 \ldots c_k)$.

Thus, for m combinations the width k of the instruction register can be calculated using the formula:

$$k = \lceil \log_2(m) \rceil.$$

The CAS generator developed is based on this heuristic and takes as parameters the N and P values and the kind of switch implementation. It provides a VHDL description of the chosen CAS, which can be synthesized with a any synthesis tool. This generator is written in the C language.

2.3.2. Wrapper Generator. A wrapper generator has also been developed. The software, designed in C language, takes as input the VHDL description of a core I/Os. It generates the corresponding wrapped core in the VHDL format. The generated wrapper is as close as possible to the IEEE P1500 wrapper, as the standard is still under development. The design of the wrapper is based on public information that can be found in [9]. In order to be easily updated, the generator is designed as a collection of specialized modules, and is based on the use of a cell library. The cells are those described by the standardization group on its web site. Some choices had to be made since some features are not completely defined yet. Currently, the generator takes as parameters the kind of boundary cell (basic or shift/update cell) and a file where are described the characteristics of the core INOUT pins. We currently use this file to solve core INOUT pins problems. This file will be replaced by the Core Test Language (CTL [9] file that will be provided with the IP core, when the standard will be completed and adopted.

2.4. *Benefits and Comparison to Other Approaches*

The CAS-BUS TAM architecture has multiple advantages over other approaches described in the literature, due to its flexible, scalable, reconfigurable and dynamic aspects:

- It supports different core test methods (Scan, BIST, parallel test),
- It can provide parallel test access to many cores, depending on N and P. If N is limited, the cores can be partially or completely daisy chained. Whatever the routing scheme is, many wrappers can be active simultaneously.
- It allows the test of hierarchical cores without degrading performances in terms of reconfigurability and flexibility,
- It can easily be reconfigured, even during test sessions, in order to optimize test performances. A good collaboration between the test designer and the test programmer leads to a good trade-off between test overhead and test time. For example, in case of scanned cores, the test programmer can balance the length of the scan chains within the test programs, in order to reduce the test time. In the same way, SoC interconnect test time can be optimized when adopting a good configuration of the test chains.
- It allows maintenance test, by testing some embedded cores while others are in normal functioning mode. This is very useful when, e.g., an embedded memory test is periodically required.

Moreover, the CAS-BUS has the advantage of being independent from any industrial proprietary SoC architecture. In addition, we propose an architecture where the TAM can be used with any kind of wrappers (either proprietary or standard to be defined, e.g. the P1500 wrapper in its current status), contrarily to approaches like [16], where the TAM and the wrapper are closely merged, leaving few freedom of decision to the system integrator. On the contrary, the CAS-BUS TAM eases the SoC test architecture design by using plug-and-play CAS modules.

Like the Adressable Test Port architecture [26], the CAS-BUS TAM is only intended to provide test access to the embedded cores, independently of the wrapper configuration. However many differences appear between the two architecures. Among these differences we can see that in the CAS-BUS TAM the same bus is used to input and output data and many different cores can be tested at the same time. The CASes are configured to determine the routing scheme contrarily to the Adressable Test Ports that are addressed for enabling purpose. However the CASes do not provide capabilities such as test generation and test compaction allowed within the Adressable Test Ports.

With its control features based on the IEEE 1149.1 standard, the CAS-BUS TAM is very flexible and easily controlled using the well known TAP controller. It allows the concurrent application of test stimuli to many wrapped cores through a scalable bus. Even if they are too different to allow any comparison, architectures like HTAP [3] or TLA [25] have only one TDI/TDO couple and do not allow to test many cores at the same time.

In the CAS-BUS architecture, Boundary Scan I/Os are used not only to control the CASes and the core wrappers, but also to apply test stimuli to cores inputs, contrarily to approaches like [22]. Indeed, even if this architecture includes Boundary Scan features at SoC level like our architecture, the TDI/TDO wires are only used to control the global SoC test, the TAM being directly connected to the SoC primary I/Os.

3. Extension of the Architecture to TAPed Cores Testing

The basic test architecture presented in Section 2 can be used when the cores within the SoC are wrapped cores. Since TAPed cores are not wrapped cores, they can only be accessed through their test I/Os (*TMS*, *TDI*, *TDO*, *TCK* and *TRST*, if present). Unlike wrapped cores, these cores include a TAP controller. Obviously this needs to define a hierarchical test architecture able to drive these TAP controllers. Moreover, configuring the TAPed cores in the chosen mode cannot be done in the same way as wrapped cores. Direct access to the wrapper shift/update instruction register (WIR) can be enabled, while the instruction register within the TAPed core can only be accessed through the TAP finite state machine. Thus TAPed cores and wrapped cores cannot be configured serially in the same test session.

Our goal being to test both wrapped and TAPed cores within the same test session and with the same TAM, some improvements must be made on the CAS-BUS architecture in order to solve the problems induced by the TAP controllers, while avoiding any IEEE 1149.1 standard rule violation.

The two main improvements consist of defining on the one hand a special Core Access Switch for TAPed cores and of modifying on the other hand the central TAP controller of the CAS-BUS architecture in order to manage the hierarchical aspect of the global control.

3.1. Core Access Switch for TAPed Cores

Since TAPed cores cannot be serially configured with the wrappers, a new Boundary Scan instruction named **TAP_CONFIG** has been added, which enables the configuration of the TAPed Cores. The configuration of the wrappers is still made by the CAS_CONFIG instruction. Fig. 7 shows a SoC example containing both kind of cores. This TAP_CONFIG instruction serially connects CASes of TAPed Cores (TAPCASes) one to the other and connects each TAPCAS to its associated core. It also connects *TDI*1 and *TDO*1 respectively to the first and the last TAPCAS.

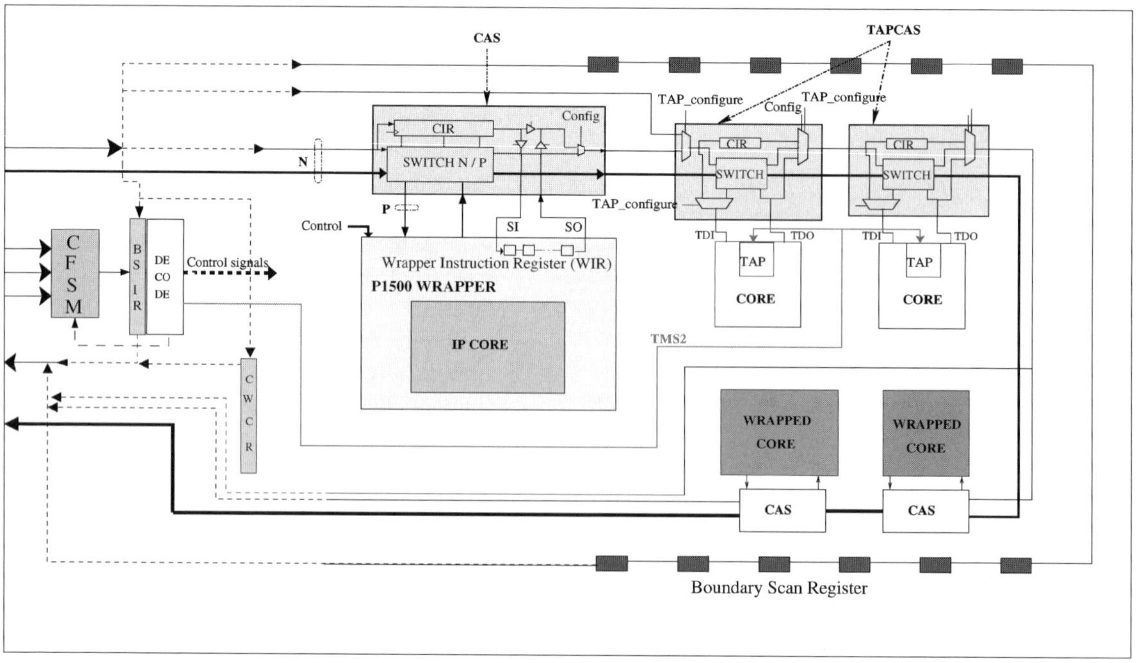

Fig. 7. SoC test architecture for TAPed and wrapped cores.

The TAPCAS is slightly different from the normal CAS. The CIR and the switch remain the same, the tristates disappear and multiplexers are added. However, two kinds of TAPCAS architectures are needed, depending on the placement of the TAPed core within the SoC. In the global chaining of all the CASes, if the TAPCAS follows a CAS then the TAPCAS must have a multiplexer at its inputs. In the other cases this multiplexer is not needed.

Obviously, for the TAPCAS, P is equal to 1: the TAPCAS is connected to *TDI*/*TDO* pair of the TAPed core. Depending on the boundary scan instruction, test data at TAPCAS inputs can be routed differently. The different multiplexers will route these data to:

- The **TAPCAS instruction register (CIR)**. This is done with the **CAS_CONFIG** instruction. The switch is configured like all the other switches.
- The **internal instruction registers of the TAPed cores**. In that case the **TAP_CONFIG** instruction must be loaded. A configuration word is shifted in the internal instruction registers of the TAPed cores, these registers being daisy-chained. The data skip the switch in the TAPCAS and are directly applied to the *TDI* input of the TAPed core. The different operations like shifting, updating, etc., are executed thanks to the TMS2 signal that drives the internal TAP controller (Fig. 8). This signal TMS2, described in the next section, is generated from the central TAP controller. Once the internal instruction registers are updated, the TAPed cores are then in the correct test mode (bypass, INTEST, etc.) and ready for test vectors application.
- The **internal data registers of the TAPed cores**. When the **CAS_TEST** instruction is loaded, test vectors are routed from one of the N wires of the TAPCAS inputs to the *TDI* input, through the switch. The test vectors are shifted in the internal test data registers (boundary scan register, bypass register...). The responses to the stimuli are then shifted out to the next core or to the SoC outputs. Here also the shift/update operations are controlled through TMS2.

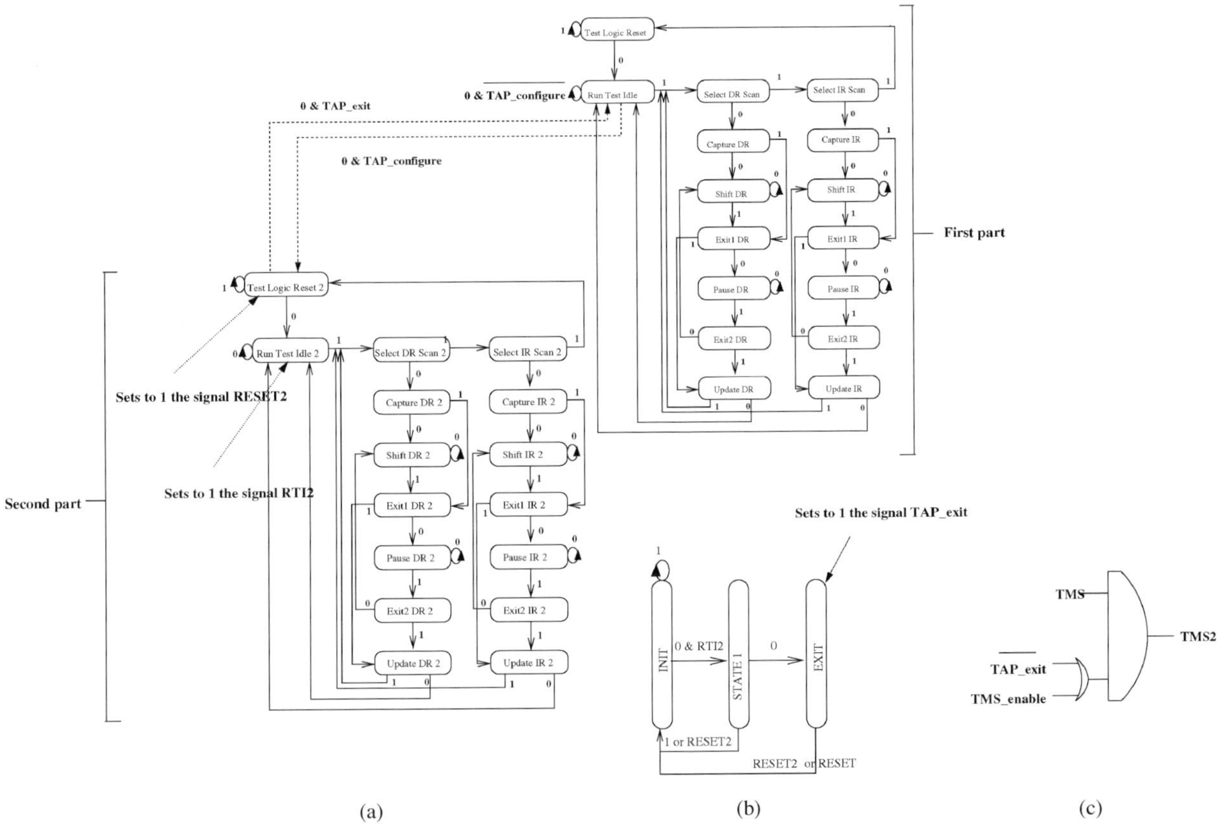

Fig. 8. SoC test architecture controller. (a) The modified state machine; (b) The counter; (c) Added logic to generate TMS2.

3.2. The New Central TAP Controller (CFSM)

The new central TAP controller (Fig. 8) is made up of a 1149.1 like TAP Finite State Machine (FSM) and a counter.

This Central FSM (CFSM) is in charge of controlling the global architecture as a classical boundary scan feature but must also drive the different Internal Finite State Machines (IFSM) of the TAPed cores. To do that, it was decided to add to the usual 16 states of a TAP, a second set of 16 new states corresponding to the states of the TAPed cores. Except Test logic 2 and Run Test Idle 2, the added states do not drive any signal and only are transition states. Depending on the instruction loaded, TMS will control the CFSM and the IFSMs or only the CFSM.

The test instructions to be loaded in the Boundary Scan instruction register should respect the following chronological sequence:

1. **CAS_WRAPPER_COUPLING** to define which CIRs and WIRs will be connected during the configuration step. This instruction operates only on CASes and wrapped cores.
2. **CAS_CONFIG** to configure the CASes and the TAPCAS. The routing scheme of the overall switches is defined.
3. **TAP_CONFIG** to configure the test mode of the TAPed cores.
4. **CAS_TEST** to apply test stimuli and shift out the test responses. The TAPed cores and the wrapped cores are tested concurrently through the same bus.

The following subsections describe more in details what happens when the TAP_CONFIG and the CAS_TEST instructions are loaded.

3.2.1. TAP_CONFIG Instruction Effect. This instruction is needed to shift values in the instruction registers of the TAPed cores.

When this instruction is loaded and updated, the signal TAP_configure is set to 1. Run Test Idle state (Fig. 8(a)) is left to enter in the second part of the CFSM in the Test Logic Reset 2 state. This state sets to 1 the RESET2 signal that sets the counter (Fig. 8(b)) in the state INIT. The counter is initialized, the signal TAP_exit is set to 0. TMS2 (Fig. 8(c)) will then have the same value as TMS. To synchronize the CFSM and the IFSM, which states must be equivalent to those of the second part of the CFSM, TMS must be held at 1 during 5 cycles to reset the IFSM.

When entering in the Run Test Idle 2 state, the signal RTI2 is set to 1, Reset2 is set to 0 and if TMS is held at 0 during 2 cycles, the counter is put in the state EXIT. The TAP_exit signal is set to 1. This sets TMS2 to 0 and the IFSMs are no more controlled by TMS. The IFSMs are and stay in the Run Test Idle state. The counter is used not only to enable and disable IFSMs control but also to leave the second part of the CFSM.

If TMS is not held at 0 during 2 cycles, when the CFSM is in the Run Test Idle 2 state, the counter stays in the INIT state. The IFSMs are controlled by TMS and their current state is equivalent to the current state of the second part of the CFSM. The instruction words can be shifted in the TAPed core as they would be if placed on a boundary scan board.

3.2.2. CAS_TEST Instruction Effect. This instruction is used to shift in, apply and shift out test vectors to and from all the cores within the SoC. With the right configuration of the CASes and the TAPCAS, all the test data registers are connected as multiple scan chains. Internal test data registers of the TAPed cores (boundary scan or bypass registers) can be considered as internal scan chains of the P1500 wrapped cores. When updated this CAS_TEST instruction sets to 1 the signal TMS_enable and sets to 0 the signal TAP_configure. TMS controls the IFSMs through TMS2 (Fig. 8(c)). During this test step the current state of the CFSM cannot be one of those of the second part.

When leaving the Update IR state (Fig. 8(a)), the next state must be Run Test Idle. The configuration step of the TAPed cores having left the IFSMs in the Run Test Idle state, the synchronization of the CFSM and the IFSMs is done thanks to this state. This way, the current state of all the finite state machines of the SoC being exactly the same, shifting and applying test vectors within wrapped and TAPed cores can be made at the same time.

The IFSMs remain controlled by the CFSM until a new instruction is loaded in the Boundary Scan Instruction Register (BS IR) of the SoC. This new instruction will set to 0 the signal TMS_enable. This is equivalent to setting TMS2 to 0 and the IFSMs are left in the state Run Test Idle. However, while shifting this new instruction in the BS IR, unknown values are concurrently shifted in the internal instruction registers of the TAPed cores. This is not important because the test responses have already been shifted out from the internal test data registers and because the next test session will reset these instruction registers. The next TAP configuration step will load these registers with the right values.

3.3. Benefits and Limitations

The main benefit of this new CAS-BUS architecture is that TAPed cores do not need to be tested apart from the other wrapped cores using added test resources, hence decreasing the test area overhead. TAPed cores can be tested with the CAS-BUS TAM taking advantage of the possible optimizations provided by this TAM. Test time can still be optimized through correct configuration of all CASes and TAPCASes.

With this architecture it is also possible to test only TAPed cores with the TAP_CONFIG instruction. The cores can be tested as stand-alone chips on a boundary scan board, controlled by the CFSM, because direct test access is provided to their I/Os, skipping the wrapped cores. However BIST testing can not be processed within these TAPed cores. During the TAP_CONFIG step the second part of the CFSM is active. If the RUN_BIST instruction is loaded in the TAPed core, to execute the self test, the current state must be Run Test Idle 2 during many cycles, with TMS held at 0. With this CFSM, this cannot be done. In that case, after TMS is held at 0 during two cycles, the next state is Run Test Idle, in the first part of the CFSM, and then the IFSMs stop being controlled by the CFSM. If really needed, a solution could be to add a register that would be concurrently loaded with the internal instruction registers of the TAPed cores. If the RUN_BIST instruction is detected in this new register, the current state remains in the Run test Idle 2 state without exiting to the main part of the CFSM.

As for the previous CAS-BUS architecture, interconnect testing can be performed, between wrapped cores and TAPed cores, assuming that the EXTEST instructions are loaded.

The CAS generator has been extended and a new parameter has been added. The user can also generate a VHDL description of the TAPCAS with the desired kind of switch, with or without the input multiplexer. Some results are discussed in Section 6.1.

This extension of the CAS-BUS TAM architecture does not degrade the characteristics of the previous, basic architecture, while it offers a solution for testing SoCs with TAPed cores. If the SoC does not have any TAPed cores included, the integrator should obviously prefer the basic CAS-BUS architecture.

4. Addressing the I/O Bandwidth Issue

Some constraints may degrade the TAM performances. The one we focus on in this paper arises when the TAM width is greater than the SoC test pins number (Fig. 9). In such a case, the expansion of one or more external input and output pair of serial test pins is necessary in order to connect all test bus wires. This will obviously increase test time and then degrade test performances, since we have to expand one test pin pair into two or more serial test data chains (Fig. 10).

To cope with this problem, we propose an expansion mechanism, based on a compression/decompression method, allowing to limit test time increase. This novel compression/decompression method is primarily dedicated to our CAS-BUS architecture but it can be adapted to other architectures. The compression gain reached with this method is completely independent from test data correlation. The choice of other compression/decompression statistical coding methods like those based on Huffman or similar coding [11] or test vector decompression via cyclical scan chains techniques [12] was studied, but their dependence on test data correlation represents an important drawback.

4.1. Theory

The method we propose is lossless and treats n bit-length samples of the test data. The choice of the optimal bit-length value will be explained later.

To ease the explanation of this method, let us take the expansion example shown in Fig. 11, with 4 bit-length samples ($n = 4$). It consists of three scan chains

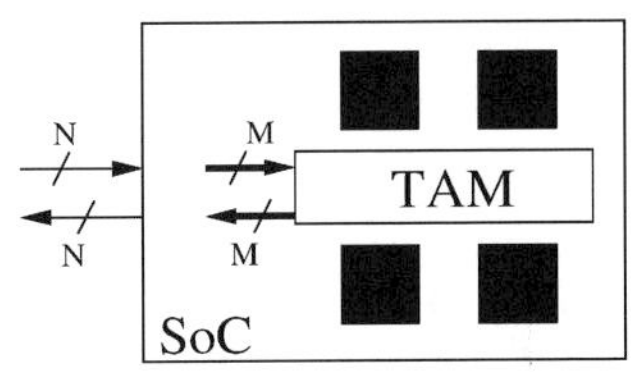

Fig. 9. Problem: N < M.

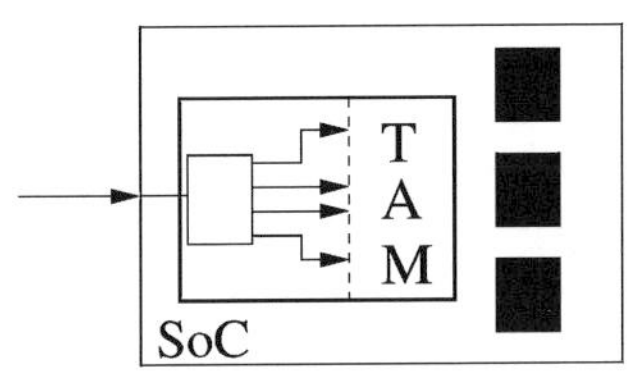

Fig. 10. Expansion.

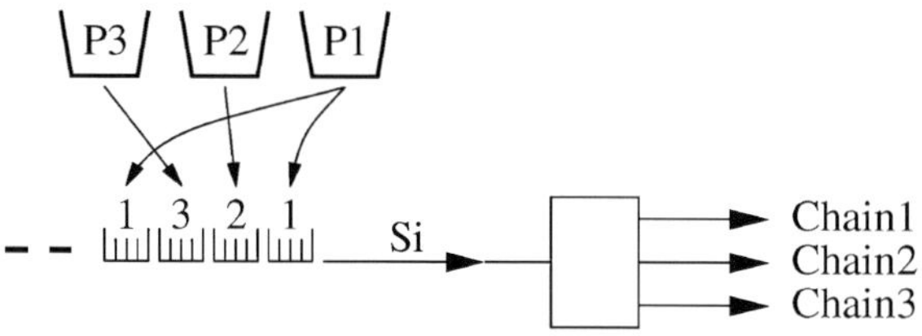

Fig. 11. Example of decompression/expansion.

which must be accessed through only one serial input (**si**). $P1$, $P2$ and $P3$ respectively represent the global length of test patterns for chain1, chain2 and chain3. We assume a cyclic transfer of the 4 bit-length samples coming on **si** to the three chains. Without any compression of test data, it is clear that the total test time needed to transfer all test patterns is equal to $(P1 + P2 + P3)$ cycles. This represents a large test time, knowing that if we had 3 serial inputs, the total test time would have been reduced to $\max(P1, P2, P3)$ cycles (parallel testing of the three chains).

For 4 bit-length samples, the number of possible combinations is equal to 16. We arbitrary classify the 16 combinations in 4 groups of 4 combinations each. Fig. 12(a) shows a possible example of such classification.

Within one specific group, each combination is normally coded using 2 bits (offset of the combination in the group). Groups are also coded using 2 bits, but we consider another way to select them. Indeed, we consider that they are cyclically chained (Fig. 12(b)), so there are three possibilities to change from one group to another:

- One forward jump in the cyclic group chain,
- One backward jump,
- Two backward (or forward) jumps.

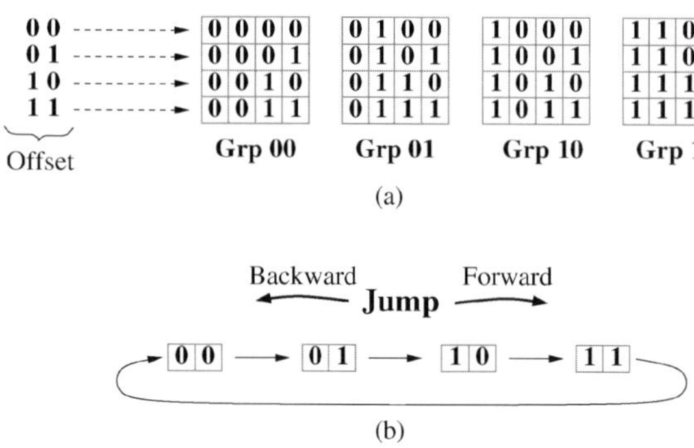

Fig. 12. Groups example.

Thus, the idea is to start from one chosen group (for example group 00) and to code samples using 2 bits. If any group modification is necessary, a dedicated signal (for example **Grp_chg**) must indicate it. When the **Grp_chg** signal is active, the value at the input **si** indicates the direction of the jump (1: forward jump, 0: backward jump). In case of two jumps, the signal **Grp_chg** must stay active during two clock cycles while the value at the input **si** must be also stable (at 0 or at 1).

The gain for each 4bit-length sample can be estimated as follows:

(i) case of no jump: 2 bits (2 cycles) for the coding $\rightarrow$ 50% gain,
(ii) case of one jump: 2 bits (2 cycles) for the coding and one cycle for the jump indication $\rightarrow$ 25% gain,
(iii) case of two jumps: 2 bits (2 cycles) for the coding and two cycles for the jumps indication $\rightarrow$ 0% gain.

When there is no special correlation in the test data, the percentage of jumps can be statically evaluated as: 25% of no jumps, 25% of one forward jump, 25% of one backward jump and 25% of two jumps. So the global average gain can be calculated as: $(0.25 \times 50\%) + (0.25 \times 25\%) + (0.25 \times 25\%) + (0.25 \times 0\%) = 25\%$. Of course this gain is not static and essentially depends on the number of jumps within a test process. The maximum gain is reached when no jump is necessary and is equal to 50%.

One way to limit the number of jumps, is to modify the distribution of the 16 main combinations among the 4 groups. For each test process, an optimal distribution can be found.

To limit the number of possible jumps, the number of groups must be small. Suppose now that we use a 6 bit-length samples, the number of possible combinations will be 64. Distributed among four groups (for example), we will obtain 16 combination per group: 4 bit for coding offsets. In case of no jumps, the gain will be equal to $(6 - 4/6) = 33\%$ which is less than the gain obtained for 4 bit-length samples. When distributed among more groups, 16 for example, offsets will be coded using 2 bits but the number of jumps will be very important. Suppose for example that we must make a jump of 8 groups, so 8 clock cycles must be added to the offset 2 cycles: then we have 10 cycles to code a 6 bit-length sample (-66% gain!). The same problems occur for larger samples (8, 10, ...).

4.2. *Integration in the CAS-BUS Architecture*

As said in the introduction, the compression-decompression/expansion method we present is first dedicated for the CAS-BUS architecture controlled using the 1149.1 TAP. This method has been integrated considering 4 bit-length samples. The serial input is obviously TDI, and in order to avoid additional signals we decide to use the TMS signal to indicate a jump (**Grp_chg**). An expansion/decompression module is added to the architecture as well as an MISR from which a signature can be shifted out through TDO (Fig. 13).

In addition to the three new instructions presented in Section 2.2, a fourth instruction called **DECOMPRESS_test** replaces **CAS_test** when transferring test vectors. This instruction connects the TDI serial input to be expanded to the expansion/decompression module and the TDO output to the MISR.

It is important to note that only one clock cycle is allowed to indicate a jump (2 cycles for two jumps). Thus, using the TMS signal to indicate group jumps with a classic standard TAP controller is impossible. Indeed, when TMS is activated within a test vector transfer, it is impossible to go back to the **shift-DR** state before 3 clock cycles, which does not satisfy constraints.

The solution we propose is to use a modified TAP which we call MTAP (Fig. 14). This TAP behaves exactly like a classic standard 1149.1 controller except when a **DECOMPRESS_test** is shifted in. In that case, two additional states (**ONE-jump** and **TWO-jumps**) are added to the data register control part of the TAP. This is automatically done when the instruction register decoder decodes the **DECOMPRESS_test** instruction.

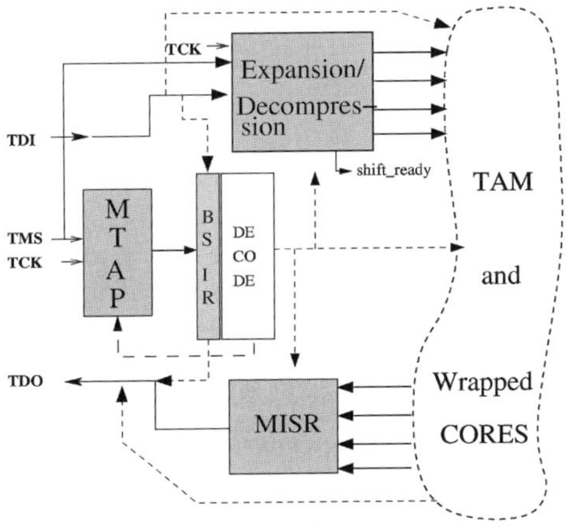

Fig. 13. CAS-BUS/TAP architecture with expansion/decompression module.

The main advantage of MTAP is that the constraint of one clock cycle for one jump is satisfied. Indeed when TMS signal is activated within a test vector transfer, MTAP moves to the state **ONE-jump** (the direction of the jump depends on the value of TDI). If another jump is necessary, TMS must stay at one and MTAP moves to the state **TWO-jumps**, otherwise it goes back to a shift state: **shift-DR**.

The expansion module takes as inputs TDI, TMS, TCK and control signals generated from the MTAP controller. It outputs the N chains corresponding to the TAM width, and a control signal (**shift_ready**) specifying that decompressed data are ready to be applied to the TAM. This module includes four bit-length registers and a table containing the four group samples. As we have N chains, N registers are needed to store successively the decompressed samples. Coded samples and jump informations are shifted in, decompressed and affected to the correct register, continuously, without waiting delay. This is possible since TMS is directly connected to the module. This process is stopped if we are not in one of the states (**ONE-jump**, **TWO-jumps** and **shift-DR**). When the N registers are filled the module asserts the signal **shift_ready**. Decompressed and expanded data can then be shifted out and be applied to the TAM.

The main advantage of this implementation is that TMS is both used for controlling the MTAP and for giving jump informations to the decompression/expansion module.

Fig. 15 shows an expansion/decompression sequence of one TDI input to three serial chains. **CHi-VALj** represents a 4 bit-length sample which is part to chain i. **CHi-CDj** indicates the compressed code of **CHi-VALj**. The four groups are named A, B, C and D and the register **Current-group** is initialized to B. Once three codes (ex: CH1-CD1,CH2-CD1 and CH3-CD1) are shifted in, the decompressed values (CH1-VAL1,CH2-VAL1 and CH3-VAL1) are shifted out through the three chains.

Since 12 clock cycles are needed to shift three 4 bit-length samples without any compression/decompression process, the gain obtained for each sequence is: 3 cycles for CHi-VAL1 and 2 cycles for CHi-VAL2.

4.3. *Special Cases*

In this section we present two special cases which we think are the most important ones.

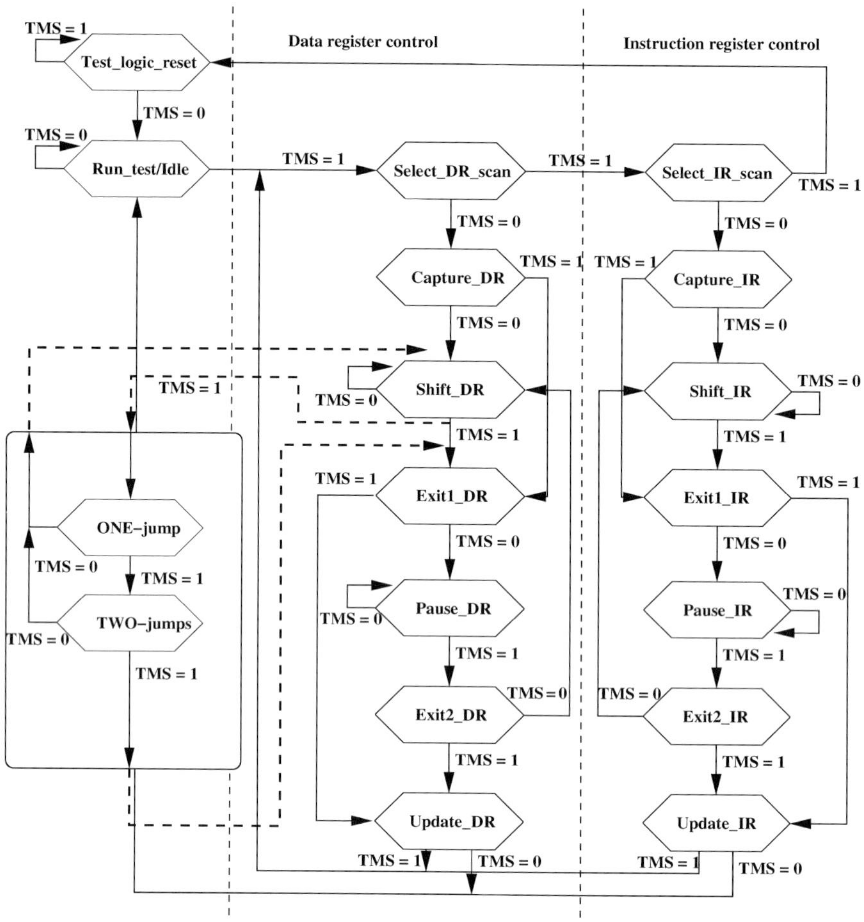

Fig. 14. Modified TAP state machine.

First special case corresponds to no needed expansion (one serial input and one internal chain). This is still possible, but two test clocks are necessary: one slow clock for the compressed patterns (tester) and the second, a fast clock, for the decompressed chain (Fig. 16). The same technique is used in [11]. Note that this is not necessary in case of an expansion, since the time needed to input compressed codes is at least equal to $2N$ cycles (N being the number of target internal chains, with $N \geq 2$), which is enough to shift in parallel 4 bit-length samples.

Second special case corresponds to using the method when more than one serial input are available. In such a case, a simple rule is to minimize the number of chains to be connected to the expansion/decompression module. So, for N serial inputs and M internal chains, we can connect $(N-1)$ inputs to $(N-1)$ chains and use the last input to expand the remaining $(M-(N-1))$ chains. The distribution of chains on the direct or expanded serial inputs must be a result of a global trade-off taking into account test chain length and relative test data amount. Using more than one decompression module is possible, but this necessitates more than one MTAP mechanism.

4.4. Developed Tools

A compression software module has been developed. It first serializes the N test patterns at its inputs in one test pattern, then it makes its compression and finally include the values giving the jump directions. The compression is made according to the partitioning of the 4 groups of 4 bits. To enhance the compression degree we decided to slightly modify the proposed theory. We affect the two bits coming on TDI not to the offset value, as previously described, but to the place of the 4 bits pattern in the group. This change gives us complete

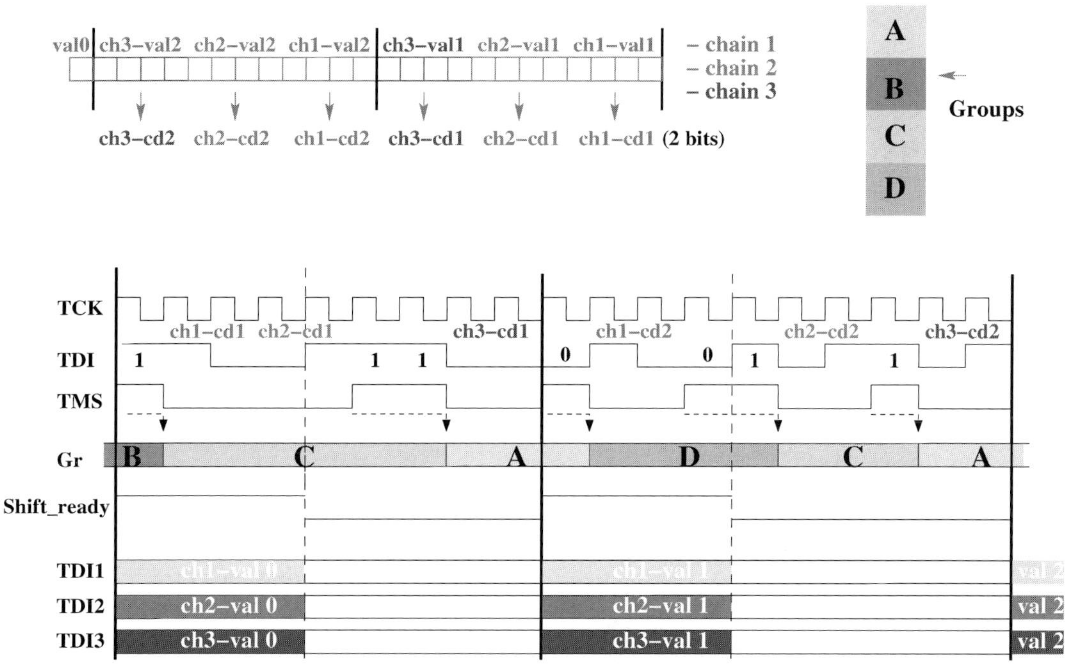

Fig. 15. Expansion/decompression sequence.

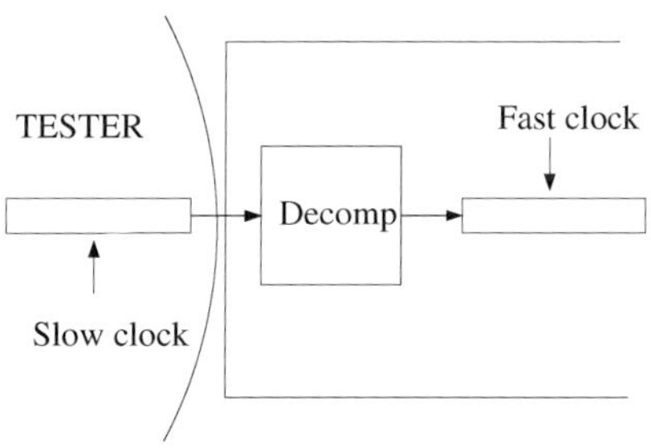

Fig. 16. Case of no expansion.

freedom to partition the 16 combinations of 4 bits in 4 groups. A software module processes the serialized test pattern and gives the corresponding optimized partitioning of the 16 samples. For a given test pattern, the optimization is made in order to reduce the global number of jumps. The overall processing time needed to serialize, to calculate the best partitioning and to compress the *N* patterns is less than one minute. Obviously processing a test pattern in order to calculate the best partitioning corresponds to the study of the natural correlation between samples, and then use this correlation to save test time. Even if the proposed method has been developed for patterns with loose correlation, this optimization provide an overall gain on the global test time.

However when using the CAS-BUS TAM with the expansion/decompression module, before each test pattern processing, the test engineer must initialize the four groups with the optimized partition. This requires 64 extra cycles, which is very few compared to the global test length.

An application of this software to a real example is presented in Section 6.2.

4.5. *Discussion on Sample Length*

The length of the sample data obviously influences the compression rate. If n is the sample's length, 2^a the number of groups and 2^b the number of samples within a group, we have $n = a + b$. The average compression gain is given by the formula (not detailed in this paper):

$$\text{gain}(\%) = 100 \times (1 - ((b + 2^a/4)/(a + b)))$$

The optimal gain value is obtained with $b = 1$ and $a = 2$. Thus the optimal average gain is obtained with 3 bit-length samples. In this case the average gain is equal to 33% and the maximum gain is equal to 66%.

The 33% minimum average gain is completely independent from any test data correlation.

A version of the compression software, processing 3 bit-length samples (optimal value), has been developed. In this case the processing time is less than a second and the test engineer must initialize the four groups with 24 extra cycles only.

5. Comments on the Proposed Instances of the CAS-BUS TAM

Depending on the SoC design constraints, a choice must be made between the two test architecture instances. If the SoC is made up of TAPed and wrapped cores, then the SoC integrator should use the test architecture presented in Section 3, controlled by the CFSM. If few SoC test pins are available, the architecture described in Section 4, controlled by the MTAP, well fits the test integration. Merging the two architectures into a single one enabling both test pin expansion process and TAPed core testing is currently under study. As they have been defined till now, the CFSM and MTAP controllers cannot be merged into a single controller because the TAP controllers of the TAPed cores do not include the two states *ONE-jump* and *TWO-jumps*. Thus, when the instruction DECOMPRESS_test is loaded, when expanding and shifting test data, the current state of the MTAP cannot be the same as the ones of the internal controllers of the TAPed cores. This leads to define a new test controller able to drive, in the same SoC, test pin expansion modules, the CAS-BUS TAM, wrapped cores and TAPed cores.

6. Experimental Results

To validate the three CAS-BUS architectures and the associated tools, simulations and synthesis have been done. Considering the lack of SoC application availability, the complete integration of the global architectures in a real SoC have not been done yet.

6.1. Architecture Synthesis

6.1.1. CAS and TAPCAS Generation. In our experiments, we have simulated the different features of the CAS-BUS architecture. We also have synthesized various configurations of CASes using the Synopsys synthesis tool [21]. Some results are presented in Table 1. CAS (a) corresponds to a CAS which switch is implemented by N decoders and CAS (b) a CAS with a switch including only one decoder. The CAS Instruction Register width (k) and the area in terms of transistors (tr) are presented. A trade-off must be made between CAS (a) and CAS (b) depending on the

Table 1. CAS synthesis results.

		CAS (a)		CAS (b)		TAPCAS (a)		TAPCAS (b)	
N	P	k	tr	k	tr	k	tr	k	tr
5	1	5	696	3	626	5	684	3	614
5	3	10	1.224	6	2.026	–	–	–	–
5	4	15	1.636	7	2.634	–	–	–	–
8	1	8	1.084	4	912	8	1.072	4	900
8	3	16	1.952	9	7.286	–	–	–	–
8	4	24	2.548	11	39.990	–	–	–	–
8	6	24	2.948	15	>100.000	–	–	–	–
10	1	10	1.338	4	1.044	10	1.326	4	1.032
10	4	30	3.236	13	>100.000	–	–	–	–

area constraints and the scan cycles overhead needed for CAS configuration. For high values of N and P (great number of combinations) CAS (b) cannot be used because the area grows quickly with N and P.

These two architectures allows the SoC integrator to make a trade-off between the total k bits word control and the area overhead induced by the decoder size. In the first case, the total k bit word is longer but the total number of gates is reduced. In the second case, synthesizing one decoder for the whole N inputs requires more gates but fewer control bits. A trade-off can be done to optimize the CIRs length and the area overhead. However when summing up all CIRs and WIRs length, we found out that it is far smaller than a core scan chain length. In the same way, the area overhead is not significant when compared to a core area.

Concerning the architecture dealing with TAPed cores testing, we can see in Table 1 that TAPCASes (b) (one decoder per switch) are more interesting than TAPCASes (a) (N decoders per switch) since they are smaller and need fewer control bits. The area overhead induced for TAPed cores testing is not significant since the TAPCASes are very small and the new TAP controller is implemented with 1264 transistors. However configuring TAPed cores, in comparison with wrapped cores, needs 27 extra scan cycles (loading the TAP_CONFIG instruction, entering and leaving the second part of the central controller).

6.1.2. Wrapper Generation. Some experiments have been made with the wrapper generator, using the AMD2901 microprocessor as a core. The wrapped core has been simulated and synthesized with ALLIANCE, a home-designed CAD suite [15]. Although the obtained wrapper area seems a very large overhead with the chosen core example, the absolute area values are very reasonable: the area of one boundary cell is equal to 98 transistors (this value may slightly vary depending on the kind of cell), and the area of the 3-bits Wrapper instruction Register (WIR) is equal to 240 transistors.

6.2. *Implementation of the Compression Method*

To check the feasibility of the method, we implemented a synthesizable description of the decompression/expansion module, the MISR, the MTAP and all the IEEE 1149.1 remaining parts. The modular descriptions was synthesized with the Synopsys synthesis tool [21]. Simulations coincide well with expected results.

The decompression/expansion module implemented deals with 4 bit-length samples and is available in generic VHDL format. An update of this module, processing 3 bit samples is under development. Its area depends on the TAM width (N). For $N = 3$ the module's area is equal to 10488 transistors, 12842 transistors for $N = 10$ and 31528 transistors for $N = 100$. Another advantage of the presented method over some coding/decoding approaches is that it requires few hardware to be implemented.

In order to experiment the compression method with a real test example, we decided to apply it to the test patterns of a real circuit designed in our laboratory. The circuit, named PCIDDC, is a 200 K transistor, Network interface Component for PCI bus including 4 scan chains with a 1028 total scan length [24].

Using our method, when using 4 bit-length samples, we easily reach 25% of compression rate with a random distribution of combinations within groups. When using 3 bit-length samples we reach 35% of compression rate. Calculating the optimized distribution adds 2% to this compression rate. For the same patterns, we calculated the percentage of appearance of each combination in order to prepare a Huffman coding:

(i) with 3 bit-length samples: all the 8 combinations have a close appearance degree (between 10% and 20%).
(ii) with 4 bit-length samples: all the 16 combinations have a close appearance degree (between 5.3% and 8%).
(iii) with 6 bit-length samples: all the 64 combinations have a close appearance degree (between 1% and 1.9%).

This means that correlation between samples is very loose and no important compression can be reached with a Huffman coding. Obviously, for all the methods based on similar coding, the problem remains the same, since they all necessitate a good correlation between samples. This kind of problems are avoided with the expansion/decompression method detailed in this paper.

7. Conclusion

The basic CAS-BUS TAM described in this paper is flexible, scalable and dynamically reconfigurable. It allows multiple trade-off regarding the choice of Nmax, the number of TDI/TDO couples, the kind of switch implementation and the CASes configurations. The architecture is both IEEE P1500-compliant (in its current development status) at core level and 1149.1-compliant at SoC level. By correctly configuring the CASes, the test programmer can choose during each test session which core scan chains must be serialized in order to optimize their total length. The goal is to have for each of the N wires the same test length. The control of the global architecture is easy through a simple test access port.

Two instances of SoC test architecture based on the use of the CAS-BUS TAM have been presented in this paper.

The first one offers a complete solution for testing both wrapped and TAPed cores within a SoC, concurrently if needed. With the same test access mechanism, wrapped and TAPed cores can be tested, thanks to a hierarchical test control. In order to manage the TAPed cores testing, new CASes have been designed and the central TAP controller has been modified. The area overhead induced by the TAPed core testing is not significant. The SoC integrator has the choice of using this architecture or the previous CAS-BUS TAM depending on the presence of TAPed cores in the SoC.

The second instance has been developed to integrate a new compression-decompression/expansion method within the CAS-BUS TAM architecture. Even if it does not provide spectacular compression performances with correlated test patterns like Huffman/similar or run-length coding methods, it is very useful with non correlated test patterns. Compression degree varies between 33% (minimum average) and 66% (maximum). Particularly, we propose this method to enhance test performances in term of test time when expanding a test serial input to more than one SoC internal test chain. An expansion/decompression sequence example has been presented. A simple modification has been introduced in the IEEE 1149.1 TAP controlling the CAS-BUS architecture. The modified TAP (MTAP) behaves exactly as a classic TAP, except when it is an expansion/decompression process. The method has been tested on the real test patterns of a circuit made in our laboratory (PCIDDC: a Network Interface Component) and 35% of compression degree is reached. The use of other compression methods was studied for the same patterns, but due to their feeble correlation, compression degrees were not significant.

Independent and modular test architecture generators have been developed in the C language. Depending on the SoC features and constraints, the user can generate the appropriate TAM architecture in synthesizable format. These pieces of software make available a toolbox environment for automatic design of plug-and-play components, thus easing the design of SoC test architectures based on the CAS-BUS TAM and its developed instances. This kind of architecture is compliant with existing or under development IEEE standards, independently from any proprietary architecture.

Acknowledgments

Authors would like to thank Le Yin Keu for the development of the wrapper generator, in the framework of his DEA project, and Zakaryae Bekkouri and Gilles Richard for the development of the compression-decompression/expansion software, in the framework of their final undergraduate study project.

References

1. M. Benabdenbi, W. Maroufi, and M. Marzouki, "Cas-Bus: A Scalable and Reconfigurable Test Acces Mechanism for Systems on a Chip," in *IEEE Design Automation and Test in Europe (DATE)*, Paris, France, March 2000, pp. 141–145.
2. M. Benabdenbi, W. Maroufi, and M. Marzouki, "Testing Taped Cores and Wrapped Cores with the Same Test Access Mechanism," in *IEEE Design Automation and Test in Europe (DATE)*, Munich, Germany, March 2001, pp. 150–155.
3. D. Bhattacharya, "Hierarchical Test Access Architecture for Embedded Cores in an Integrated Circuit," in *IEEE VLSI Test Symposium (VTS)*, Dana Point (CA), USA, April 1998, pp. 8–14.
4. K. Chakrabarty, "Design of System on a Chip Test Access Architectures Under Place-and-Route and Power Constraints," in *IEEE/ACM Design Automation Conference (DAC)*, Los Angeles (CA), USA, June 2000, pp. 432–437.
5. K. Chakrabarty, "Optimal Test Access Architectures for System on a Chip," *ACM Trans. Design Automation of Electronic Systems*, vol. 6, no. 1, pp. 26–49, Jan. 2001.
6. B. Dervisoglu and J. Swamy, "A Novel Approach for Designing a Hierarchical Test Access Controller for Embedded Core Designs in an SoC Environment," in *4th IEEE International Workshop on Testing Embedded Core-Based System-Chips (TECS)*, Montreal (QC), Canada, May 2000, pp. 1.4.1–1.4.7.
7. C. Feige and C. Wouters, "Integration of Structural Test Methods into an Architecture Specific Core-Test Approach," in *2nd IEEE International Workshop on Testing Embedded Core-Based System-Chips (TECS)*, Washington (DC), USA, Oct. 1998, pp. 5.2.1–5.2.8.
8. S.K. Goel and E.J. Marinissen, "TAM Architectures and their Implication on Test Application Time," in *IEEE International Workshop on Testing Embedded Core-Based Systems (TECS)*, Marina del Rey (CA), USA, May 2001, pp. 3.3.1–3.3.10.
9. IEEE P1500 Standard for Embedded Core Test Web Site, http://grouper.ieee.org/groups/1500. Public Informations.
10. V. Iyengar, K. Chakrabarty, and E.J. Marinissen, "Test Wrapper and Test Acces Mechanism Co-Optimization for System on a Chip," in *IEEE International Test Conference (ITC)*, Baltimore, MD, USA, Oct. 2001, pp. 1023–1032.
11. A. Jas, J. Ghosh-Dastidar, and N.A. Touba, "Scan Vector Compression/Decompression Using Statistical Coding," in *17th IEEE VLSI Test Symposium (VTS)*, San Diego (CA), USA, April 1999.
12. A. Jas and N.A. Touba, "Test Vector Decompression Via Cyclical Scan Chains and its Application to Testing Core-Based Designs," in *IEEE International Test Conference (ITC)*, Washington (DC), USA, 1998, pp. 458–464.
13. E. Larsson and Z. Peng, "An Integrated System on a Chip Test Framework," in *IEEE Design Automation and Test in Europe Conference (DATE)*, Munich, Germany, March 2001, pp. 138–144.
14. D.A. Lelewer and D.S. Hirschberg, "Data Compression," *ACM Computing Surveys (CSUR)*, vol. 19, no. 3, pp. 261–296, Sept. 1987.
15. LIP6-ASIM, Alliance CAD System. http://www-asim.lip6.fr/alliance/. University Paris 6-France.
16. E.J. Marinissen, R. Arendsen, G. Bos, H. Dingemanse, M. Lousberg, and C. Wouters, "A Structured and Scalable Mechanism for Test Access to Embedded Reusable Cores," in *IEEE International Test Conference (ITC)*, Washington (DC), USA, Oct. 1998.
17. E.J. Marinissen and M. Lousberg, "The Role of Test Protocol in Testing Embedded Core-Based System ICs," in *IEEE European Test Workshop (ETW)*, Constance, Germany, May 1999, pp. 70–75.
18. E.J. Marinissen, Y. Zorian, R. Kapur, T. Taylor, and L. Whetsel, "Towards a Standard for Embedded Core Test: An Example," in *IEEE International Test Conference (ITC)*, Atlantic City (NJ), USA, Sept. 1999.
19. W. Maroufi, M. Benabdenbi, and M. Marzouki, "Solving the I/O Bandwith Problem in System on a Chip Testing," in *XIII Symposium on Integrated Circuits and System Design (SBCCI)*, Manaus (AM), Brazil, Sept. 2000, pp. 9–14.
20. W. Maroufi, M. Benabdenbi, and M. Marzouki, "Controlling the CAS-BUS TAM with IEEE 1149.1 TAP: A Solution for Systems on a Chip Testing," in *4th IEEE International Workshop on*

Testing Embedded Core-Based System-Chips (TECS), Montreal (QC), Canada, May 2000, pp. 4.5.1–4.5.6.

21. Synopsys, Synopsys Design Compiler Family. http://www.synopsys.com/products/logic/logic.html. Synopsys, Inc.
22. J. van Beers and H. van Herten, "Test Features of a Core-Based Co-Processor Array for Video Applications," in *IEEE International Test Conference (ITC)*, Atlantic City (NJ), USA, Sept. 1999, pp. 638–647.
23. P. Varma and S. Bhatia, "A Structured Test Reuse Methodology for Core-Based System Chips," in *IEEE International Test Conference (ITC)*, Washington (DC), USA, Oct. 1998.
24. F. Wajsbürt, J.L. Desbarbieux, C. Spasevski, S. Penain, and A. Greiner, "An Integrated PCI Component for IEEE 1355," in *European Multimedia Microprocessor Systems and Electronic Commerce Conference and Exhibition*, Florence, Italy, Nov. 1997.
25. L. Whetsel, "An IEEE 1149.1 Based Test Acces Architecture for ics with Embedded Cores," in *IEEE International Test Conference (ITC)*, Washington (DC), USA, Nov. 1997, pp. 69–78.
26. L. Whetsel, "Addressable Test Ports an Approach to Testing Embedded Cores," in *IEEE International Test Conference (ITC)*, Atlantic City (NJ), USA, Sept. 1999, pp. 1055–1064.
27. L. Whetsel and M. Ricchetti, "Tapping into IEEE P1500 Domains," in *5th IEEE International Workshop on Testing Embedded Core-Based Systems (TECS)*, Los Angeles (CA), USA, May 2001.
28. I.H. Witten, R.M. Neal, and J.G. Cleary, "Arithmetic Coding for Data Compression," *Communications of the ACM*, vol. 30, no. 6, pp. 520–540, June 1987.
29. Y. Zorian, "Testing the Monster Chip," in *IEEE Spectrum*, July 1999, pp. 54–60.

Mounir Benabdenbi was born in 1971 in Meknes, Morocco. He received the Masters Degree in Electronics and the D.E.A Degree in Microelectronics from the University of Montpellier. He is currently working towards his Ph.D. degree at LIP6 Laboratory under the direction of Meryem Marzouki. His research interests include Boundary Scan and SoC Testing.

Walid Maroufi was born in 1970 in Monastir, Tunisia. He received his Engineering degree from the Faculty of Electrical Engineering of Monastir, Tunisia and his Ph.D. degree in Computer Science from the University Pierre et Marie Curie in Paris, France, where he spent afterwards one year of post-doctoral research. He is currently with Nortel Networks, Ottawa, Canada.

Meryem Marzouki was born in 1961 in Tunis, Tunisia. She received the D.E.A Degree and the Ph.D. Degree, both in Computer Science, from the National Polytechnical Institute of Grenoble in 1987 and 1991 respectively. Before that, she received the Engineering Degree in Computer Science too, from the University of Tunis, in 1986.

From 1987 to 1992, she was mainly involved in prototype validation of ICs in the framework of electron-beam testing. Since 1992, she has been a researcher with the CNRS (French National Research Center), in charge of the *Diagnosis of Complex System* Group at TIMA Laboratory in Grenoble, France, with artificial intelligence approaches for test and diagnosis as main research interests, until 1997, when she joined the LIP6 Laboratory in Paris. Since then, her research interests are related to high-level synthesis for testability and system on a chip testing.

JOURNAL OF ELECTRONIC TESTING: Theory and Applications 18, 475–485, 2002

An Integrated Approach to Testing Embedded Cores and Interconnects Using Test Access Mechanism (TAM) Switch*

SUBHAYU BASU
Princeton University, NJ, USA

INDRANIL SENGUPTA AND DIPANWITA ROY CHOWDHURY
Department of Computer Science and Engineering, Indian Institute of Technology, Kharagpur 721302, India

SUDIPTA BHAWMIK
Agere Systems, Holmdel, NJ, USA

Received September 2, 2001; Revised January 17, 2002

Editor: Krishnendu Chakrabarty

Abstract. The present paper introduces a new strategy for testing embedded cores using Test Access Mechanism (TAM) switches. An algorithm has been proposed for testing the cores using the TAM switch architecture. In addition, a scheme for testing the interconnections between cores in parallel is also presented. Experiments have been carried out on several synthetic SOC benchmarks. Results show significant optimization of area overhead as well as test time.

Keywords: system-on-chip, TAM switch, interconnect testing

1. Introduction

Embedded cores are becoming popular day by day due to their design-reuse, greater on-chip functionality and shorter production cycle [10]. But the testing of such cores in SoC has become a major challenge. Due to non-accessibility of the core inputs and outputs some special access mechanisms are used to test them at the chip level. Using some suitable test access mechanism (TAM), the test data are transported from the chip level to the core inputs, and conversely from the core outputs to the chip level. A number of TAM architectures have been proposed, like macro-test, core-transparency [7], dedicated test-bus, multiplexed access, and a test architecture based on test rail [8]. Test rail provides a flexible and scalable architecture, and helps to trade off between test time and area overhead by varying test data width. The recent research emphasis is on test scheduling for core based system [1, 3, 6]. In [3], it is assumed that TAM has one external test bus and the cores are assigned to the test bus. The author treated the problem as an m-processor open shop scheduling problem and used Mixed Integer Linear Programming to solve the problem. In [6], a fixed number of test buses has been assumed instead of one and each core is assign to a test bus. An Integer Linear Programming model is proposed to minimize the test time. The same problem has been solved by considering a cluster of cores into a module and test scheduling has been done to reduce the total test time. Most of these works have restricted themselves to time scheduling only [1, 2, 6]. In the present

*This work is supported by Lucent Technologies, Bell Labs, USA.

paper, we present a new strategy for optimally using the test bandwidth and controlling the whole testing mechanism using TAM switches. It attempts to optimize both the area overhead and the test application time.

2. Proposed Work

The given core based system is represented by a graph $G = (V, E)$, where the set of vertices V represents the cores in the SoC, and the set of edges E represents the interconnections between the cores. Multiple interconnections between two cores are labeled as a single edge. Hyper edges are replaced by complete graphs between the associated cores. It is assumed that for the given core-based system, information like the time cycles required to test each core, number of scan chains for each core, size of each scan chain, and the test bandwidth provided by the chip, are available. Using these information an algorithm is proposed to generate an optimal number of Test Access Mechanism (TAM) switches, which can deliver tests to the embedded cores, based on the given inputs and user defined constraints. At the same time, an attempt is made to determine the optimal size of the TAM switches, keeping the total area overhead due to these switches minimum while keeping the test application time the shortest possible. Finally, in this algorithm it is proposed to generate test schedules (consisting of several test sessions) for optimal application time. A strategy is also designed to test interconnects between the cores to cover faults like stuck-at-0, stuck-at-1, coupling, etc and to ensure full testability. The following subsection introduces the Test Access Mechanism (TAM) switch. The assumptions undertaken in these algorithms are:

(a) every core is to be scan-tested;
(b) there are no BIST enabled cores;
(c) due to the partition of test input pins into smaller test rails, several TAM switches that are on different test rails may be tested in parallel;
(d) all cores have a common clock; and
(e) the wire cost for the entire TAM is not considered.

2.1. *Test Access Mechanism (TAM) Switch*

The TAM switch is a synthesizable RTL core that can be instantiated in an ASIC design to provide test data access to embedded cores. The test can be either deterministic vectors (generated by an ATPG) for a scan based core, or pseudo-random vectors generated by a BIST test pattern generator for a BIST ready core. The TAM switch allows the SoC test integrator to maximally utilize the test data bandwidth available.

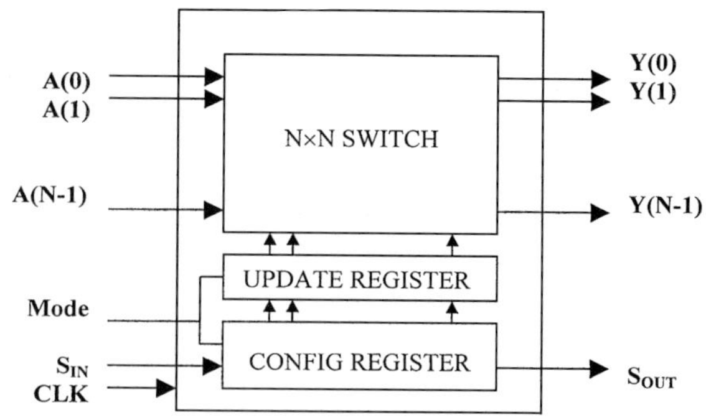

Fig. 1. A TAM switch.

It is a programmable crossbar switch (Fig. 1) that allows a flexible way to deliver test vectors to embedded cores at varying bandwidth. It consists of an $N \times N$ switch matrix where N is the number of input and output lines. The switch configuration can be programmed by serially loading the configuration information into a *Configuration Register* ($N * \log(N)$ bits wide), which decides which input will be connected to which output, under the control of a *Mode* signal (Mode = high for shifting in). There is an *Update Register* that loads the configuration when *Mode* is low.

3. Generation of Optimal Number and Size of TAM Switches

In the proposed scheme, several cores can be under the control of a TAM switch. In essence, every TAM switch represents a *cluster*, which is a set of one or more cores and the interconnections between them. This is illustrated in Fig. 2. Basically the TAM provides a way for transporting test patterns to/from the core-under-test and the chip periphery. The core test wrapper, on the other hand, interfaces between the core and the rest of the chip (including TAM). The wrapper possesses the capability to switch between three modes: normal mode, core test mode, and interconnect test (bypass) mode.

For the purpose of applying the test vectors to the SoC, several test I/O pins must be provided. The

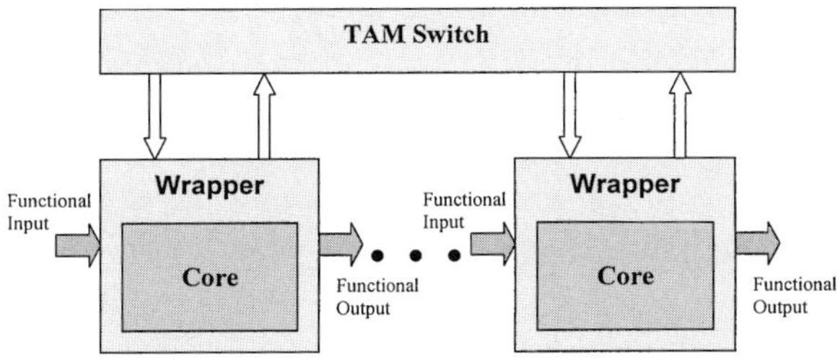

Fig. 2. A TAM cluster and the core wrappers.

number of such I/O pins is called *test bandwidth*. Since there may be several TAM switches inside the chip, it is necessary to have a *test bandwidth division*, for the purpose of maximum utilization of the test hardware. Different test bandwidth divisions may lead to solutions, some of them being better than the others with regards to hardware overhead and/or test application time. For example, a test bandwidth of 7 can be variously partitioned as (7), (6,1), (5,2), (3,2,2), (4,3), etc.

Definition 1. The *deficiency* of a cluster C_i is defined as

$$d_i = (\text{Maximum Bandwidth Division}) - \left(\sum_j \text{number of scan chains in core } j\right)$$

for all cores j in cluster i.

In other words, the deficiency for a cluster is the difference between the *highest available bandwidth division* and *the sum of the total number of scan chains in the cluster*. We try to minimize the deficiency of a core as much as possible. In the proposed algorithm, we consider only clusters whose deficiency values are positive. Thus if the bandwidth partition is (8,0) and we have a cluster which has two cores of 80 and 10 scan chains respectively, this cluster will have a negative deficiency ($8 - 90 = -82$) and will not be considered. Such a cluster is assigned a TAM switch of proper size and dealt with in the latter part of the algorithm.

For the sake of comparison between several solutions, an objective function has been defined based on the hardware overhead and the test application time. The hardware overhead clearly depends on the number of TAM switches and their widths. Although there can be several choices, the following function has been used for the purpose of evaluating a solution.

$$\text{Objective function} = \text{Hardware Overhead} * \text{Test Application Time}$$

3.1. *Algorithm*

The inputs to the algorithm are:

1. Number of cores to be tested
2. Number of time cycles required to test each core
3. Number of scan chains for each core
4. Size of each scan chain
5. Number of I/O pins required to test the SoC
6. A graph representing the SoC

The algorithm generates the following outputs:

1. Number of TAM switches
2. Size of each TAM switch
3. A netlist showing the connection between the TAM switches and the cores
4. Partition of input bandwidth between TAM switches

3.1.1. Initial Allotment of Cores. Initially the number of TAM switches is set to be equal to the number of cores. For a particular partition of bandwidth, the algorithm would assign a core to the bandwidth partition, based on the following three cases:

1. If the number of scan chains of a core is equal to a bandwidth division then the core is serviced by that partition.
2. If the number of scan chains of a core is more than the highest division possible then the core is initially allotted to the highest bandwidth division.
3. For all other cases, assign the core to the division that has scan chains nearest (less) with respect to the bandwidth.

Example 1. For a partition (5,3), if there is a core with 3 scan chains, then the initial allotment is done according to case 1 by allotting it to the test bandwidth division of 3. If there was a core with 8 scan chains, allotment will be done according to case 2 by stitching the scan chains to 5 (assuming that all scan chains can be stitched). And if the core has 4 scan chains it will be allotted to test bandwidth division of 5 according to case 3.

3.1.2. Algorithmic Steps

```
For all bandwidth partition
  {
      initial_allotment_cores( );
      while (deficiency > 0 or cores in the neighborhood of the cluster > 0)
        {     assign_cluster( );
              expand_cluster( );
              find_deficiency of cluster generated by expand_cluster( );
              if (deficiency < 0)
              {   find_deficiency of cluster generated by assign_cluster( );
                  break;
              }
        }
      while (deficiency < 0)
        {     stitch_scanchains( );
              build_scanchains_partitions( );
              assign_scanchains_partitions( );
        }
      identify_max_test_time_bandwidth_division( );
      identify_min_test_time_bandwidth_division( );
      while (max_test_time_bandwidth_division < max_bandwidth_division)
        {      if (max_test_time_bandwidth_division < min_test_time_bandwidth_division)
                  {       select min_test_time_cluster in max_test_time_bandwidth;
                          place this cluster in min_test_time_bandwidth;
                          if (max_test_time_bandwidth_division == max_bandwidth_division)
                              break;
                  }
               else
                      shift to next min_test_time_bandwidth_division    ;
        }
      find_objective_func.( );
  }
find_min_objective func.( );
find_total_testing_time( );
```

Here the function **assign_cluster( )** allots the cluster to the bandwidth division which is greater than or equal to the total number of scan chains in the cluster. In the function **expand_cluster( )**, a cluster is expanded by including cores that are neighbors to the cores of the cluster. That neighbor is chosen which decreases the deficiency the most. If there are two or more such neighbor cores that decrease the deficiency by the same amount, the core that has the least testing time is chosen. The **find_deficiency( )** function finds the deficiency and if the deficiency becomes negative after expanding cluster then it breaks from the while loop. The deficiency of the assigned cluster is then calculated. For negative deficiency, the function **stitch_scanchains( )** stitches the lesser order scan chains to form one large scan chain, the size of which is smaller than the size of the largest scan chain of the core. This step is rigorously done using exhaustive search (feasible because of the limited problem size) so as to lessen the number of scan chains for a core as much as possible. Only those scan chains are stitched which are on the same clock domain. Now **build_scanchains_partitions( )** partition the lessened number of scan chains into uniform partitions based on several strategies. First it finds the number of partitions by dividing the total number of scan chains by the available test bandwidth or some multiple of it. Then for each partition dedicate one TAM switch. Offer a choice to the designer, a tradeoff between the number of TAM switches and the size of each TAM switch. The designer may now wish to test this core with large number of scan chains with lesser number of TAM switches but increased complexity of each TAM switch, or alternatively with greater number of TAM switches leading to lesser complexity of each core. Ultimately **assign_scanchains_partitions( )** assign the partitions to the appropriate bandwidth division. **identify_max_test_time_bandwidth_division( )** identifies the bandwidth division which has the maximum test time (Test time in each bandwidth division is found out by adding the testing time of each TAM switches in that particular bandwidth division). **identify_min_test_time_bandwidth_division( )** similarly identifies the bandwith division which has the minimum testing time. After this for a particular configuration of bandwidth the objective function is calculated by **find_objective_func.( )** The function **find_min_objective func.( )** selects the bandwidth partition configuration which has minimum objective function. Finally **find_total_testing_time( )** calculates the total testing time. The testing time is the maximum of the testing times for each bandwidth partition.

4. Testing of Interconnects Between Cores

Here we assume that the test vectors are fed through the TAM switch by changing the switch configuration. At the wrapper, from the test inputs it is bypassed and fed to the wrapper cells of the outputs, and then it is fed to the interconnects. These interconnects feed to the input wrapper cells and bypassed to feed the TAM. But this is beyond the scope of the present work, as it needs the configuration of the wrapper configuration register.

We use the *walking zeroes & walking ones* and *all zeroes & all ones* tests for detecting all single and multiple stuck-at faults and also coupling faults in the interconnects between cores.

4.1. Algorithm

The inputs to the algorithm are:

1. Number of cores
2. Number of scan chains in each core
3. Number of input/output pins required to test the SoC
4. Number of functional input/output pins present in each core
5. A graph representing the SoC

The outputs generated by the algorithm are:

1. The interconnects that will be tested parallel
2. The overall test architecture

4.1.1. Algorithmic Steps

1. In the first step, the interconnects which are fed in parallel to the same core are considered. For example, if two interconnects from different cores feed the same core, then they may be bunched at some part and they are to be tested parallely to cover all the faults like stuck-at-0, stuck-at-1, bridge fault, cross-talk, etc. Fig. 3 illustrates the idea.
2. Next, all such cores are considered the outputs of which are feeding to different cores (Fig. 4). The interconnects at the inputs of both the cores being

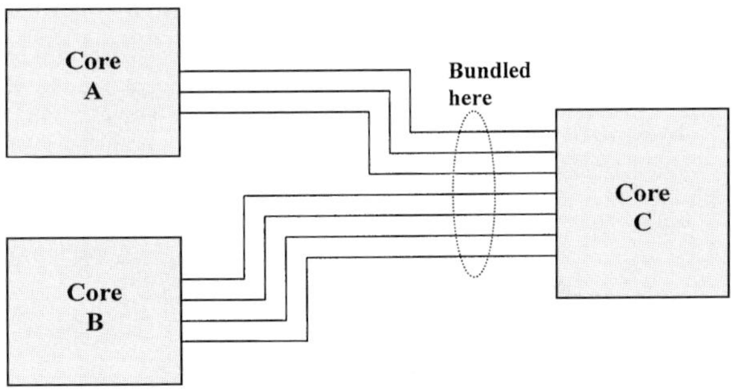

Fig. 3. Two cores feeding to a third core.

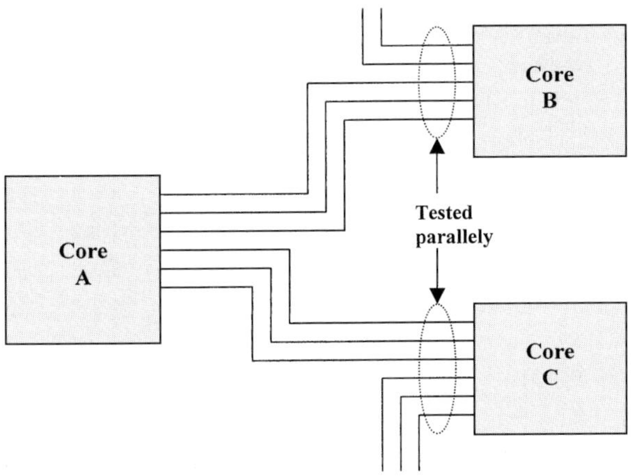

Fig. 4. Core A feeding Core B and Core C.

fed can be tested parallely. During this step, all other output lines of the core are fed with either all zeroes or all ones.

5. Results

The proposed algorithms have been tested by running them on five synthetic SoC designs (Appendix-A), which use ISCAS sequential benchmark circuits (Appendix-B) as cores. We have developed different cases either giving more stress on the hardware function or on the total test time. In the following tables, two of the cases are highlighted. In case-1 all the clusters and the partitions are assigned separately to a particular TAM switch. In case-2, The TAM switches feeding to different clusters in each bandwidth partition, are combined into a single bigger TAM switch. Also the cores that have more scan chains than the highest test bandwidth partition, are partitioned into bigger blocks. Then each block is assigned to a TAM switch.

Tables 1 to 5 depict the results of the test runs. In the tables, TAM Switch no. indicates the serial number of the TAM switch, BDIV indicates the width of the test bandwidth partition, no. of Scan Chn indicates the number of scan chains connected to that TAM Switch, Cores Conn is the serial number of the cores connected

Table 1. *System-on-Chip-1*: Test input pins: 3, test output pins: 3, test bandwidth partition: (1,2).

Case-1						Case-2					
TAM switch no.	BDIV	No. of scan chn	Cores conn	Size	Time	TAM switch no.	BDIV	No. of scan chn	Cores conn	Size	Time
0	1	1	0	2 × 2	3	0	2	3	2	5 × 5	12
1	1	1	1	2 × 2	6	1	2	3	3	5 × 5	14
2	2	2	2	4 × 4	6	2	2	3	4	5 × 5	12
3	2	1	2	3 × 3	6	3	1	2	0,1,2	3 × 3	11
4	2	2	3	4 × 4	7						
5	2	1	3	3 × 3	7						
6	2	2	4	4 × 4	6						
7	2	1	4	3 × 3	6						

Table 2. *System-on-Chip-2*: Test input pins: 4, test output pins: 4, test bandwidth partition: (2,2).

Case-1						Case-2					
TAM switch no.	BDIV	No. of scan chn	Cores conn	Size	Time	TAM switch no.	BDIV	No. of scan chn	Cores conn	Size	Time
0	2(1)	2	0,1	4 × 4	3	0	2(2)	3	4	5 × 5	12
1	2(1)	2	2	4 × 4	7	1	2(2)	4	5	6 × 6	14
2	2(1)	2	3	4 × 4	8	2	2(2)	4	5	6 × 6	14
3	2(2)	2	4	4 × 4	6	3	2(2)	3	5	5 × 5	12
4	2(2)	1	4	3 × 3	6	4	2(1)	6	0,1,2,3	8 × 8	18
5	2(2)	2	5	4 × 4	6						
6	2(2)	2	5	4 × 4	7						
7	2(2)	2	5	4 × 4	7						
8	2(2)	2	5	4 × 4	7						
9	2(2)	2	5	4 × 4	7						
10	2(2)	1	5	3 × 3	7						

Table 3. *System-on-Chip-3*: Test input pins: 3, test output pins: 3, test bandwidth partition: (1,2).

Case-1						Case-2					
TAM switch no.	BDIV	No. of scan chn	Cores conn	Size	Time	TAM switch no.	BDIV	No. of scan chn	Cores conn	Size	Time
0	2	2	0,1	4 × 4	6	0	2	3	4	5 × 5	14
1	1	1	2	2 × 2	6	1	2	4	0,1,3	6 × 6	16
2	2	2	3	4 × 4	10	2	1	1	2	2 × 2	6
3	2	2	4	4 × 4	7						
4	2	1	4	3 × 3	7						

Table 4. *System-on-Chip-4*: Test input pins: 4, test output pins: 4, test bandwidth partition: (2,2).

Case-1						Case-2					
TAM switch no.	BDIV	No. of scan chn	Cores conn	Size	Time	TAM switch no.	BDIV	No. of scan chn	Cores conn	Size	Time
0	2(1)	1	0	3 × 3	5	0	2(2)	3	2	5 × 5	12
1	2(1)	2	1	4 × 4	10	1	2(2)	3	3	5 × 5	14
2	2(2)	2	2	4 × 4	6	2	2(2)	4	4	6 × 6	14
3	2(2)	1	2	3 × 3	6	3	2(2)	4	4	6 × 6	14
4	2(2)	2	3	4 × 4	7	4	2(2)	3	4	5 × 5	12
5	2(2)	1	3	3 × 3	7	5	2(2)	4	5	6 × 6	18
6	2(2)	2	4	4 × 4	6	6	2(2)	4	5	6 × 6	16
7	2(2)	2	4	4 × 4	7	7	2(2)	4	5	6 × 6	16
8	2(2)	2	4	4 × 4	7	8	2(2)	4	5	6 × 6	16
9	2(2)	2	4	4 × 4	7	9	2(2)	4	5	6 × 6	16
10	2(2)	2	4	4 × 4	7	10	2(2)	2	5	4 × 4	16
11	2(2)	1	4	3 × 3	7	11	2(1)	3	0,1	5 × 5	15
12	2(2)	2	5	4 × 4	8						
13	2(2)	2	5	4 × 4	8						
14	2(2)	2	5	4 × 4	8						
15	2(2)	2	5	4 × 4	8						
16	2(2)	2	5	4 × 4	8						
17	2(2)	2	5	4 × 4	8						
18	2(2)	2	5	4 × 4	8						
19	2(2)	2	5	4 × 4	8						
20	2(2)	2	5	4 × 4	8						
21	2(2)	2	5	4 × 4	9						
22	2(2)	2	5	4 × 4	9						

to a particular TAM switch, Size is the number of inputs/outputs in the TAM switch, and Time is the total testing time required for the cluster identified by that TAM switch. As can be seen from the tables, the scan chains for a core may be connected to more than one TAM switch.

Table 6 summarizes the results for all the five examples.

Table 5. *System-on-Chip-5*: Test input pins: 3, test output pins: 3, test bandwidth partition: (3) in case-1, (1,2) in case-2.

Case-1						Case-2					
TAM switch no.	BDIV	No. of scan chn	Cores conn	Size	Time	TAM switch no.	BDIV	No. of scan chn	Cores conn	Size	Time
0	3	3	0,1	6 × 6	7	0	2	3	2	5 × 5	14
1	3	3	2	6 × 6	7	1	2	3	3	5 × 5	14
2	3	3	3	6 × 6	7	2	2	3	4	5 × 5	14
3	3	3	4	6 × 6	7	3	1	1	0	2 × 2	5
						4	2	2	1	4 × 4	7

Table 6. Summary of results for the five SoC designs.

	Case-1				Case-2			
SoC no.	Hardware overhead	Total test time	Objective function	Number of TAM switches	Hardware overhead	Total test time	Objective function	Number of TAM switches
1	83	44	3652	8	84	41	3444	4
2	162	61	9882	11	186	56	10416	5
3	61	34	2074	5	65	32	2080	3
4	344	178	61232	23	368	166	61088	12
5	144	32	4608	4	97	53	5141	5

Table 7. Results of running the interconnect test algorithm.

SoC number	The interconnects to be tested in parallel and in this order (between)	SoC number	The interconnects to be tested in parallel and in this order (between)
SoC-1	Core 0 and Core 2	SoC-2	Core 0 and Core 2
	Core 1 and Core 2		Core 1 and Core 2
	Core 2 and Core 3		Core 1 and Core 4
	Core 0 and Core 4		Core 2 and Core 3
	Core 1 and Core 4		Core 3 and Core 4
	Core 3 and Core 4		Core 4 and Core 5
SoC-3	Core 0 and Core 5	SoC-4	Core 0 and Core 1
	Core 0 and Core 1		Core 0 and Core 2
	Core 1 and Core 2		Core 1 and Core 3
	Core 1 and Core 4		Core 2 and Core 3
	Core 2 and Core 5		Core 3 and Core 4
	Core 3 and Core 4		Core 0 and Core 4
	Core 4 and Core 5		
SoC-5	Core 0 and Core 1		
	Core 0 and Core 2		
	Core 0 and Core 3		
	Core 1 and Core 4		
	Core 2 and Core 4		

The hardware overhead is the sum of the switch sizes of all the TAM switches. Some of the TAM switches in case-1 for a SoC combine to form bigger TAM switches in case-2. Therefore the hardware overhead increases in case-2, except for SoC-5. The number of clock cycles required for initialization of the TAM switches decreases with the decrease in the number of switches. Thus the total test time is less in case-2 than case-1, with the exception of SoC-5. Here we have got minimum objective function for the test bandwidth partition of 3 in case-1. But the minimum objective function in case-2 is achieved by the test bandwidth partition of 2 and 1. In case-1, for that particular configuration, the test bandwidth is utilized to the maximum; this decreases the total test time drastically.

Table 7 shows the results obtained by running the interconnect test algorithm on the five SoC benchmarks.

6. Conclusion

We have presented a strategy based on TAM switches for optimal test application strategy using the test bandwidth available at the chip level. In the first algorithm we have optimized the area overhead and the test application time, and in the second algorithm we have given a strategy for testing the interconnects of the SoC. Results obtained on several synthetic SoC benchmarks are promising. Here we have also kept an option to the designer where one can choose between less numbers of TAM switches with more complexity or more number of TAM switches with less complexity. In our algorithm we have not considered the interconnection wire cost. The wire cost decreases with the decrease in the number of TAM switches. A proper cost estimation would be required if the wire costs are taken into consideration.

Appendix A

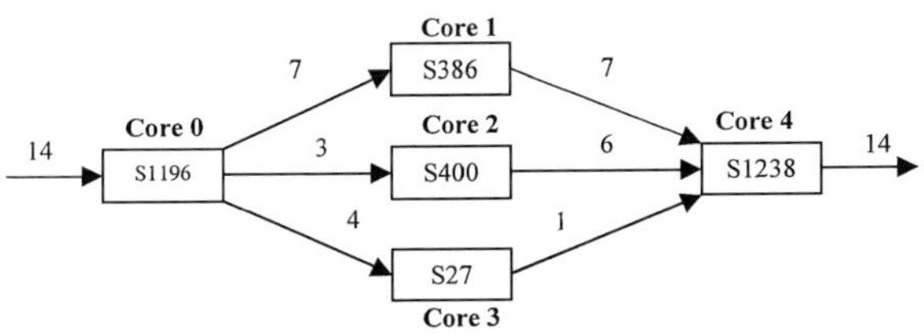

System On Chip-1

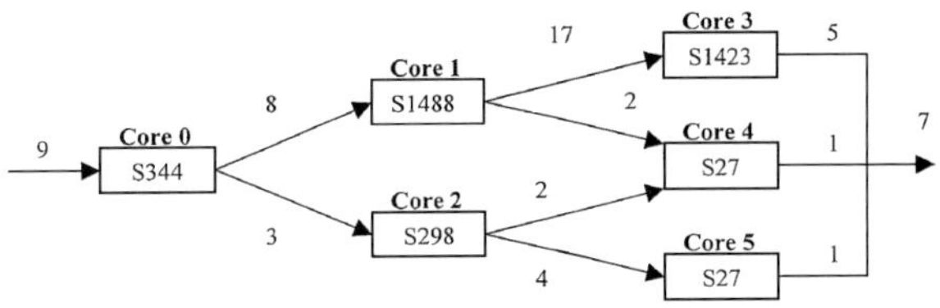

System On Chip-2

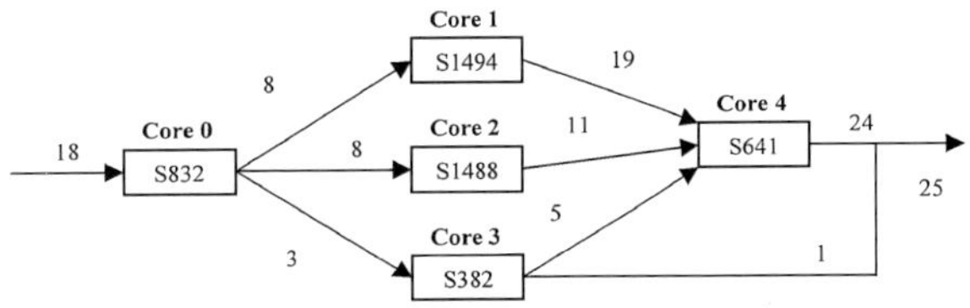

System On Chip-3

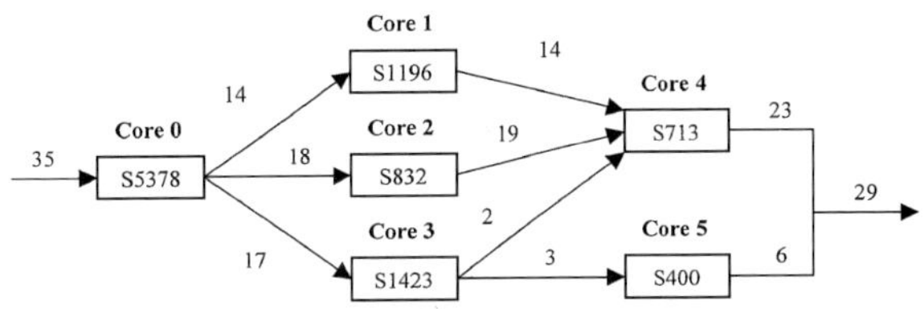

System On Chip-4

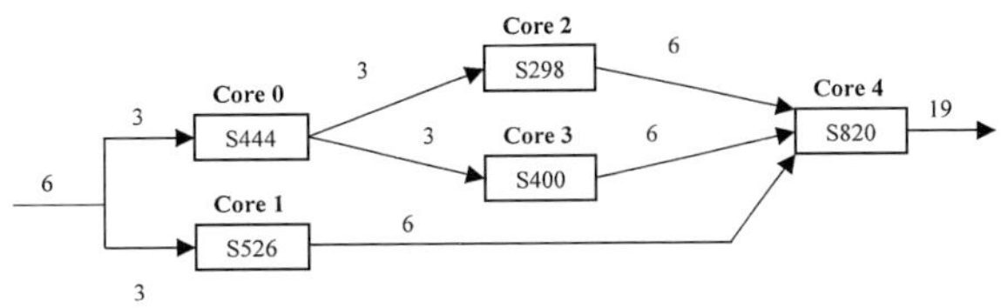

System On Chip-5

The number beside the arrow determines the width of the interconnection bus. The arrow in the edge determines the direction of the data flow.

Appendix B

Each sequential benchmark circuit is inserted with scan chains, and the length depends on the number of flip flops in the circuit. The library used is Synopsys' own design library.

Name of the sequential benchmark core	Functional inputs	Functional outputs	Number of D-flip flops	Number of scanchains	Length of each scanchain
S1196	14	14	18	3	6×3
S1238	14	14	18	3	6×3
S1423	17	5	74	11	$7 \times 8, 6 \times 3$
S1488	8	19	6	1	6
S1494	8	19	6	1	6
S27	4	1	3	1	3
S298	3	6	14	2	7×2
S344	9	11	15	2	7,8
S349	9	11	15	2	7,8
S382	3	6	21	3	7×3
S386	7	7	6	1	6
S400	3	6	21	3	7×3
S444	3	6	21	3	7×3
S526	3	6	21	3	7×3
S5378	35	49	179	22	$8 \times 19, 9 \times 3$
S641	35	24	19	2	10,9
S713	35	23	19	2	10,9
S820	18	19	5	1	5
S832	18	19	5	1	5

References

1. D. Bagchi, D.R. Chowdhury, and J. Mukherjee, "A Novel Strategy to Test Core Based Designs," *VLSi Design*, pp. 122–127, 2001.
2. K. Chakrabarty, "Test Scheduling for Core-Based Systems," in *Proc. Intl. Conf. on Computer Aided Design (ICCAD)*, Nov. 1999, pp. 391–394.
3. K. Chakrabarty, "Design of a System-on a Chip Test Access Architectures Using Integer Linear Programming," *VTS 2000*, pp. 127–134.
4. K. Chakrabarty, "Design of System-on-a-Chip Test Access Architectures Using Integer Linear Programming," in *Proc. IEEE VLSI Test Symposium (VTS)*, Montreal, Canada, April 2000, pp. 127–134.
5. K. Chakrabarty, "Design of System-on-a-Chip Test Access Architectures Under Place-and-Route and Power Constraints," in *Proc. ACM/IEEE Design Automation Conference (DAC)*, Los Angeles, CA, June 2000, pp. 432–437.
6. K. Chakrabarty, "Test Scheduling for Core-Based Systems Using Mixed Integer Linear Programming," *IEEE TCAD*, vol. 19, pp. 1163–1174, Oct. 2000.
7. I. Ghosh, S. Dey, and N.K. Jha, "A Fast and Low Cost Technique for Core-Based System-on-Chip," in *Proc. Design Automation Conference*, San Francisco, CA, June 1998.
8. E.J. Marinissen et al., "A Structured and Scalable Mechanism for Test Access to Embedded Reusable Cores," in *Proc. Intl. Test Conference*, Washington DC, Oct. 1998.
9. E.J. Marinissen, S.K. Goel, and M. Lousberg, "Wrapper Design for Embedded Core Test," in *Proc. IEEE Intl. Test Conference*, Oct. 2000.
10. Y. Zorian, E.J. Marinissen, and S. Dey, "Testing Embedded-Core Based System Chips," in *Proc. Intl. Test Conference*, Washington DC, Oct. 1998.

Subhayu Basu holds B-Tech (Hons) degree in Electrical Engineering from Indian Institute of Technology Kharagpur, India. From 2000 to 2001, he was graduate research consultant in Indian Institute of Technology working on a project sponsored by Agere Technologies, USA on VLSI Testing. At present he is undergoing the Ph.D. degree in Computer Engineering at Princeton University, New Jersey, USA. His research interests include VLSI testing, VLSI design and Bio Engineering with special emphasis on programming cell behaviors and cell-to-cell communications.

Indranil Sengupta is an Associate Professor in the Department of Computer Science and Engineering, Indian Institute of Technology, Kharagpur, India. He is also the Associate Head of the Computer and Informatics Center, which is in charge of one of the largest ATM/Gigabit Ethernet networking infrastructure in Asia, with more than 5000 computing nodes. His current interests are in the field of VLSI design and testing, fault tolerance, and computer networks. Dr. Sengupta received his B.Tech., M.Tech. and Ph.D. degrees in Computer Science from the University of Calcutta, and is in the teaching profession for more than 16 years.

Dipanwita Roy Chowdhury received the B.Tech and M.Tech degrees in Computer Science from the Department of Computer Science, University of Calcutta, India in 1987 and 1989 respectively, and the PhD degree from the Department of Computer Science and Engineering, Indian Institute of Technology, Kharagpur, India in 1994. She received the prestigious Indian National Science

Academy (INSA) Young Scientist Award in 1994. Currently, Dr. Roy Chowdhury is an Assistant Professor in the Department of Computer Science and Engineering, Indian Institute of Technology, Kharagpur. Her research interests include VLSI design and testing, cellular automata, error correcting code and encryption.

Sudipta Bhawmik is a member of technical staff in the SoC Design, Verification and DFT group, at Agere Systems, Holmdel, NJ. During his tenure at the Bell Labs, Dr. Bhawmik has made significant contributions towards the development and application of several generations of Built-In Self Test technology and tools currently in use within Lucent and Agere. His current interests are in the field of System Level Design and System-on Chip test methods. He has several technical publications and patents to his credit. Dr. Bhawmik is an active member of the IEEE P1500 working group and chairs the task force on mergeable core test requirements. Dr. Bhawmik received his B.Tech (Hons) and M.Tech degrees in Electronics and Electrical Communication Engineering and Ph.D degree in Computer Science and Engineering from the Indian Institute of Technology, Kharagpur, India.

JOURNAL OF ELECTRONIC TESTING: Theory and Applications 18, 487–501, 2002

Design for Consecutive Testability of System-on-a-Chip with Built-In Self Testable Cores

TOMOKAZU YONEDA AND HIDEO FUJIWARA
Graduate School of Information Science, Nara Institute of Science and Technology, 8916-5 Takayama, Ikoma, Nara, 630-0101, Japan
tomoka-y@is.aist-nara.ac.jp
fujiwara@is.aist-nara.ac.jp

Received August 28, 2001; Revised January 8, 2002

Editor: Krishnendu Chakrabarty

Abstract. This paper introduces a new concept of testability called consecutive testability and proposes a design-for-testability method for making a given SoC consecutively testable based on integer linear programming problem. For a consecutively testable SoC, testing can be performed as follows. Test patterns of a core are propagated to the core inputs from test pattern sources (implemented either off-chip or on-chip) consecutively at the speed of system clock. Similarly the test responses are propagated to test response sinks (implemented either off-chip or on-chip) from the core outputs consecutively at the speed of system clock. The propagation of test patterns and responses is achieved by using interconnects and consecutive transparency properties of surrounding cores. All interconnects can be tested in a similar fashion. Therefore, it is possible to test not only logic faults but also timing faults that require consecutive application of test patterns at the speed of system clock since the consecutively testable SoC can achieve consecutive application of any test sequence at the speed of system clock.

Keywords: consecutive testability, consecutive transparency, test access mechanism, system-on-a-chip, design for testability, built-in self test

1. Introduction

A fundamental change has taken place in the way digital systems are designed. It has become possible to design an entire system, containing millions of transistors, on a single chip. In order to cope with the growing complexity of such modern systems, designers often use pre-designed, reusable megacells knows as cores. Core-based systems-on-a-chip (SoC) design strategies help companies significantly reduce the time-to-market and design cost for their new products.

However, SoCs are difficult to test after fabrication [16]. In order to make SoC testable, the following three conditions have to be satisfied. (1) *There exist test pattern source (TPS) and test response sink (TRS) for each core*. The TPS generates the test patterns for the embedded core and the TRS observes the test responses. TPS as well as TRS can be implemented either off-chip or on-chip. (2) *There exists test access mechanism for each core*. The test access mechanism propagates test patterns and responses. It can be used for on-chip propagation of test patterns from a TPS to the core-under-test, and for on-chip propagation of test responses from the core-under-test to a TRS. (3) *Interconnects that exist between cores are testable*.

A major difficulty to make SoC testable concerns accessibility of embedded cores. Several techniques of design-for-testability (DFT) have been proposed. There are three main approaches to achieve accessibility of embedded cores. The first approach is based on *test bus architectures* by which the cores are isolated from each other in test mode using a dedicated bus [2–4, 10, 13] or flexible *TESTRAIL* [8] around the cores to propagate test data. Test time reduction is the main objective in the majority of these methods. For example, [3] used an integer linear programming formulation to find the best test assignment and optimize the bandwidth distribution among various test buses to minimize time. The second approach uses *boundary scan architectures* [12, 14] to isolate the core during test. The third approach uses *transparency* [6, 7, 11] or *bypass* [9] mode for embedded cores to reduce the problem to one of finding paths from TPS to core inputs and from core outputs to TRS.

Under the design environment for SoCs, pre-computed test sets are provided for every core. These test sets may contain functional vectors, scan vectors or ordered test sequences for non-scan sequential circuits. They may be for logic faults such as stuck-at faults or timing faults such as delay faults. Moreover, some cores may be able to be at-speed testable in order to increase the coverage of non-modeled and performance-related defects. For that reason, it is necessary to be capable of applying any test sequence to each core and observing any response sequence from the core consecutively at the speed of system clock. We call such a test access *consecutive test access*. Although test bus approach is consecutively test accessible for cores, it is difficult to perform consecutive test access for interconnects. On the other hand, boundary scan, transparency, and bypass mode approaches can test interconnects. However, they are not consecutively test accessible.

There have been reported two works [5, 15] to realize the consecutive test accessibility for both cores and interconnects. In [15], assuming that TPS and TRS are implemented only off-chip (i.e., embedded cores are tested by using external automatic test equipment), we proposed a new testability of SoCs called *consecutive testability*. A consecutively testable SoC consists of *consecutively transparent* cores and can achieve consecutive test access to all cores and all interconnects. Consecutive transparency of a core guarantees consecutive propagation of any test/response sequence from the core input to the core output with some latency. In [5], a synthesis-for-transparency approach was presented to make cores *single-cycle transparent* by embedding multiplexers. This single-cycle transparency is a special case of consecutive transparency of [15] such that the latency of the consecutive transparency is restricted to zero, i.e., single-cycle transparency is the consecutive transparency with zero latency. Therefore, area overhead for making cores consecutively transparent with some latency is generally lower than that for making cores single-cycle transparent (i.e., transparent with zero latency).

In this paper, we consider SoCs that include *BISTed* (Built-In Self Tested) cores and *opaque* cores as well as non-BISTed cores and consecutively transparent cores, and extend the concept of *consecutive testability* of SoCs so that TPS and TRS implemented both on-chip and off-chip can be dealt with. Then, we present a DFT method to make a given SoC consecutively testable. Consecutive testability of an SoC guarantees that, for each core (for each interconnect), by using interconnects and consecutive transparency properties of surrounding cores, test patterns can be fed into the core (the interconnect, respectively) from TPS and the responses can be propagated to TRS consecutively at the speed of system clock. Therefore, consecutively testable SoCs can achieve high quality of test since any test sequence for a core can be applied to the core from TPS and any response sequence can be observed at TRS consecutively at the speed of system clock.

This paper is organized as follows. We introduce an SoC model in Section 2. In Section 3, we introduce the consecutive transparency, the consecutive testability, and present a new test methodology for testing SoCs. We present a graph model for an SoC in Section 4. In Section 5, we present a DFT method for consecutive testability. The experimental results are discussed in Section 6. Finally, Section 7 concludes this paper.

2. System-on-a-Chip

An SoC consists of *cores, primary inputs, primary outputs* and *interconnects* (Fig. 1). For the sake of uniformity, user-defined logic can be considered as another core. Each individual core is testable by either external test or built-in self test. In case a core is testable by external test, a pre-computed test set is available for the core which, if applied to the core, will result in a very high fault coverage. We introduce *ports* of each core as interface points in a natural fashion: signals enter into a core through its *input ports*, and exit through its *output ports*. An interconnect connects an output port with an input port, a primary input with an input port, or an output ports with a primary output. Any number of interconnects can connect to the same output port (i.e., fanout is allowed), but only one

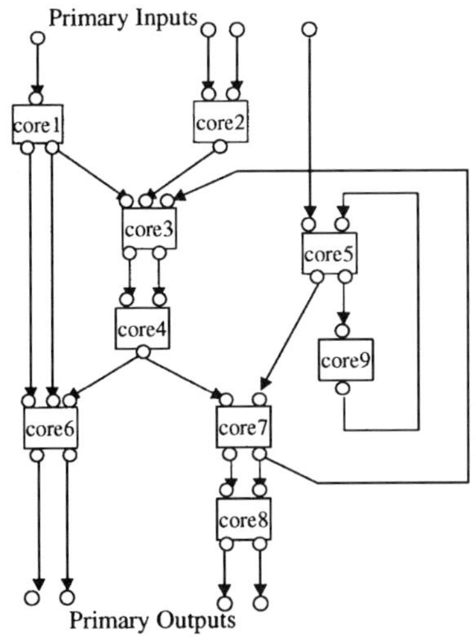

Fig. 1. System-on-a-chip.

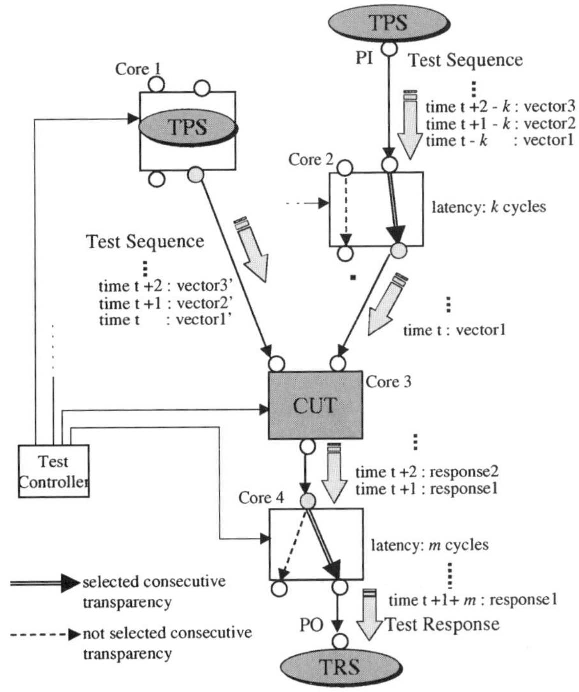

Fig. 2. Consecutive test access.

interconnect can connect to the same input port. It is not necessary that interconnects are of the same bit width.

3. A Test Methodology for System-on-a-Chip Based on Consecutive Testability

We present a new test methodology for SoCs based on *consecutive testability*. Fig. 2 illustrates a consecutively testable SoC and the consecutive test access to Core 3. A control signal is provided for each core by a test controller (either off-chip or on-chip). Each control signal of a core determines the current test mode of the core called a *configuration*. The types of configurations are consecutive transparencies and functions as a TPS and a TRS. Core 1 works as a TPS for Core 3. Core 2 realizes a consecutive transparency of shaded output port and Core 4 realizes a consecutive transparency of shaded input port. Consecutive transparency of an input port of a core guarantees that any input sequence applied to the input port can propagate to some output ports of the core consecutively at the speed of system clock. Consecutive transparency of an output port of a core guarantees that any output sequence that appears at the output port can propagate from some input ports of the core consecutively at the speed of system clock. Consecutive testability of an SoC guarantees that, for each core (for each interconnect) in the SoC, by selecting configurations of other cores, any test sequence can be consecutively fed into the core (the interconnect, respectively) from TPSs and any response sequence can be consecutively propagated to TRSs through consecutive transparencies of other cores and interconnects. We define the consecutive transparency of a core and the consecutive testability of an SoC in the following subsections.

3.1. *Consecutive Transparency of a Core*

Definition 1 (Consecutive transparency of a core). Let $I(i)$ be the ith bit of an input port I, and $O(j)$ be the jth bit of an output port O. Suppose that there exists a configuration of a core which can realize a path P between $I(i)$ and $O(j)$. P is called a *consecutively transparent path* if any input sequence applied to $I(i)$ can be consecutively observed at $O(j)$ after some latency, and then $I(i)$ and $O(j)$ are said to be *consecutively transparent*. Moreover, a core is called to be *consecutively transparent* if, for each port of the core, there exists a configuration that can make all bits of the port consecutively transparent.

Fig. 3 illustrates various configurations of a consecutively transparent core. A consecutively transparent core has generally several configurations, and each configuration can be identified by an ID number. By

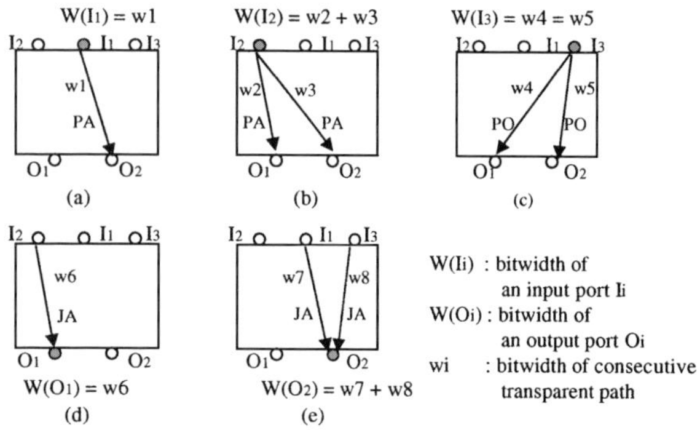

Fig. 3. Various configurations of a consecutively transparent core. (a) Configuration ID 1; (b) Configuration ID 2; (c) Configuration ID 3; (d) Configuration ID 4; (e) Configuration ID 5.

selecting a configuration of a core, consecutively transparent paths of an I/O port are realized and the I/O port can be made consecutively transparent. For each configuration, all consecutively transparent paths between an input port and an output port are represented as one consecutively transparent path.

We classify consecutively transparent paths into three types, *PA* (Propagation AND), *PO* (Propagation OR), and *JA* (Justification AND). *PA* is a type for a consecutively transparent path of an input port that propagates part of bit-width of test responses applied to the input port. On the other hand, *PO* is a type for a consecutively transparent path of an input port that propagates all bit-width of test response applied to the input port. For an input port, all consecutively transparent paths of type *PA* are necessary to make the input port consecutively transparent. However, only one consecutively transparent path of type *PO* is sufficient to make the input port consecutively transparent. *JA* is a type for a consecutively transparent path of an output port that propagates all or part of bit-width of test sequence which appears at the output port. For an output port, all consecutively transparent paths are necessary to make the output port consecutively transparent.

Fig. 3(a) illustrates type *PA* such that any input sequence applied to an input port I_1 propagates to only one output port O_2. Fig. 3(b) illustrates type *PA* such that any input sequence applied to an input port I_2 propagates to two output ports (O_1 and O_2), where any input sequence of bit width $W(I_2)$ is bit-sliced ($W(I_2) = w2 + w3$) and observed at two output ports (O_1 and O_2). Fig. 3(c) illustrates type *PO* such that any input sequence applied to I_3 propagates to two output ports (O_1 and O_2), where any input sequence of bit width $W(I_3)$ is fanouted ($W(I_3) = w4 = w5$) and observed at two output ports (O_1 and O_2). Fig. 3(d) illustrates type *JA* such that any output sequence that appears at the output port O_1 is propagated from only one input port I_2. Fig. 3(e) illustrates type *JA* such that any output sequence that appears at the output port O_2 is propagated from two input ports (I_1 and I_3), where any output sequence of bit width $W(O_2)$ is constructed by the two input sequences ($W(O_2) = w7 + w8$).

3.2. Test Pattern Source and Test Response Sink

The test pattern source (TPS) generates test patterns for cores and interconnects, and the test response sink (TRS) observe the test responses. TPS and TRS can be implemented either off-chip or on-chip. In this paper, we classify TPS and TRS into the following three types (Fig. 4).

1. S_{BIST}. S_{BIST} is a type of TPS and TRS implemented inside of a core (i.e., on-chip) and used for testing the core itself (Fig. 4(c)). A core which has this type of TPS and TRS can be self-testable.
2. S_{off}. S_{off} is a type of TPS and TRS implemented off-chip by external automatic test equipment (ATE) (Fig. 4(a)). TPS of type S_{off} can generate any test sequence of any length, and TRS of type S_{off} can observe any response sequence of any length consecutively at the speed of ATE system clock, which is usually slower than SoC system clock.
3. S_{on}. S_{on} is a type of TPS and TRS implemented inside of a core (i.e., on-chip) and used for testing other cores and interconnects (Fig. 4(b)). Since TPS and TRS of type S_{on} are implemented on-chip, memory spaces for them are limited. Therefore, TPS and TRS of type S_{on} cannot deal with arbitrary long sequences like TPS and TRS of type S_{off}. However,

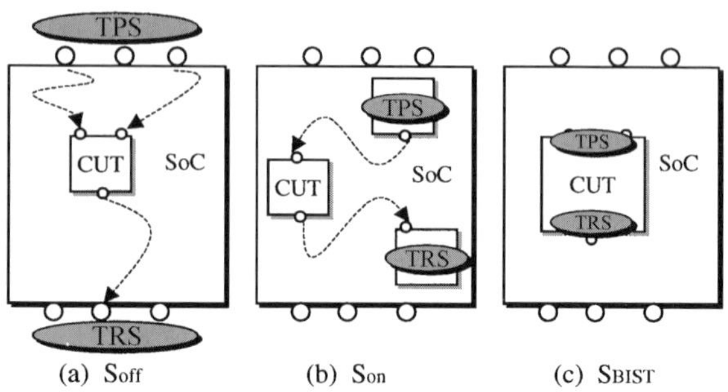

Fig. 4. Types of TPS and TRS.

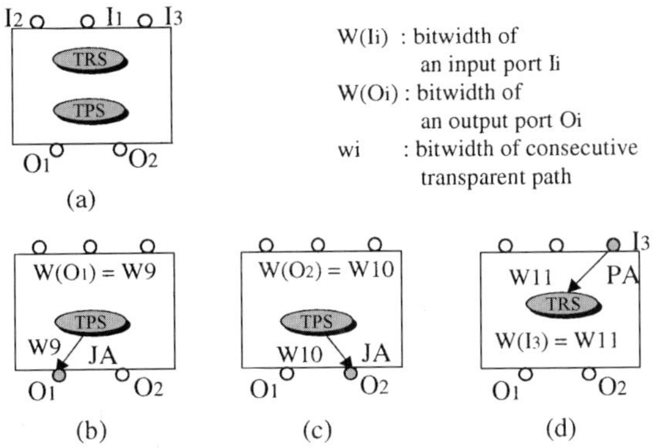

Fig. 5. Various configurations of a core that has TPS and TRS of type S_{on}. (a) Son; (b) Configuration ID 6; (c) Configuration ID 7; (d) Configuration ID 8.

within the limited memory spaces, TPS of type S_{on} can generate any test sequence and TRS of type S_{on} can observe any response sequence consecutively at the speed of system clock. A core which can be tested by TPS and TRS of type S_{on} can also be tested by TPS and TRS of type S_{off}. A core which has TPS and TRS of type S_{on} has several configurations (Fig. 5), and each configuration can be identified by an ID number. By selecting a configuration of the core, the core can realize functions as a TPS and a TRS.

3.3. *Consecutive Testability of a System-on-a-Chip*

In this subsection, we introduce a new testability of an SoC called *consecutive testability*. In this paper, we assume that the following informations are given as an SoC.

- Connectivity information between cores
- Test informations of each core
  - type of TPS/TRS that can test the core (S_{BIST} or S_{off} or S_{on})
  - configurations if the core is consecutively transparent
  - configurations if the core has TPS/TRS of type S_{on}

The length of test sequence required to test an interconnect is usually much shorter than that required to test a core. Hence, we assume all interconnect can be tested by TPS/TRS of type S_{on}. In order to test a core, it is necessary to apply test patterns consecutively to all input ports of the core simultaneously. On the other hand, it is not necessary to observe all output ports of the core simultaneously. It is sufficient only to observe one output port at a time. Therefore, we define the consecutive test accessibility of a core and the consecutive test accessibility of an interconnect as follows.

Definition 2 (Consecutive test accessibility of a core). A core C is said to be *consecutively test accessible* if the following two conditions are satisfied at the same time for each output port O of C.

1. Any test sequence generated by the TPS required to test C can be applied to all input ports of C consecutively at the speed of system clock (*consecutive controllability of C for TPS*).
2. Any response sequence appeared at O can be propagated to the TRS required to test C consecutively at the speed of system clock (*consecutive observability of O for TRS*).

Definition 3 (Consecutive test accessibility of an interconnect). For an interconnect E that connects an output port O with an input port I, E is said to be *consecutively test accessible* if O and I satisfies the following two conditions at the same time.

1. Any test sequence generated by the TPS required to test E can be applied to O consecutively at the speed of system clock (*consecutive controllability of E for TPS*).
2. Any response sequence appeared at I can be propagated to the TRS required to test E consecutively at the speed of system clock (*consecutive observability of I for TRS*).

Then, we define the consecutive testability of an SoC as follows.

Definition 4 (Consecutive testability of an SoC). An SoC is said to be *consecutively testable* if all cores and all interconnects in the SoC are consecutively test accessible.

4. Graph Modeling

In this section, we define a core connectivity graph to represent an SoC, and consider the consecutive testability on the graph.

Definition 5 (Core connectivity graph). We define a *core connectivity graph* $G = (V, E, \lambda)$ as a following directed graph to represent an SoC.

- $V = V_{PI} \cup V_{PO} \cup V_{in} \cup V_{out} \cup V_{source} \cup V_{sink}$ where V_{PI} is the set of all PIs of the SoC, V_{PO} is the set of all POs of the SoC, V_{in} is the set of all input ports of cores in the SoC, and V_{out} is the set of all output ports of cores in the SoC. V_{source} is the set of all TPSs of type S_{on} in the SoC. V_{sink} is the set of all TRSs of type S_{on} in the SoC.
- $E = E_{core} \cup E_{net}$ where $E_{core} = \{(x, y) \in V_{in} \times V_{out} \mid$ input port x is connected to output port y by a consecutively transparent path$\}$, and $E_{net} = \{(y, x) \in V_{out} \times V_{in} \mid$ output port y is connected to input port x by an interconnect$\}$.
- Labeling function $\lambda : E \to 2^{C \times I \times T \times W}$ where C is the set of all cores in the SoC, I is the set of all ID numbers of configurations, $T = \{JA, JO, PA, PO \mid$ types of consecutively transparent path (JO is for fanouted interconnects)$\}$, and W is the set of all bit widths of $e \in E$. Especially for $e \in E_{net}$, $\lambda(e) = \{\{\phi, \phi, JO,$ bit width of $e\}, \{\phi, \phi, PO,$ bit width of $e\}\}$.

Fig. 6 illustrates a core connectivity graph G which corresponds to the SoC of Fig. 1. Fig. 7 illustrates edges labeled by λ which correspond to the core of Figs. 3 and 5.

We refer to a vertex that has no input edge as a *source*, and a vertex that has no output edge as a *sink*. For a core connectivity graph G, selecting a configuration of a core is to leave edges which have labels of the configuration and to remove other edges from the core.

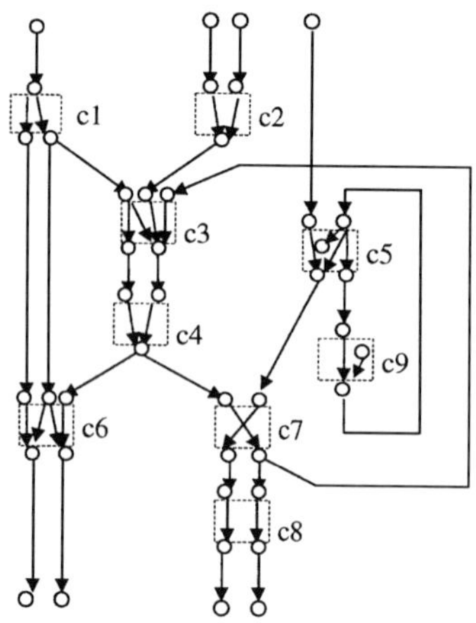

Fig. 6. Core connectivity graph.

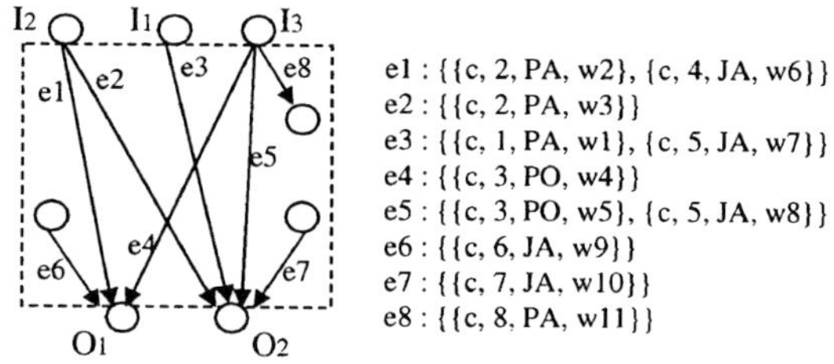

Fig. 7. Label by λ.

Then, we define a justification subgraph of a core, a justification subgraph of an interconnect and a propagation subgraph of a port as subgraphs of a core connectivity graph.

Definition 6 (Justification subgraph of a core). Let $G = (V, E, \lambda)$ be a core connectivity graph of an SoC and $G_J = (V_J, E_J, \lambda)$ be an acyclic subgraph of G. For a core $c \in C$, G_J is called a *justification subgraph of c* if G_J satisfies all the following conditions.

1. All input ports of c are sinks in G_J and there exists no sink except for all input ports of c in G_J.
2. For each edge $u \in E_J$, u has a label of either JO or JA.
3. Let $G' = (V', E', \lambda)$ be a subgraph of G obtained by selecting a configuration for each core. For each edge $u \in E_J$,
   (a) u contains all input edges of u in G', and
   (b) u contains only one output edge of u in G' when output edges have labels of JO in G'.

Lemma 1. *Let V_S be the set of all source vertices in G_J of core c. Then c is consecutively controllable for V_S.*

Proof: By Definition 5 and condition 2 of Definition 6, all edges in G_J can be used to apply test patterns consecutively at the speed of system clock since each edge in G_J represents either a consecutively transparent path or an interconnect, and has a label of either JO or JA. By condition 1 of Definition 6, there exist simple paths from more than one element in V_S to each input port of c. By condition 3 of Definition 6, all edges in the same core have the same ID number of configuration since only one configuration is selected for each core. Let v be the vertex in V_{out} (i.e., v is an output port of a core). If a configuration to realize a consecutive transparency of v is selected, all consecutively transparent paths for v exist in G_J (condition 3(a) of Definition 6). If a configuration to realize a consecutive transparency of v is not selected, v is a source vertex in G_J. By condition 3(b) of Definition 6, it is possible to apply any test sequence for all simple paths at the same time.

Therefore, we can see that any test sequence generated at V_S can be applied to all input ports of c along all simple paths in G_J consecutively at the speed of system clock. This completes the proof. □

Definition 7 (Justification subgraph of an interconnect). Let $G = (V, E, \lambda)$ be a core connectivity graph of an SoC and $G_J = (V_J, E_J, \lambda)$ be an acyclic subgraph of G. For an interconnect $e = (y, x) \in E_{net}$, G_J is called a *justification subgraph of e* if G_J satisfies all the following conditions.

1. Only y is a sink in G_J.
2. For each edge $u \in E_J$, u has a label of either *JO* or *JA*.
3. Let $G' = (V', E', \lambda)$ be a subgraph of G obtained by selecting a configuration for each core. For each edge $u \in E_J$,
   (a) u contains all input edges of u in G', and
   (b) u contains only one output edge of u in G' when output edges have labels of *JO* in G'.

Lemma 2. *Let V_S be the set of all source vertices in G_J of interconnect e. Then e is consecutively controllable for V_S.*

Proof: The proof is similar to the proof of Lemma 1. □

Definition 8 (Propagation subgraph of a port). Let $G = (V, E, \lambda)$ be a core connectivity graph of an SoC and $G_P = (V_P, E_P, \lambda)$ be an acyclic subgraph of G. For a vertex $v \in V$, G_P is called *propagation subgraph of v* if G_P satisfies all the following conditions.

1. Only v is a source in G_P.
2. For each edge $u \in E_P$, u has a label of either *PO* or *PA*.
3. Let $G' = (V', E', \lambda)$ be a subgraph of G obtained by selecting a configuration for each core. For each edge $u \in E_P$,
   (a) u contains all output edges of u in G' when the output edges have labels of *PA*, and
   (b) u contains more than one output edge of u in G' when the output edges have labels of *PO* in G'.

Lemma 3. *Let V_E be the set of all sink vertices in G_P of vertex v. Then v is consecutively observable for V_E.*

Proof: By Definition 8 and condition 2 of Definition 6, all edges in G_P can be used to propagate test responses consecutively at the speed of system clock since each edge in G_P represents either a consecutively transparent path or an interconnect, and has a label of either *PO* or *PA*. By condition 1 of Definition 8, there exist simple paths from v to each element in V_E. By condition 3 of Definition 8, all edges in the same core have the same ID number of configuration since only one configuration is selected for each core. Let v' be the vertex in V_{in} (i.e., v' is an input port of a core). If a configuration to realize a consecutive transparency of v' is selected and the consecutively transparent paths for v' are type *PA*, all consecutively transparent paths for v' exist in G_P (condition 3(a) of Definition 8). If a configuration to realize a consecutive transparency of v' is selected and the consecutively transparent paths for v' are type *PO*, there exist at least one consecutively transparent path for v' G_P (condition 3(b) of Definition 8). If a configuration to realize a consecutive transparency of v' is not selected, v' is a sink vertex in G_P.

Therefore, we conclude that any response sequence appeared at v can be propagate to V_E along all simple paths in G_P consecutively at the speed of system clock. This completes the proof. □

Theorem 1. *Let $G = (V, E, \lambda)$ be a core connectivity graph of an SoC. An SoC is said to be* consecutively testable *if the SoC satisfies the following two conditions.*

1. *For each output port $v \in V_{out}$ of each core $c \in C$, there exist one* justification subgraph *G_J of c and one* propagation subgraph *G_P of v where G_J and G_P are disjoint and satisfy the following conditions.*
   - *if TPS/TRS type required to test c is S_{BIST} $G_J = G_P = \phi$*
   - *if TPS/TRS type required to test c is S_{off} $V_S \subseteq V_{PI}$, $V_E \subseteq V_{PO}$*
   - *if TPS/TRS type required to test c is S_{on} $V_S \subseteq (V_{PI} \cup V_{source})$, $V_E \subseteq (V_{PO} \cup V_{sink})$.*
2. *For each interconnect $e = (y, x) \in E_{net}$, there exist one* justification subgraph *G_J of e and one*

propagation subgraph G_P *of* x *where* G_J *and* G_P *are disjoint and satisfy the following conditions.*

- $V_S \subseteq (V_{PI} \cup V_{source})$, $V_E \subseteq (V_{PO} \cup V_{sink})$.

Proof: The proof follows from Definitions 2–4 and Lemmas 1–3. □

5. DFT for Consecutive Testability

This section presents a *design-for-testability* (DFT) method that makes a given SoC consecutively testable. We assume that each individual core is testable by either external test or built-in self test. In case a core is testable by external test, a pre-computed test set is available for the core which, if applied to the core, will result in a very high fault coverage. Additionally, we assume (i) the internal design of the cores cannot be modified by DFT due to IP (Intellectual Property) protection and (ii) control signals for configurations can be controlled independently of normal operations. In the rest of this paper, we consider the DFT under such assumptions.

5.1. Problem Formulation

Each core (interconnect) in a consecutively testable SoC is consecutively controllable for the required TPS and consecutively observable for the required TRS. In other words, for each output port v of each core $c \in C$, a core connectivity graph G that represents a consecutively testable SoC has one justification subgraph G_J of c and one propagation subgraph G_P of v where G_J and G_P are disjoint and satisfy the condition 1 of Theorem 1. Similarly, for each interconnect $e = (y, x) \in E_{net}$, there exist one justification subgraph G_J of e and one propagation subgraph G_P of x where G_J and G_P are disjoint and satisfy the condition 2 of Theorem 1.

When a core (an interconnect) in a given SoC is not consecutively controllable for the required TPS, paths from the TPS are added by using *test multiplexers* (MUXes) in the proposed DFT (Fig. 8(a)). Similarly, when a core (an interconnect) in a given SoC is not consecutively observable for the required TRS, paths to the TRS are added by using test MUXes (Fig. 8(a)). When an interconnect-under-test is directly connected to an input or output port of a core which is not consecutively transparent, it is necessary to isolate the interconnect from the core in order to make the interconnect

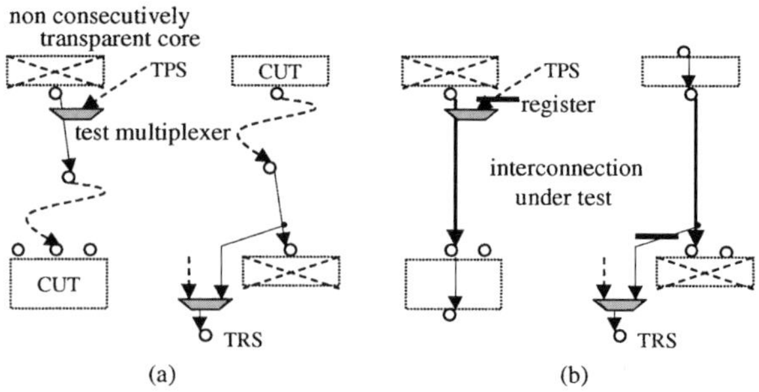

Fig. 8. DFT elements. (a) DFT for consecutive test access; (b) DFT for isolation of interconnection under test.

consecutively test accessible. This isolation is implemented by using test MUXes and *registers* (Fig. 8(b)). Assuming that any SoC includes enough number of TPSs and TRSs to make each core (each interconnect) consecutively controllable and observable, we formulate a DFT for making the SoC consecutively testable as the following optimization problem.

Definition 9. DFT for consecutive testability
Input: An SoC (a core connectivity graph)
Output: A consecutively testable SoC
Optimization: Minimizing hardware overhead (i.e., total bit width of added MUXes and registers).

5.2. DFT Algorithm

We propose a DFT algorithm for consecutive testability. The algorithm consists of the following four stages.

Stage 1. Augment a given SoC so that all cores are consecutively controllable for the required TPS.
Stage 2. Augment a given SoC so that all cores are consecutively observable for the required TRS.
Stage 3. Augment a given SoC so that all interconnects are consecutively controllable for the required TPS.
Stage 4. Augment a given SoC so that all interconnects are consecutively observable for the required TRS.

5.2.1. DFT for Consecutive Controllability of Cores (Stage 1). The objective of the first stage is to modify a given SoC with minimum hardware overhead so that all cores are consecutively controllable for the required TPS (i.e., each core $c \in C$ has a justification subgraph G_J of c where G_J satisfies the condition 1 of Theorem 1). The strategy of the algorithm is that, for each core, it first creates *control initial graph*, and then, it creates *control middle graph*. After that, it induces conditions such that the control middle graph satisfies the Definition 6 and the core is consecutively controllable for the required TPS. Finally, the algorithm formulates the DFT in this stage as *an integer linear programming problem*. All cores are made consecutively controllable with minimum hardware overhead by solving the integer linear programming problem.

5.2.1.1. Step 1: Creation of Control Initial Graph. The *control initial graph* G_{J_c} of a core $c \in C$ is created from a core connectivity graph G as follows.

1. Remove the edges which have labels of c and let the vertices which correspond to the input ports of c be sinks.
2. Remove the edges which have labels of neither *JA* nor *JO*.
3. We define the control initial graph G_{J_c} as the set of vertices and edges reachable to sinks.

Fig. 9 illustrates a control initial graph $G_{J_{c6}}$. Each edge in $G_{J_{c6}}$ has a label of either *JO* or *JA* and the number beside $e \in E_{core}$ represents a label of configuration ID.

Let A_{J_c} be the set of cores that exist in G_{J_c}. Here, a core $c' \in C$ that exists in G_{J_c} means that there exists more than one edge which has a label of c' in G_{J_c}. For each $a \in A_{J_c}$, let B_{J_a} be the set of all configuration IDs of a. We define K_{J_c} as the following equation.

$$K_{J_c} = \prod_{a \in A_{J_c}} B_{J_a}$$
$$= B_{J_{a1}} \times B_{J_{a2}} \times B_{J_{a3}} \times \dots .$$

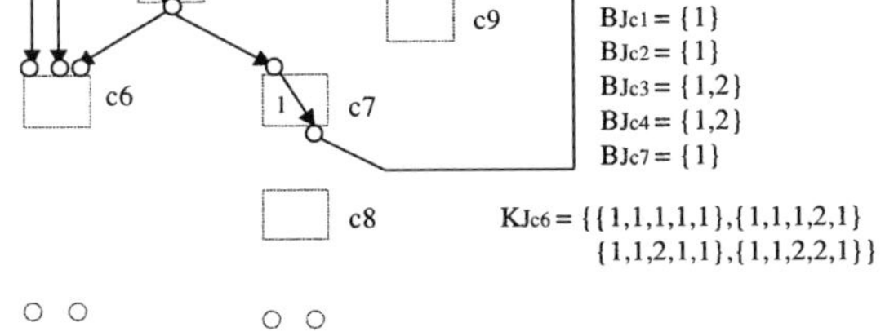

Fig. 9. Control initial graph $G_{J_{c6}}$.

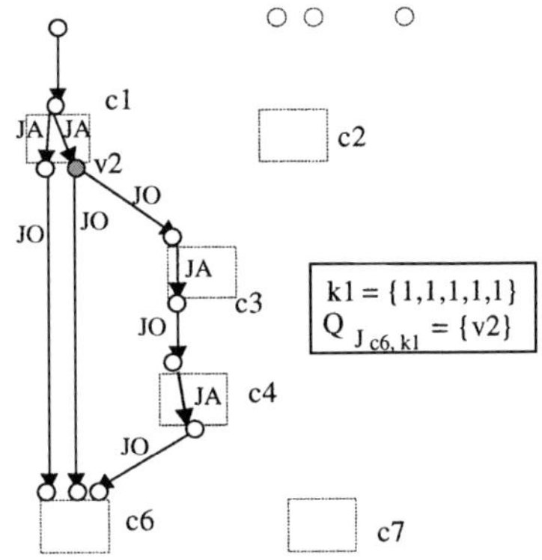

Fig. 10. Control middle graph $G_{J_{c6,k1}}$.

A control initial graph G_{J_c} contains several configurations for each core $a \in A_{J_c}$, and consecutive transparency of each core $a \in A_{J_c}$ is not realized.

5.2.1.2. Step 2: Creation of Control Middle Graph. For each $k \in K_{J_c}$, the *control middle graph* $G_{J_{c,k}}$ is created from a control initial graph G_{J_c} as follows.

1. For each $a \in A_{J_c}$, select a configuration that corresponds to k.
2. We define the control middle graph $G_{J_{c,k}}$ as the set of vertices and edges reachable to sinks.

Fig. 10 illustrates a control middle graph $G_{J_{c6,k1}}$. *JO* and *JA* beside $e \in E$ represent types of consecutively transparent path e. A control middle graph $G_{J_{c,k}}$ contains only one configuration for each core $a \in A_{J_c}$, and consecutive transparency of each core $a \in A_{J_c}$ is realized.

For $G_{J_{c,k}}$, let $Q^1_{J_{c,k}}$, $Q^2_{J_{c,k}}$ and $Q^3_{J_{c,k}}$ be the sets of all vertices $q \in G_{J_{c,k}}$ that satisfies the following conditions respectively (Fig. 11).

1. $Q^1_{J_{c,k}}$: q is a source.
2. $Q^2_{J_{c,k}}$: q has more than two output edges which have labels of *JO*.
3. $Q^3_{J_{c,k}}$: There exist cycles which contain q.

We define $Q_{J_{c,k}}$ as follows.

$$Q_{J_{c,k}} = Q^1_{J_{c,k}} \cup Q^2_{J_{c,k}} \cup Q^3_{J_{c,k}}$$

$Q^2_{J_{c,k}}$ and $Q^3_{J_{c,k}}$ are the sets of all vertices that do not satisfy the Definition 6. Moreover, we define $Q^{1,PI}_{J_{c,k}}$ and

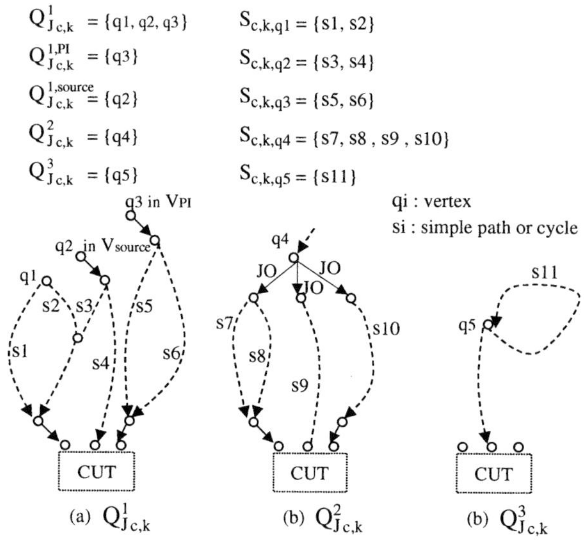

Fig. 11. $Q^1_{J_{c,k}}$, $Q^2_{J_{c,k}}$, $Q^3_{J_{c,k}}$.

$Q^{1,source}_{J_{c,k}}$ as follows.

$$Q^{1,PI}_{J_{c,k}} = Q^1_{J_{c,k}} \cap V_{PI}, \quad Q^{1,source}_{J_{c,k}} = Q^1_{J_{c,k}} \cap V_{source}$$

Then, let $S_{c,k,q}$ be the set of all simple paths from q in $Q_{J_{c,k}}$ to each sink vertex in $G_{J_{c,k}}$.

5.2.1.3. Step 3: Integer Linear Programming Formulation. We define the following variables as integer linear programming variables.

$$y_c = \begin{cases} 1 & \text{core } c \text{ is consecutively controllable for TPS} \\ 0 & \text{otherwise} \end{cases}$$

$$a_{c,k} = \begin{cases} 1 & G_{J_{c,k}} \text{ is consecutively controllable for TPS} \\ 0 & \text{otherwise} \end{cases}$$

$$d_{c,k,q} = \begin{cases} 1 & G_{J_{c,k}} \text{ is consecutively controllable for vertex } q \\ 0 & \text{otherwise} \end{cases}$$

$$z_{q,r} = \begin{cases} 1 & \text{output edge } r \text{ of vertex } q \text{ is consecutively controllable for } q \\ 0 & \text{otherwise} \end{cases}$$

$$m_s = \begin{cases} 1 & \text{if MUX is inserted to simple path } s \\ 0 & \text{otherwise} \end{cases}$$

$$x_e = \begin{cases} 1 & \text{if MUX is inserted to interconnect } e \\ 0 & \text{otherwise} \end{cases}$$

The following integer linear programming formulation minimizes the test overhead (i.e., total bit width of MUXes) while making all cores consecutively controllable.

Minimize

$$\sum_{e \in E_{net}} x_e \cdot width(e) \tag{1}$$

Subject to:

1. for each core $c \in C$,

$$y_c \geq 1 \tag{2}$$

2. For each core $c \in C$ which can be tested by either S_{off} or S_{on},

$$\sum_{k \in K_{J_c}} a_{c,k} \geq y_c \tag{3}$$

3. For each element $k \in K_{J_c}$,

$$\sum_{q \in Q_{J_{c,k}}} d_{c,k,q} \geq |Q_{J_{c,k}}| \cdot a_{c,k} \tag{4}$$

$|Q_{J_{c,k}}|$ is a constant value which represents the number of elements in $Q_{J_{c,k}}$.

4. (a) In case TPS type required to test c is S_{off} for each vertex $q \in (Q^1_{J_{c,k}} - Q^{1,PI}_{J_{c,k}})$,

$$\sum_{s \in S_{c,k,q}} m_s \geq |S_{c,k,q}| \cdot d_{c,k,q} \tag{5}$$

   (b) In case TPS type required to test c is S_{on}, for each vertex $q \in (Q^1_{J_{c,k}} - (Q^{1,PI}_{J_{c,k}} \cup Q^{1,source}_{J_{c,k}}))$,

$$\sum_{s \in S_{c,k,q}} m_s \geq |S_{c,k,q}| \cdot d_{c,k,q} \tag{6}$$

5. For each vertex $q \in Q^2_{J_{c,k}}$, let $R_{c,k,q}$ be the set of all output edges of q. For each element $r \in R_{c,k,q}$, let $S^r_{c,k,q}$ be the set of all simple paths between r and all sink vertices in $G_{J_{c,k}}$. Then, for each vertex $q \in Q^2_{J_{c,k}}$,

$$\sum_{r \in R_{c,k,q}} z_{q,r} \geq d_{c,k,q} \tag{7}$$

$$\sum_{s \in (S_{c,k,q} - S^r_{c,k,q})} m_s \geq \left|S_{c,k,q} - S^r_{c,k,q}\right| \cdot z_{q,r} \tag{8}$$

6. For each vertex $q \in Q^3_{J_{c,k}}$,

$$\sum_{s \in S_{c,k,q}} m_s \geq |S_{c,k,q}| \cdot d_{c,k,q} \tag{9}$$

7. For each simple path $s \in S_{c,k,q}$,

$$\sum_{e \in E_s} x_e \geq m_s \tag{10}$$

E_s represents the set of all edges which correspond to interconnects in simple path s.

Equation (2) guarantees that all cores are consecutively controllable for the required TPS. If TPS type required to test a core is either S_{off} or S_{on}, more than one $G_{J_{c,k}}$ must be consecutively controllable for the TPS. This is guaranteed by equation (3). In order to make $G_{J_{c,k}}$ consecutively controllable for the TPS, all vertices in $Q_{J_{c,k}}$ must be consecutively controllable for the TPS. This is guaranteed by Eq. (4). In order to make q in $Q^1_{J_{c,k}}$ consecutively controllable for the TPS, all simple paths in $S_{c,k,q}$ must be inserted MUXes and paths from TPS must be added. However, if q is a vertex that represents the TPS required to test the core, q is already consecutively controllable for the TPS. This is guaranteed by Eqs. (5) and (6). Each vertex q in $Q^2_{J_{c,k}}$ has more than two output edges which have label of JO, and all the edges propagate only the same sequence. In order to make q in $Q^2_{J_{c,k}}$ consecutively controllable for the TPS, MUXes must be inserted to all simple paths which include all the output edges except one output edge. This is guaranteed by Eqs. (7) and (8). In order to make q in $Q^3_{J_{c,k}}$ consecutively controllable for the TPS, all cycles which contain q must be broken by MUXes and paths from TPS must be added. This is guaranteed by Eq. (9). Let s be a simple path and let E_s be the set of all edges which correspond to interconnects in s, insertion of MUX to s means that MUX is inserted to more than one element in E_s. This is guaranteed by Eq. (10).

Test MUXes are inserted to the edges obtained by solving the above integer linear programming problem (Fig. 12), and all cores can be made consecutively controllable with minimum hardware overhead. However, in case TPS type required to test core c is S_{off}, it is necessary to add TPSs of type S_{off} (i.e., add vertices to V_{PI}) if the sum bit width of edges that must be inserted MUXes to make c consecutive controllable is larger than that of available TPSs (i.e., vertices in $V_{PI} - Q^{1,PI}_{J_{c,k}}$). Similarly, in case TPS type required to test core c is S_{on}, it is necessary to add TPSs of type S_{on} (i.e., add vertices to V_{source}) if the same condition as above is satisfied.

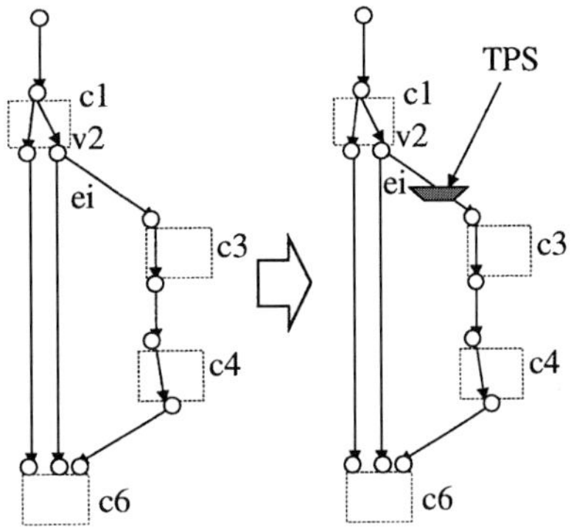

Fig. 12. Insertion of a MUX to an edge e_i for consecutive controllability.

5.2.2. DFT for Consecutive Observability of Cores (Stage 2).

The objective of the second stage is to modify a given SoC with minimum hardware overhead so that all cores are consecutively observable for the required TRS (i.e., each output port $v \in V_{out}$ of cores has a propagation subgraph G_P of v where G_P satisfies the condition 1 of Theorem 1). The strategy of the algorithm is that, for each core,

Step 1: remove consecutively controllable paths for the core.
Step 2: create observation initial graph.
Step 3: create observation middle graph.
Step 4: formulate as integer linear programming problem.

In Step 1, the algorithm first selects each configuration for the core in order to realize the consecutively controllable paths which are already decided in Stage 1. Then, it removes the consecutively controllable paths from the core connectivity graph G in order to guarantee that consecutively controllable paths and consecutively observable paths are disjoint. Procedures for Step 2, Step 3, and Step 4 are similar to Step 1, Step 2, and Step 3 in Stage 1, respectively. In Step 2, it creates *observation initial graph* for each output port of the core. After that, it creates *observation middle graph* in Step 3. In Step 4, the algorithm induces conditions such that the observation middle graph satisfies the Definition 8 and the output port of the core is consecutively observable for the required TRS. Finally, the algorithm formulates the DFT in this stage as *an integer linear programming problem*. All cores can be made consecutively observable with minimum hardware overhead by solving the integer linear programming problem.

5.2.3. DFT for Consecutive Controllability of Interconnects (Stage 3). The objective of the third stage is to modify a given SoC with minimum hardware overhead so that all interconnects are consecutively controllable for the required TPS (i.e., each interconnect $e \in E_{net}$ has a justification subgraph G_J of e where G_J satisfies the condition 2 of Theorem 1). The algorithm for this stage is similar to that of Stage 1 except for addition of test registers. When an interconnect-under-test is directly connected to the output port of a core which is not consecutively transparent (i.e., control initial graph $G_{J_{e,k}}$ for the interconnect e is empty), it is necessary to isolate the interconnect from the core in order to make the interconnect consecutively test accessible (Fig. 8(b)). Therefore, this stage can be formulated as the following integer linear programming problem exchanging the objective function (Eq. (1)) and constraint 2 (Eq. (3)) as follows.

Minimize

$$\sum_{e \in E_{net}} (x_e + x_{e,reg}) \cdot width(e) \tag{11}$$

Subject to:

2. (a) if $G_{J_{e,k}}$ is empty,

$$x_e \geq 1 \tag{12}$$

$$x_{e,reg} \geq 1 \tag{13}$$

(b) otherwise,

$$\sum_{k \in K_{J_e}} a_{e,k} \geq y_e \tag{14}$$

Here, $x_{e,reg}$ is the integer linear programming variable, and is equal to one if a test register is inserted to interconnect e, otherwise, is equal to zero.

All interconnects can be made consecutively controllable with minimum hardware overhead by solving the integer linear programming problem.

5.2.4. DFT for Consecutive Observability of Interconnects (Stage 4). The objective of the fourth stage is to modify a given SoC with minimum hardware overhead so that all interconnects are consecutively observable for the required TRS (i.e., each interconnect $e \in E_{net}$ has a propagation subgraph G_P of e where G_P satisfies the condition 2 of Theorem 1). The algorithm for this stage is similar to that of Stage 2 except for addition of test registers. The procedure for addition of test registers can be presented in a similar fashion to Stage 3. Therefore, all interconnects can be made consecutively observable with minimum hardware overhead by solving the integer linear programming problem.

After above four stages, we can modify a given SoC so that all cores and all interconnect are consecutively controllable and observable (i.e., consecutively testable).

6. Experimental Results

In this section, we present experimental results of the proposed method. We applied the method to three SoC examples shown in Fig. 13. System S_1 consists of four consecutively transparent cores and two non-consecutively transparent cores, and it contains one on-chip TPS inside of core5 and one on-chip TRS inside of core1. System S_2 consists of six consecutively transparent cores and it has the same connectivity information and on-chip TPS/TRS as system S_1. System S_3 consists of eight consecutively transparent cores and have no on-chip TPS/TRS.

We used the *lp_solve* package from Eindhoven University of Technology [1]. Assuming that all interconnects are of the same bit-width, the running time is negligible (less than 0.01 second) for each stage of all examples on a SUN Ultra 5 workstation. The results of the running SoC examples are shown in Table 1. Column "systems" denotes system name. Columns "Stage 1" and "Stage 2", "Stage 3", "Stage 4" and "Total" denote the number of DFT elements added in Stage 1, Stage 2, Stage 3, Stage 4 and all four stages, respectively. e_i shows the edge to which DFT element is added.

Table 2 shows the estimations of the area overhead. Column "interconnects" denotes the bit width of interconnects in each system. Columns "0.1" and "0.5", "1" and "10" of each system denote the area overhead when we assume that the system contains 0.1 million gates, 0.5 million gates, 1 million gates and 10 million gates, respectively. For example, "0.61" in the first row of the column "0.1" of system S_1 shows the area overhead of the DFT element (5 MUXes and 4 registers) when we assume that S_1 contains 0.1 million

Table 1. Results of the running SoC examples.

	# DFT elements				
Systems	Stage 1	Stage 2	Stage 3	Stage 4	Total
S_1	MUX: 2 (e_3, e_5)	MUX: 0	MUX: 1 (e_7) Reg: 2 (e_5, e_7)	MUX: 2 (e_3, e_8) Reg: 2 (e_3, e_8)	MUX: 5 Reg: 4
S_2	MUX: 1 (e_1)	MUX: 0	MUX: 0 Reg: 0	MUX: 0 Reg: 0	MUX: 1 Reg: 0
S_3	MUX: 2 (e_4, e_5)	MUX: 3 (e_3, e_4, e_5)	MUX: 2 (e_1, e_9) Reg: 0	MUX: 1 (e_9) Reg: 0	MUX: 8 Reg: 0

Table 2. Estimation of area overhead (%).

	System S_1 (million gates)				System S_2 (million gates)				System S_3 (million gates)			
Interconnects (bit)	0.1	0.5	1	10	0.1	0.5	1	10	0.1	0.5	1	10
32	0.61	0.12	0.06	0.01	0.10	0.02	0.01	0.00	0.77	0.15	0.08	0.01
64	1.22	0.24	0.12	0.01	0.19	0.04	0.02	0.00	1.54	0.31	0.15	0.02
128	2.44	0.48	0.24	0.02	0.38	0.08	0.04	0.00	3.08	0.62	0.31	0.03
256	4.87	0.97	0.49	0.05	0.77	0.15	0.08	0.01	6.15	1.23	0.62	0.06
512	9.73	1.94	0.97	0.10	1.54	0.31	0.15	0.02	12.29	2.46	1.23	0.12
1024	19.46	3.89	1.95	0.20	3.07	0.61	0.31	0.03	24.58	4.92	2.46	0.25

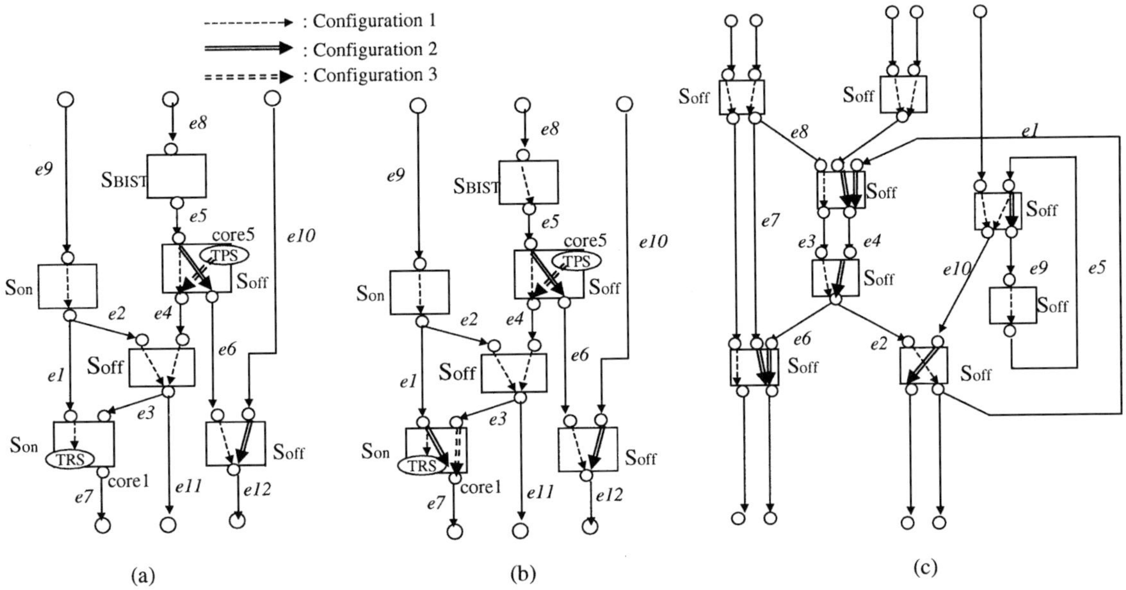

Fig. 13. SoC examples. (a) System S_1; (b) System S_2; (c) System S_3.

gates and all interconnects in S_1 are of 32 bit width.

In our proposed method, the delay overhead is negligible since at most only one multiplexer is inserted to each interconnect.

7. Conclusions

In this paper, we introduced a new testability called consecutive testability. For a consecutively testable SoC, testing can be performed as follows. Test patterns of a core are propagated to all input ports of the core from TPS, and the test responses appeared at an output port of the core are propagated to TRS consecutively at the speed of system clock. The propagation of test patterns and responses is achieved by using interconnects and consecutively transparent paths of surrounding cores. All interconnects can be tested in a similar fashion. Therefore, it is possible to apply any test sequence and observe any response sequence consecutively at the speed of system clock. We also proposed a design-for-testability method for making a given SoC consecutively testable based on integer linear programming problem. Our future work is to propose a DFT method for making cores consecutively transparent with minimum hardware overhead.

Acknowledgments

This work was sponsored in part by NEDO (New Energy and Industrial Technology Development Organization) through the contract with STARC (Semiconductor Technology Academic Research Center) and supported in part by Foundation of Nara Institute of Science and Technology under the Grant for Activity of Education and Research. Authors would like to thank Toshimitsu Masuzawa (Osaka University), Michiko Inoue and Satoshi Ohtake (Nara Institute of Science and Technology) for their valuable discussion.

References

1. M. Berkelaar, *lp_solve*, version 3.2, Eindhoven University of Technology, The Netherlands, *ftp://ftp.ics.ele.tue.nl/pub/lp_solve*.
2. S. Bhatia, T. Gheewala, and P. Varma, "A Unifying Methodology for Intellectual Property and Custom Logic Testing," in *Proc. 1996 Int. Test Conf.*, Oct. 1996, pp. 639–648.
3. K. Chakrabarty, "Design of System-on-a-Chip Test Access Architectures Using Integer Linear Programming," in *Proc. 18th VLSI Test Symp.*, May 2000, pp. 127–134.
4. K. Chakrabarty, "Design of System-on-a-Chip Test Access Architectures Under Place-and-Route and Power Constraints," in *Proc. 37th Design Automation Conf.*, June 2000, pp. 432–437.
5. K. Chakrabarty, R. Mukherjee, and A. Exnicios, "Synthesis of Transparent Circuits for Hierarchical and System-on-a-Chip Test," in *Proc. IEEE International Conference on VLSI Design*, Jan. 2001, pp. 431–436.
6. I. Ghosh, S. Dey, and N.K. Jha, "A Fast and Low Cost Testing Technique for Core-Based System-Chips," *IEEE Trans. on CAD*, vol. 19, no. 8, pp. 863–877, Aug. 2000.
7. I. Ghosh, N.K. Jha, and S. Dey, "A Low Overhead Design for Testability and Test Generation Technique for Core-Based Systems-on-a-Chip," *IEEE Trans. on CAD*, vol. 18, no. 11, pp. 1661–1676, Nov. 1999.
8. E. Marinissen, R. Arendsen, G. Bos, H. Dingemanse, M. Lousberg, and C. Wouters, "A Structured and Scalable Mechanism for Test Access to Embedded Reusable Cores," in *Proc. 1998 Int. Test Conf.*, Nov. 1998, pp. 284–293.
9. M. Nourani and C.A. Papachristou, "Structural Fault Testing of Embedded Cores Using Pipelining," *Journal of Electronic Testing: Theory and Applications*, vol. 15, pp. 129–144, 1999.
10. T. Ono, K. Wakui, H. Hikima, Y. Nakamura, and M. Yoshida, "Integrated and Automated Design-for-Testability Implementation for Cell-Based ICs," in *Proc. 6th Asian Test Symp.*, Nov. 1997, pp. 122–125.
11. S. Ravi, G. Lakshminarayana, and N.K. Jha, "Testing of Core-Based Systems-on-a-Chip," *IEEE Trans. on CAD*, vol. 20, no. 3, pp. 426–439, March 2001.
12. N.A. Touba and B. Pouya, "Testing Embedded Cores Using Partial Isolation Rings," in *Proc. 15th VLSI Test Symp.*, May 1997, pp. 10–16.
13. P. Varma and S. Bhatia, "A Structured Test Re-Use Methodology for Core-Based System Chips," in *Proc. 1996 Int. Test Conf.*, Oct. 1998, pp. 294–302.
14. L. Whetsel, "An IEEE 1149.1 Based Test Access Architecture for ICs with Embedded Cores," in *Proc. 1997 Int. Test Conf.*, Nov. 1997, pp. 69–78.
15. T. Yoneda and H. Fujiwara, "A DFT Method for Core-Based Systems-on-a-Chip Based on Consecutive Testability," in *Proc. 10th Asian Test Symp.*, Nov. 2001, pp. 193–198.
16. Y. Zorian, E.J. Marinissen, and S. Dey, "Testing Embedded-Core Based System Chips," in *Proc. 1998 Int. Test Conf.*, Oct. 1998, pp. 130–143.

Tomokazu Yoneda received the B.E. degree in information systems engineering from Osaka University, Osaka, Japan, in 1998, and M.E. degree in information science from Nara Institute of Science and Technology, Nara, Japan, in 2001. Presently he is a Ph.D. candidate in Graduate School of Information Science, Nara Institute of Science and Technology. His research interests are VLSI CAD, design for testability and SoC testing.

Hideo Fujiwara received the B.E., M.E., and Ph.D. degrees in electronic engineering from Osaka University, Osaka, Japan, in 1969, 1971, and 1974, respectively. He was with Osaka University from 1974 to 1985 and Meiji University from 1985 to 1993, and joined Nara Institute of Science and Technology in 1993. In 1981 he was a Visiting Research Assistant Professor at the University of Waterloo, and in 1984 he was a Visiting Associate Professor at McGill University, Canada. Presently he is a Professor at the Graduate School of Information Science, Nara Institute of Science and Technology, Nara, Japan.

His research interests are logic design, digital systems design and test, VLSI CAD and fault tolerant computing, including high-level/logic synthesis for testability, test synthesis, design for testability, built-in self-test, test pattern generation, parallel processing, and computational complexity. He is the author of Logic Testing and Design for Testability (MIT Press, 1985). He received the IECE Young Engineer Award in 1977, IEEE Computer Society Certificate of Appreciation Award in 1991, 2000 and 2001, Okawa Prize for

Publication in 1994, IEEE Computer Society Meritorious Service Award in 1996, and IEEE Computer Society Outstanding Contribution Award in 2001. He is an advisory member of IEICE Trans. on Information and Systems and an editor of IEEE Trans. on Computers, J. Electronic Testing, J. Circuits, Systems and Computers, J. VLSI Design and others. Dr. Fujiwara is a fellow of the IEEE, a Golden Core member of the IEEE Computer Society, a fellow of the IEICE (the Institute of Electronics, Information and Communication Engineers of Japan) and a member the Information Processing Society of Japan.

JOURNAL OF ELECTRONIC TESTING: Theory and Applications 18, 503–514, 2002

Deterministic Test Vector Compression/Decompression for Systems-on-a-Chip Using an Embedded Processor

ABHIJIT JAS AND NUR A. TOUBA
Computer Engineering Research Center, Department of Electrical and Computer Engineering, Engineering Science Building, University of Texas at Austin, Austin, TX 78712-1084, USA
touba@ece.utexas.edu

Received August 31, 2001; Revised January 15, 2002

Editor: Krishnendu Chakrabarty

Abstract. A novel approach for using an embedded processor to aid in deterministic testing of the other components of a system-on-a-chip (SOC) is presented. The tester loads a program along with compressed test data into the processor's on-chip memory. The processor executes the program which decompresses the test data and applies it to scan chains in the other components of the SOC to test them. The program itself is very simple and compact, and the decompression is done very rapidly, hence this approach reduces both the amount of data that must be stored on the tester *and* reduces the test time. Moreover, it enables at-speed scan shifting even with a slow tester (i.e., a tester whose maximum clock rate is slower than the SOC's normal operating clock rate). A procedure is described for converting a set of test cubes (i.e., test vectors where the unspecified inputs are left as X's) into a compressed form. A program that can be run on an embedded processor is then given for decompressing the test cubes and applying them to scan chains on the chip. Experimental results indicate a significant amount of compression can be achieved resulting in less data that must be stored on the tester (i.e., smaller tester memory requirement) and less time to transfer the test data from the tester to the chip.

Keywords: system-on-chip testing, test data compression, deterministic testing

1. Introduction

Printed circuit board (PCB) based systems are being replaced by widespread use of systems-on-a-chip (SOC). All the functionality found in a complete system is put together on a single chip. A wide variety of electronic devices ranging from embedded processors, embedded memories, logic elements, communication peripherals, and various analog components can be integrated together on a single piece of silicon. The high level of integration has made the manufacturing process simpler thus driving down the manufacturing costs. However, the advent of SOC technology has rapidly increased the complexity of testing these chips. One of the increasingly difficult challenges in testing SOCs is dealing with the large amount of test data that must be transferred between the tester and the chip [25]. The entire set of test vectors for all components of the SOC must be stored on the tester and transferred to the chip during testing. This poses a serious problem in terms of ATE (automated test equipment) throughput because of the cost and limitations of ATE. Testers have limited speed, channel capacity, and memory. The amount of time required to test a chip depends on how much test data needs to be transferred to the chip and how fast the data can be transferred (i.e., the test data bandwidth). This depends on the speed and channel capacity of the tester and the organization of the scan chains on the chip. Both test time and test storage are major concerns for SOCs.

There are usually quite a few computing resources available on an SOC chip, and if this computing power is harnessed effectively, it could be used to greatly reduce the overhead of testing. As test cost is becoming a significant percentage of the manufacturing cost, chip vendors are looking into sophisticated DFT techniques to contain the exploding test cost.

This paper presents a novel approach for using an embedded processor to aid in deterministic testing of the other components of the SOC (preliminary results were published in [17]). The basic idea is that the tester loads a program along with compressed test data into the processor's on-chip memory. The processor executes the program which decompresses the test data and applies it to scan chains in the other components of the SOC to test them. This approach reduces both the amount of data that must be stored on the tester *and* reduces the test time. Moreover, it can enable at-speed shifting using a slow tester (i.e., a tester whose maximum clock rate is slower than the SOC's normal operating clock rate).

The compression/decompression scheme presented in this paper is very well suited for implementation on an embedded processor. The decompression process requires very few processor instructions and thus can be done very quickly. The decompression process is pipelined to maximize the throughput of decompressed test vectors and thereby minimize the test time.

The paper is organized as follows: Section 2 discusses previous research that relates to the work presented here. Section 3 describes the proposed compression scheme. Section 4 explains the implementation details for the proposed scheme. Section 5 discusses ordering heuristics for the test vectors to maximize compression. Section 6 discusses techniques for avoiding memory overflow. Section 7 analyzes the test time reduction that can be obtained using this scheme. Section 8 discusses hardware overhead for the proposed scheme. Section 9 presents experimental results for benchmark circuits. Section 10 is a conclusion.

2. Related Work

Previous research in using embedded processors to aid in testing has focused primarily on performing memory tests [19, 20], or pseudo-random built-in self-test (BIST). Techniques for generating pseudo-random patterns and compacting test responses using simple programs have been proposed in [1, 9, 18, 21–23]. Techniques for mixed-mode BIST using embedded processors have been described in [7] and [10]. In [19], an embedded microprocessor is used to perform embedded memory test using the march algorithm and also to test other embedded function specific cores like D/A converters (DACs). A self-test methodology for bus-based programmable SOCs is described in [11]. The approach presented here is a fully deterministic test approach which supports either external testing or mixed-mode BIST. The advantage of decompressing deterministic vectors is that a targeted fault coverage can be achieved with a short test time. The number of vectors that need to be applied to the circuit is much less than that required for pseudo-random BIST. Moreover, it supports structured delay fault testing and testing of intellectual property (IP) blocks. In some cases, an IP provider may not be willing to provide any information of the internal logic of an IP block and thus fault simulation is not possible thereby precluding pseudo-random BIST.

Previous research has also been done in compressing/decompressing deterministic test data. Novel approaches for compressing test data using the Burrows-Wheeler transform and run-length coding were presented in [12, 24]. These schemes were developed for reducing the time to transfer test data from a workstation to a tester (not for use on chips). Iyengar et al. [13] presented the idea of statistically encoding test data. They described a BIST scheme for non-scan circuits based on statistical coding using comma codes (very similar to Huffman codes) and run-length coding. A scheme for compression/decompression of test data using cyclical scan chains is described in [16]. It uses careful ordering of the test set and formation of cyclical scan chains to achieve compression with run-length codes. A general technique for statistically encoding test vectors for full scan circuits using selective Huffman coding is presented in [14]. Chandra and Chakrabarty have proposed test vector compression techniques based on Golomb codes [3, 4] and frequency directed run-length (FDR) codes [5]. Test vector compression based on hybrid BIST techniques have been described in [6] and [15]. In [8], a test vector compression technique based on geometric primitives is proposed. In this technique, test vectors are optimally reordered and divided into blocks and then encoded based on geometric shapes. Although the technique achieves high compression, the authors have not discussed the implementation details of the scheme and assumed that an embedded processor will be used to decode the test vectors. However, the complexity of the decoding process in [8] is a severe limitation of the approach and may result in long test times. The compression/decompression scheme described in this paper can be very efficiently implemented using an embedded processor. The decompression program is very simple and compact and performs the decompression on the fly so that both the test data and test time are significantly reduced.

3. Proposed Scheme

The compression/decompression scheme described in this paper is based on generating the next test vector from the previous one by storing only the information about how the vectors differ. In this scheme, each test vector is divided into fixed length blocks as shown in Fig. 1. The size of the test vector blocks depends on the word size of the processor. The way this block size is determined will be explained later. The next test vector is built from the previous test vector by replacing the blocks in which they differ. For example, in Fig. 1, the blocks in which test vector $t+1$ differs from test vector t are shaded. Hence test vector $t+1$ can be built from the t-th one by replacing only the shaded blocks.

Because of the structural relationship among faults in a circuit, there will be a lot of similarity between the test vectors. The test vectors can be ordered in an optimal way such that two successive test vectors will differ in a relatively fewer number of blocks. Hence the amount of information required to store these differences will be less than that required for storing the entire test vector. These differences are represented by "replacement words" which are encoded pieces of information that tell the processor how to build the next test vector from the previous one. Each replacement word has three fields as shown in Fig. 2. A single bit field called the *last flag*, a $\log_2 N$ bit field called the *block number* (where N is the number of blocks into which the test vector is divided) and a b bit field called the *new block pattern*. The block number field contains the address of the block that is to be replaced with the new pattern contained in the new block pattern field. If two successive test vectors differ in x blocks

Last Block	Block #	New Block Pattern
1	$m = \lceil \log_2 N \rceil$	b=block size

Fig. 2. Replacement words.

Last Flag	Block #	New Block Pattern
0	001	1011010011010011010001101010
0	010	1011010011010011010001101010
0	101	1011010011010011010001101010
1	111	1011010011010011010001101010

Word Size of Processor (e.g., 32 bits)

Fig. 3. Replacement words for example in Fig. 2.

then this information is represented as a sequence of x replacement words where the last flag field of the x-th replacement word (the last replacement word in the sequence) has its last flag bit set (1). All other replacement words have their last flags turned off (0). The sequence of replacement words for the example of Fig. 1 is shown in Fig. 3. The processor reads these replacement words and then replaces the appropriate blocks with the new block patterns. When it sees the last flag bit set, then it knows that the next test vector formation has been completed. It then shifts the test vector into the scan chain(s) and applies it to the core-under-test. The block size, b, is chosen in such a way that

$$1 + \lceil \log_2 N \rceil + b = W \tag{1}$$

where W is the word size of the processor.

There are several advantages to selecting the block size in the way as explained above. This leads to the most efficient packing of the replacement words in the memory as each replacement word fits completely into one memory word. Not only does it lead to the most efficient utilization of storage but also the decoding becomes easier as the processor has to just read one word, mask the index and the last flag bit and update the appropriate memory block with the block pattern. Choosing a block size that is too small would necessitate double packing of multiple replacement words into

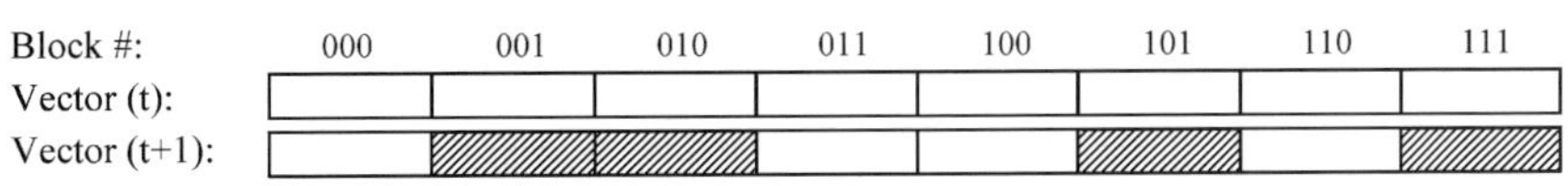

Fig. 1. Dividing test vector into blocks.

one memory location to save storage. Doing so would also complicate the decoding of the replacement words. If single replacement words are packed into each memory word then there will be waste of space as each replacement word will now be much less than W bits. On the other hand choosing a block size that is too big would necessitate a replacement word to be split across multiple memory words, which will make the decoding process more time consuming and complicated.

As mentioned before, if successive test vectors differ in a small number of blocks, then the total number of bits required for representing all the replacement words (to build the next test vector) will be less than that required for representing the entire test vector (which is the same as the number of bits in a test vector), thus resulting in compression. Note that the amount of compression depends on the ordering of the test vectors. The better the ordering, the fewer the number of blocks in which successive test vectors differ and consequently the fewer the number of replacement words. In Section 5, compression results are shown for different ordering heuristics to study the effect of ordering on the compression achieved.

4. Performing Decompression Using Embedded Processor

The compression/decompression technique described in the previous section can be implemented by using an embedded processor in an SOC. This section describes the implementation details.

The block diagram in Fig. 4 gives an overview of the architectural set-up of this scheme. The processor is used to concurrently load multiple scan chains on the chip. The on-chip memory holds the instructions that the processor executes and also the data on which the processor operates. The tester initially downloads a software program into the memory which the processor executes. The tester then supplies the processor with a stream of encoded data (replacement words) about blocks that are to be replaced with new patterns to produce the next test vector. The processor just decodes the replacement words and acts accordingly.

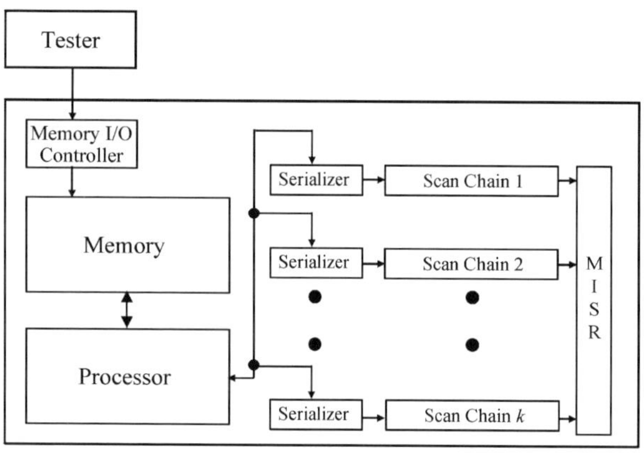

Fig. 4. Block diagram of test architecture with serializers as separate registers.

In the initialization phase (before testing begins) the tester downloads the software program (compiled machine code) into the on-chip memory. Then a fixed set of locations are reserved in the memory for the tester to continuously download the encoded data (replacement words) which the program uses. For example, let us assume that memory locations $0, 1, \ldots, M-1$ are M memory locations which are reserved for the tester to write encoded data into in a modulo M manner, i.e., the tester first writes encoded data into locations $0, 1, \ldots, M-1$ and then again starts from 0. After the initialization phase is over, the tester starts loading replacement words into the specified memory locations. The processor now starts running the program. The program is very simple. It directs the processor to routinely read the replacement words from the memory locations $0, 1, \ldots, M-1$ in a modulo M manner and act accordingly. While the processor is running the program, which results in test vectors being generated on the chip and being applied to the scan chains, the tester continues to load the memory with new replacement words for the processor to work on.

Note that, in general, there will be a memory I/O controller on the SOC for interfacing with the outside world during normal system operation. This memory I/O controller is used by the tester to load the data into the memory during testing. The memory I/O controller is typically capable of handling a slower clock rate when interfacing with the outside world. Thus, the tester can run at a slower clock rate than the normal system clock rate of the processor. Thus the processor can be operating at-speed even though the tester may be slower. Because of the tremendous cost of high-speed ATE equipment, this is a major advantage because it allows at-speed scan shifting while using slower (and cheaper) ATE equipment.

The current set of blocks is stored in the memory along with the replacement words. The program that runs on the processor causes the processor to execute a while loop until the end of the test data. In each iteration

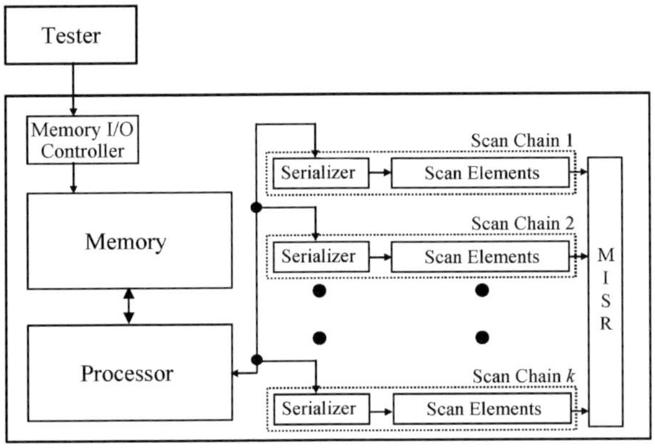

Fig. 5. Block diagram of test architecture with serializers as part of scan chains.

of the loop, it fetches a replacement word from the next memory location from which it is supposed to read the data and replaces the appropriate block. When the processor sees that the last flag bit is set, it knows that it has seen the last replacement word for that particular test vector and now needs to apply the test vector to the scan chain. The mechanism for applying a test vector to a scan chain is controlled by a *serializer*. The processor downloads each block to the appropriate serializer one at a time and the serializer shifts the block into the scan chain. The serializer is a register with a small finite state machine (FSM) controller which shifts in one bit per clock cycle into the scan chain and stops the shifting when all the bits have been shifted out. The shift register portion of each serializer can either be a separate entity that is not part of the scan chain (as shown in Fig. 4), or it can be formed from the scan elements in the scan chain (i.e., be part of the scan chain itself, as shown in Fig. 5). When the entire test vector has been loaded into the scan chain, the system clock is applied and the response is loaded back into the scan chain. The response is shifted out into a multi-input signature register (MISR) for compaction as the next test vector is shifted into the scan chain. If the serializers are part of the scan chains, then the contents of the serializer after scan capture must first be shifted out before the serializer is loaded in parallel with the next block of data.

As shown in Fig. 6, a test vector is stored as N words in the memory (where N is the number of blocks into which the test vectors have been divided). Note that there are two distinct areas (sequence of addresses) in the memory from which the processor reads/writes data. There is one area from which the processor only reads the data. This is one area into which the tester writes the replacement words and the processor reads it to update the blocks. The other area is where the patterns for the blocks of the test vector are stored. The processor both reads and writes data from and to this area. It writes data into this area when it is replacing a block with a new pattern, and it reads data from this area when it downloads them to the serializers for shifting them into the scan chains. When the processor sees the *last flag* bit set, it begins the process of applying the vector by downloading the 0th block into the serializer for the 0th scan chain and starts the serializer. The serializer then starts shifting the block into the scan chain. While it is doing so the processor continues to download the 1st block into the serializer

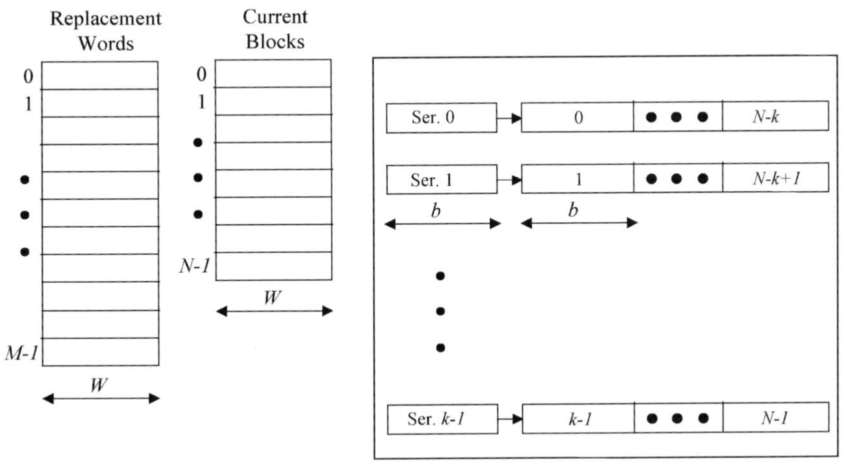

Fig. 6. Organization of data in memory.

```
void test (void)
{
  int i; // looping variable
  int k; // number of scan chains
  int N; // total number of blocks in a test vector
  unsigned int mem_index = MEM_START;
  int last_block, block_to_replace, pattern;

  // continue until read sees "end_of_test" instruction and returns FALSE
  while (read_memory(MEM_START+mem_index, &last_block, &block_to_update, &pattern))
  {
    // replace appropriate block with new block pattern
    write_memory(base_address+block_to_replace, pattern);
    // load test vector into scan chains
    if (last_block)
    {
      for (i=0; i < N; i++)
      {
        while (serializer_busy(i %k)); // wait until serializer is ready (if necessary)
        load_serializer_from_memory(&serializer[i%k], base_address+i);
        start_serializer(i%k);
      }
    }
    // point to next memory location for reading replacement word
    mem_index = (mem_index+1) % MEM_SIZE;
  }
}
```

Fig. 7. Pseudo-code for program running on processor.

for the 1st scan chain and so on and so forth. Thus if there are k scan chains and N blocks ($k < N$) then the i-th block gets downloaded into the $(i\% k)$-th serializer, where % denotes the *modulo* operator. When the processor has finished downloading the block into the k-th serializer, it comes back to the 0th serializer. Now there can be two situations. If the 0th serializer has finished shifting in the earlier block the processor can immediately download the next block and start the serializer again. Otherwise it has to wait for the serializer to finish shifting before it can download another block into it. The FSM controller associated with the serializer controls this mechanism through a "*ready*" signal, which is checked by the processor before it downloads a particular block. Once the last block has been downloaded and shifted in, the test vector is applied to the scan chains and the processor again continues to read replacement words from the memory and update the blocks.

In Fig. 7, '*C*' like pseudo-code for the decompression program that runs on the processor is given. It is a high level view of how the assembly code for the program running on the processor may look like. Note that it is just an abstraction and tries to convey the basic algorithm for the decoding. The macro calls in most cases will be just a few processor instructions. For example, the *write_memory* is actually a representative of a LOAD instruction with indirect register addressing (base and offset register). The variable *mem_index* is an offset that points to the address from which the processor should get the next data to decode. MEM_START is defined to be the starting location for reading the replacement words. It is assumed that the very last instruction of the test session causes the processor to set some kind of a flag which is abstracted by the condition of the *while* loop. MEM_SIZE is the total size of the address space reserved for the tester to write the replacement words i.e., the tester writes to locations MEM_START, (MEM_START + 1), . . . , (MEM_START + MEM_SIZE − 1). The *read_memory* macro implements reading a replacement block from a memory location and obtains the information about the address of the block to replace, pattern with which to replace, and whether it is the last block for that test vector. The *load_serializer_from_memory* macro downloads a block from the memory location addressed by the block number as offset from some base address, into the serializer for that scan chain and starts the serializer (by setting an indicator line abstracted by the *start _serializer* routine) to shift that into the scan chain.

5. Ordering Heuristics for Test Vectors to Maximize Compression

As mentioned earlier, the amount of compression achieved in the proposed scheme depends on the ordering of the test vectors (which determines the number of replacement words). There are $n!$ different ways of ordering a set of n test vectors. Hence it is impractical to examine all possible orderings for large values of n. Thus some heuristics must be used in selecting the ordering. In this section, we describe several ordering heuristics that we studied and show experimental results (number of replacement words) for each. We tried three different initial orderings, and then for each initial ordering, we tried using a greedy reordering heuristic. The first initial ordering we tried is simply the default order (which is the order produced by the ATPG tool). The second initial ordering we tried is having the test vector set ordered in ascending number of specified bits, i.e., test vectors with fewer specified bits (i.e., more don't care bits) precede those with more specified bits. The third initial ordering we tried is having the test vector set ordered in descending number of specified bits, i.e., test vectors with more specified bits precede those with fewer specified bits. For each of these three initial orderings, we computed the number of replacement words that would be required. This is shown in Table 1 under the subheading "No Reorder."

We then tried reordering the test vectors with a greedy heuristic. We started with the first test vector in the initial ordering and computed the number of replacement words that would result if each of the other $n-1$ vectors were placed after it. The test vector that resulted in the fewest number of replacement words was then placed after the first vector. For each of the non-replaced words between the first and second test vector, the don't care bits were specified as necessary so that the two vectors matched. This process was then repeated by comparing the second test vector with each of the remaining $n-2$ vectors, and so forth. The results for using this greedy reordering procedure on each of the three initial orderings are shown in Table 1 under the subheading "Greedy Reorder."

In Table 1, the lowest number of replacement words (i.e., the best compression) for each of the benchmark circuits is shown in bold. As can be seen from the results, the greedy reordering procedure improved the results in all cases. The block size b in each case was chosen according to Eq. (1) as explained earlier in Section 3.

Note that this technique is primarily targeted for stuck-at scan vectors where the ordering of the test vectors does not matter. However, for test sets where the ATPG generated ordering needs to be maintained, this technique could still be used at the cost of some loss in compression. On an average the compression produced by the "default order–no reorder" case is 6.46% worse than the corresponding best case for each of the circuits above.

6. Avoiding Memory Overflow

Since the tester is constantly transferring replacement words to the on-chip memory, one potential problem is that if the processor falls too far behind in processing the replacement words, there could be a "memory overflow." This would result in the tester overwriting a replacement word in the memory that has not yet been processed. Care must be taken to ensure that no memory overflow will occur.

Table 1. Number of replacement words for different ordering heuristics.

	Default order		Least specified first		Most specified first	
Circuit	No reorder	Greedy reorder	No reorder	Greedy reorder	No reorder	Greedy reorder
s5378	494	**398**	491	411	494	401
s9234	730	572	722	596	722	**568**
s13207	801	**737**	803	749	801	741
s15850	837	751	841	**742**	843	758
s38417	3177	2983	3206	2995	3209	**2980**
s38584	2901	2717	2901	2737	2898	**2716**

The possibility for a memory overflow depends on the relative speed of the processor to the tester. The rate at which the tester puts data into the memory depends on the tester clock rate and the number of channels. The rate at which the processor processes data in the memory depends on its clock rate, word size, and instruction set. The instruction set determines how many clock cycles are required to process the replacement words.

The tester continuously writes test data in a cyclic fashion into locations $0, 1, \ldots, M-1$ of the memory. So after every MW (where W is the word size) bits, each location of the memory gets overwritten by the tester. Hence the processor should finish processing all the replacement words in locations $0, 1, \ldots, M-1$ within the time taken by the tester to shift in MW bits to the scan chains. If the tester has n scan channels and has a clock period T_T, the time taken by the tester to shift in MW bits is $[WM/n]T_T$. Hence the processor should process all M replacement words within this time. All the processor does for each replacement word is to read the new pattern and the block address and replace the appropriate block. However for the blocks which have the last flag set, the processor has to download all the blocks into the serializers to shift them into the scan chains. This is the most time consuming part for the processor. So the speed at which the processor can process M replacement words depends on how many of them have their last flag set. Let this be denoted by e. Then the time taken by the processor to process M replacement words is $[Mu + (eNb/k)]T_P$ where T_P is the clock period of the processor, and u is the number of cycles taken by the processor to read a replacement word and replace the appropriate block. The value of u depends on the instruction set architecture of the processor. No memory overflow will occur if the following condition is satisfied:

$$[Mu + (eNb/k)]T_P < [WM/n]T_T \tag{2}$$

If the condition above is not satisfied, then for a given compressed test set, a quick check can be made to see if a memory overflow will occur. If a memory overflow would occur, then something must be done to avoid this. One solution would be to reorder the test vectors when constructing the compressed test set so that e will become smaller. This will likely result in less test data compression, but it will avoid memory overflow. Another solution may be to insert NOP's (no operation instructions) in the tester program at carefully selected locations to slow it down so that no memory overflow will occur, or to simply run the tester at a slower clock rate that ensures no memory overflow. These solutions would not reduce the amount of test data compression (and thus would still minimize tester memory requirements), however they could potentially result in less test time reduction.

Note that if the processor is sufficiently faster than the tester, then the above condition will be satisfied and the maximal reduction of both test data and test time can be achieved. Note also that as the number of scan chains (i.e., k) is increased, the condition becomes easier to satisfy. Lastly, if there is enough memory on the chip to store all the replacement words then no memory overflow will occur irrespective of all the parameters in Eq. (2) above. The results shown in the following paragraph help to understand the effect of all the parameters in the context of memory overflow.

Equation (2) can be alternatively represented as:

$$F_p/F_t > n/W[u + e/M(S/k)] \tag{3}$$

where F_p $(=1/T_p)$ is the processor frequency, F_t $(= 1/T_t)$ is the tester frequency and $S = Nb$ is the length of the scan vector. Equation (3) establishes a lower bound for the processor frequency to tester frequency ratio to prevent any memory overflow. Table 2 shows the F_p/F_t ratio for the benchmark circuits for different values of M. In each case, values of the other parameters are kept constant. The k/n value for each circuit has been carefully chosen to prevent a very high value of the F_p/F_t ratio. Also the ordered test set chosen in each case is the one that produces the best compression as shown in Table 1. The values of the parameters that are held constant for all the circuits are $W = 32$, $u = 3$. The value of e depends on the ordered test set and M. The scan length S depends on the benchmark. Note that W and u

Table 2. Variation of F_p/F_t for different values of M.

		M				
Circuit	k/n	256	512	1024	2048	4096
s5378	8/4	1.62	NA	NA	NA	NA
s9234	8/4	1.82	1.42	NA	NA	NA
s13207	8/1	2.26	1.31	NA	NA	NA
s15850	8/2	1.74	1.17	NA	NA	NA
s38417	8/4	2.10	1.90	1.62	1.34	NA
s38584	8/2	2.24	1.64	1.20	0.85	NA

are two parameters on which the system integrator will most likely have the least amount of control (as they are tightly integrated to the processor/memory micro-architecture). In Table 2 their values have been fixed for all the circuits at realistic values. The k/n ratio chosen for each circuit is also shown in the table with the actual values of k and n. The results of Table 2 illustrate how by carefully tuning the different parameters on which the system designer has control, the memory overflow can be avoided.

The entries "NA" in the above table corresponds to cases where there will be no memory overflow because of enough memory being present to store all the replacement words. Also in most cases if the processor is sufficiently faster than the tester as given by the table entries above, the memory overflow can be avoided.

7. Test Time Reduction

The total reduction in test time that can be obtained using the proposed scheme can be analytically estimated given the ordering of the test set and the other parameters that define the test architecture. In conventional scan based testing, the total test time consists of mostly the scan shifting time since the test application time (launch and capture cycle) is usually negligible compared to the scan shifting time. The scan shifting time on the other hand depends on the total amount of test data, the number of tester channels and the speed of the tester. Following the notations that are being used in this paper this time can be denoted by $T_{\text{conventional}} = (SVT_T)/n$ where V is the number of test vectors. Let T_{scheme} denote the total test time in the proposed scheme. Thus the percentage reduction in total test time using this scheme is given by

$$(1 - T_{\text{scheme}}/T_{\text{conventional}}) \times 100.$$

The following paragraphs explain how T_{scheme} can be calculated based on the test architecture. T_{scheme} consists of two components, T_{transfer} and $T_{\text{application}}$. T_{transfer} is the time taken by the tester to transfer the compressed data and $T_{\text{application}}$ is the time taken by the processor to apply the test vectors to the core-under-test. Unlike conventional scan testing in this case $T_{\text{application}}$ is not negligible and contributes significantly towards the total test application time. However, the effect is not additive because of the inherent parallelism in the operations of the tester and the processor. This means that $T_{\text{scheme}} < T_{\text{transfer}} + T_{\text{application}}$. It is obvious that $T_{\text{transfer}} < T_{\text{conventional}}$ (since the tester has to transfer less test data in this scheme) with the percentage reduction being directly proportional to the amount of test data compression. Hence the total reduction in test data is directly related to how much parallelism can be obtained between the operations of the tester and the processor. Note that T_{transfer} is affected by the tester frequency whereas $T_{\text{application}}$ is affected by the processor frequency. The following paragraph shows how the parallelism can be maximized and also derives some formulas for calculating T_{scheme}.

Let R be the total number of replacement words. Starting at time $t = 0$, let $W_1, W_2, \ldots, W_i, \ldots, W_R$ denote the times when the tester finishes writing the first, second, *i-th* and the *R-th* replacement words respectively. Maximum parallelism can be achieved if the processor can process the *i-th* replacement word as soon as the tester writes it. The processor can process a replacement word only after the tester has written that replacement word and the processor has finished processing all the previous replacement words. The time taken by the processor to process a replacement word is uT_p if the replacement word does not have the last flag bit set (because in this case all the processor has to do is to read it and update the appropriate block in memory), and $[u + S/k]T_p$ if the replacement word has the last flag bit set (because in this case besides updating the necessary block in memory it has to download the blocks into the serializer and apply them). Let this be denoted by c, a constant for a given test architecture. Starting at time $t = 0$, let $P_1, P_2, \ldots, P_i, \ldots, P_R$ denote the earliest possible times at which the processor can start processing a replacement word. The P_i's can be easily computed using the following iterative formula.

$$P_1 = (S/k)T_p \tag{4}$$

$$\text{and} \quad P_i = \max\{W_i, P_{i-1} + c\} \tag{5}$$

Equation (5) basically means that the processor can start processing the *i-th* replacement word only after the tester has written it AND it has finished processing all the previous replacement words. Equation (4) is determined by the fact that the processor can start processing the first replacement word after the very first test vector has been applied to the core-under-test. It is very easy to compute the W_i's as it depends on n, T_T and W and is given by

$$W_1 = (S/n)T_t \tag{6}$$

$$\text{and} \quad W_i = W_{i-1} + (W/n)T_t \tag{7}$$

Equation (6) is determined by the fact that the tester has to first write the memory block patterns required for the very first test vector and subsequently a constant amount of time for every other replacement word determined by the number of tester channels and its frequency. Note that each replacement word is W bits (this is how the block size b is chosen, so that each replacement word completely fits within a memory word). Assuming that both the processor and the tester is ready to start operation at $t = 0$, the time at which the testing process is completed is given by $P_R + c$, which is when the final replacement word is processed by the processor. Hence, $T_{\text{scheme}} = P_R + c$. Once T_{scheme} has been determined, the percentage reduction in test time can be determined from the formula mentioned above.

8. Hardware Overhead of the Scheme

In terms of the extra hardware that is required, this scheme is very efficient requiring only a small area overhead due only to the serializers. As mentioned earlier, the serializers consist of two components. One is a shift register and the other is a finite state machine controller that controls the operation of the shift registers along with the clocking mechanism. Part of the scan chains can be configured as the shift register (as shown in Fig. 5). The complexity of the finite state controller is independent of the size of the design or the size of the test vector set. The complexity of the controller is totally determined by the block size. This is because the controller has to keep track of the number of bits being shifted out and hence has a counter associated with it whose size depends on the block size. This means that the percentage of hardware overhead due to the serializers becomes less as the design size increases. Note that the processor and the memory are not being considered as part of the test hardware as they are assumed to be already present in the system-on-chip. As mentioned earlier, the main objective of the paper is to show how existing hardware in a system-on-chip (processor and memory) can be utilized to reduce the cost of testing other cores on the chip.

To give an idea of the hardware overhead, Table 3 shows the area overhead as a percentage of the area of the benchmark circuits. The serializers were designed in verilog and synthesized along with the benchmark circuits. The resulting area (in terms of standard library gate equivalents) was then compared to the synthesized area of the benchmarks without the serializers. The resulting area overhead is shown in Table 3 below. The area overhead is calculated as $(A_{\text{scheme}}/A_{\text{benchmark}} - 1) * 100$ where A_{scheme} is the area of the benchmark circuits with the serializers and $A_{\text{benchmark}}$ is the area of the benchmark circuits without the serializers. Also the amount of memory required to store all the replacement words for the best results obtained in each case (based on Table 1) is shown. Since each replacement word is W bits and in all the experiments W has been assumed to be 32, the memory required is four times the number of replacement words (in bytes).

Table 3. Hardware requirements of the proposed scheme.

Circuit	Percentage overhead due to serializer	Memory required for replacement words (Kbytes)
s5378	5.15	1.592
s9234	3.96	2.272
s13207	1.60	2.948
s15850	1.69	2.968
s38417	0.53	11.980
s38584	0.58	10.948

As can be seen from the results in Table 3, the area overhead of a single serializer is very small compared to the benchmark circuits. In fact, the area overhead becomes almost negligible for the largest benchmark circuits.

9. Experimental Results

The proposed scheme was used to compress test sets for the largest ISCAS 89 [2] circuits. A commercial ATPG tool was used to generate test cubes that provided 100% coverage of detectable faults in each circuit. Unspecified input assignments were left as X's to enable better compression. The block size in each case was derived according to Eq. (1). The results obtained in Table 4 are for $W = 32$.

For each circuit, Table 4 shows the size of the scan chain for each circuit and the number of bits in each block. The number of test vectors and total number of bits of test data is shown for the original test set. For the compressed test data, the number of replacement words that are required is shown along with the total number of bits of test data. The second to last column shows the percentage reduction in test data (amount of compression) that is achieved which is computed as:

$$(1 - \text{Compressed Bits/Original Bits}) \times 100$$

Table 4. Compression obtained for ISCAS benchmark circuits using proposed scheme.

Circuit	Scan size	Block size (bits)	Original test data		Compressed test data		% Reduction of test data	% Reduction of test time
			Num. vectors	Num. bits	Rep. words	Num. bits		
s5378	214	28	119	25466	398	12950	49.15	48.75
s9234	247	27	147	36309	568	18423	49.26	48.87
s13207	700	26	239	167300	737	24284	85.48	85.48
s15850	611	26	120	73320	742	24355	66.78	66.46
s38417	1664	24	95	158080	2995	97024	38.62	37.99
s38584	1464	25	131	191784	2737	88376	53.92	53.60

The last column shows the percentage reduction in test time that can be achieved using this scheme. The results are based on the calculations shown in Section 7. The test architecture used to compute the test time reduction is given by the parameters $M = 3072$ (12 KB Memory), $F_P/F_T = 2$, $u = 3$, and k/n ratios based on Table 2.

The results in Table 4 are the best results obtained from the ordering heuristics as shown in Table 1. As can be seen from the results, a significant amount of compression can be achieved with the proposed scheme. Also note that it is possible to achieve an almost equal amount of test time reduction by tuning the test architecture.

In order to see the effect of block size on the percentage of compression, experiments were done with varying block sizes on the four largest benchmark circuits. In each case the same test set was used as the one in Table 4 above. The results are shown in Table 5. As can be seen, there is a clear trend of decreasing compression with increasing block size. This is expected. With increasing block size, the number of blocks that are similar between two test vectors decreases. Also from the results in Table 4 it is evident that there is not much loss in compression in our present scheme due to the restrictions imposed on the block size because of implementation.

To give an idea of how the amount of test data compression for the proposed scheme compares with other schemes, results are shown in Table 6 comparing the final test data volume for the proposed scheme with the results in [5] (compression obtained using T_d). Note that this is not really a direct comparison as different test sets were used. As can be seen, the final results are fairly similar. The advantage of the proposed scheme is that it can be efficiently implemented using an embedded processor.

Table 5. Variation of percentage compression with block size.

Circuit	Block size					
	8	16	24	32	40	48
s13207	87.20	87.06	85.67	84.32	82.53	82.16
s15850	69.68	68.52	66.92	64.49	63.63	61.72
s38417	41.86	42.86	38.62	38.77	34.34	29.81
s38584	53.30	53.22	53.44	50.39	48.61	46.95

Table 6. Comparison of final test data volume.

Circuit	Test data in [5] (bits)	Test data in proposed scheme (bits)
s5378	12306	12950
s9234	22152	18423
s13207	30880	24284
s15850	26000	24355
s38417	93466	97024
s38584	77812	88376

10. Conclusion

The proposed approach supports external testing of embedded cores using deterministic test vectors. It harnesses the computational power of an embedded processor (already present for functional purposes) to perform test data compression/decompression in software. The decompression processes is pipelined so that the latency is minimized. Hence, it reduces both the amount of test storage *and* test time, thereby reducing the tester memory and channel capacity requirements. Such techniques are needed to keep down the cost of the ATE equipment needed to test future SOC's.

This paper presents the basic framework for how an embedded processor can be used for decompressing test data. One specific compression/decompression algorithm, which gives good results, is described. However, there is a lot of scope for future research in other compression/decompression algorithms for test data. The ability to use a processor to perform the decompression in software opens the door to many possible techniques.

Acknowledgments

This material is based on work supported in part by the National Science Foundation under Grant No. MIP-9702236, and in part by the Texas Advanced Technology Program under Grant No. 003658-0644-1999.

References

1. D. Bakalis, D. Nikolos, and X. Kavousianos, "Test Response Compaction by an Accumulator Behaving as a Multiple Input Non-Linear Feedback Shift Register," in *Proc. of International Test Conference*, 2000, pp. 804–811.
2. F. Brglez, D. Bryan, and K. Kozminski, "Combinational Profiles of Sequential Benchmark Circuits," in *Proc. of International Symposium on Circuits and Systems*, 1989, pp. 1929–1934.
3. A. Chandra and K. Chakrabarty, "Test Data Compression for System-on-a-Chip Using Golomb Codes," in *Proc. of VLSI Test Symposium*, 2000, pp. 113–120.
4. A. Chandra and K. Chakrabarty, "Efficient Test Data Compression and Decompression for System-on-a-Chip Using Internal Scan Chains and Golomb Coding," in *Proc. of Design, Automation, and Test in Europe (DATE)*, 2001.
5. A. Chandra and K. Chakrabarty, "Frequency-Directed Run-Length Codes with Application to System-on-a-Chip Test Data Compression," in *Proc. of VLSI Test Symposium*, 2001, pp. 42–47.
6. D. Das and N.A. Touba, "Reducing Test Data Volume Using External/LBIST Hybrid Test Patterns," in *Proc. of International Test Conference*, 2000, pp. 115–122.
7. R. Dorsch and H.-J. Wunderlich, "Accumulator Based Deterministic BIST," in *Proc. of International Test Conference*, 1998, pp. 412–421.
8. A. El-Maleh, S. al Zahir, and E. Khan, "A Geometric-Primitives-Based Compression Scheme for Testing Systems-on-a-Chip," in *Proc. of VLSI Test Symposium*, 2001, pp. 54–59.
9. S. Gupta, J. Rajski, and J. Tyszer, "Test Pattern Generation Based on Arithmetic Operations," in *Proc. of International Conference on Computer-Aided Design (ICCAD)*, 1994, pp. 117–124.
10. S. Hellebrand, H.-J. Wunderlich, and A. Hertwig, "Mixed-Mode BIST Using Embedded Processors," in *Proc. of International Test Conference*, 1996, pp. 195–204.
11. J.-R. Huang, M.K. Iyer, and K.-T. Cheng, "A Self-Test Methodology for IP Cores in Bus-Based Programmable SOCs," in *Proc. of VLSI Test Symposium*, 2001, pp. 198–203.
12. M. Ishida, D.S. Ha, and T. Yamaguchi, "COMPACT: A Hybrid Method for Compressing Test Data," in *Proc. of VLSI Test Symposium*, 1998, pp. 62–69.
13. V. Iyengar, K. Chakraborty, and B.T. Murray, "Built-in Self Testing of Sequential Circuits Using Precomputed Test Sets," in *Proc. of VLSI Test Symposium*, 1998, pp. 418–423.
14. A. Jas, J. Ghosh-Dastidar, and N.A. Touba, "ScanVector Compression/Decompression Using Statistical Coding," in *Proc. of VLSI Test Symposium*, 1999, pp. 114–120.
15. A. Jas, C.V. Krishna, and N.A. Touba, "Hybrid BIST Based on Weighted Pseudo-Random Testing: A New Test Resource Partitioning Scheme," in *Proc. of VLSI Test Symposium*, 2001, pp. 114–120.
16. A. Jas and N.A. Touba, "Test Vector Decompression Via Cyclical Scan Chains and Its Application to Testing Core-Based Designs," in *Proc. of International Test Conference*, 1998, pp. 458–464.
17. A. Jas and N.A. Touba, "Using an Embedded Processor for Efficient Deterministic Testing of Systems-on-a-Chip," in *Proc. of International Conference on Computer Design*, 1999, pp. 418–423.
18. J. Rajski and J. Tyszer, "Accumulator-Based Compaction of Test Responses," *IEEE Transactions on Computers*, vol. 42, no. 6, pp. 643–650, June 1993.
19. R. Rajsuman, "Testing a System-on-a-Chip with Embedded Microprocessor," in *Proc. of International Test Conference*, 1999, pp. 499–508.
20. J. Saxena, P. Ploicke, K. Cyr, A. Benavides, and M. Malpass, "Test Strategy for TI's TMS320AV7100 Device," in *IEEE Int. Workshop on Testing Embedded Core Based Systems*, 1998.
21. A.P. Stroele, "A Self-Test Approach Using Accumulators as Test Pattern Generators," in *Proc. of International Symposium on Circuits and Systems*, 1995, pp. 2010–2013.
22. A.P. Stroele, "Test Response Compaction Using Arithmetic Functions," in *Proc. of VLSI Test Symposium*, 1996, pp. 380–386.
23. A.P. Stroele, "Bit Serial Pattern Generation and Response Compaction Using Arithmetic Functions," in *Proc. of VLSI Test Symposium*, 1998, pp. 78–84.
24. T. Yamaguchi, M. Tilgner, M. Ishida, and D.S. Ha, "An Efficient Method for Compressing Test Data," in *Proc. of International Test Conference*, 1996, pp. 191–199.
25. Y. Zorian, "Test Requirements for Embedded Core-Based Systems and IEEE P1500," in *Proc. of International Test Conference*, 1996, pp. 191–199.

Abhijit Jas received the B.E. degree in computer science and engineering from Jadavpur University, Calcutta, India in 1996. He received the university gold medal for standing first in the college of engineering. He received the M.S. and Ph.D. degrees in electrical engineering from the University of Texas at Austin in 1999 and 2001, respectively. He currently works as a Sr. CAD Engineer at Intel Corporation in Austin, TX. His research interests are in VLSI testing, computer-aided design, and formal hardware verification.

Nur A. Touba received the B.S. degree in electrical engineering from the University of Minnesota where he graduated summa cum laude in 1990, and the M.S. and Ph.D. degrees in electrical engineering from Stanford University in 1991 and 1996, respectively. He is currently an Associate Professor at the University of Texas at Austin. His research interests are in VLSI testing, computer-aided design, and fault-tolerant computing. He received a National Science Foundation (NSF) Early Faculty CAREER Award in 1997. He serves on the Technical Program Committees of the International Test Conference, International Conference on Computer Design, International Test Synthesis Workshop, and International On-Line Testing Workshop.

JOURNAL OF ELECTRONIC TESTING: Theory and Applications 18, 515–527, 2002

Diagnostic Data Compression Techniques for Embedded Memories with Built-In Self-Test

JIN-FU LI, RUEY-SHING TZENG AND CHENG-WEN WU
Laboratory for Reliable Computing (LARC), Department of Electrical Engineering, National Tsing Hua University, Hsinchu, Taiwan 30013, ROC
cww@ee.nthu.edu.tw

Received August 30, 2001; Revised January 11, 2002

Editor: Krishnendu Chakrabarty

Abstract. A system-on-chip (SOC) usually consists of many memory cores with different sizes and functionality, and they typically represent a significant portion of the SOC and therefore dominate its yield. Diagnostics for yield enhancement of the memory cores thus is a very important issue. In this paper we present two data compression techniques that can be used to speed up the transmission of diagnostic data from the embedded RAM built-in self-test (BIST) circuit that has diagnostic support to the external tester. The proposed syndrome-accumulation approach compresses the faulty-cell address and March syndrome to about 28% of the original size on average under the March-17N diagnostic test algorithm. The key component of the compressor is a novel syndrome-accumulation circuit, which can be realized by a content-addressable memory. Experimental results show that the area overhead is about 0.9% for a 1Mb SRAM with 164 faults. A tree-based compression technique for word-oriented memories is also presented. By using a simplified Huffman coding scheme and partitioning each 256-bit Hamming syndrome into fixed-size symbols, the average compression ratio (size of original data to that of compressed data) is about 10, assuming 16-bit symbols. Also, the additional hardware to implement the tree-based compressor is very small. The proposed compression techniques effectively reduce the memory diagnosis time as well as the tester storage requirement.

Keywords: built-in self-test (BIST), data compression, Hamming syndrome, Huffman code, March test, memory diagnostics, memory testing, system-on-chip

1. Introduction

In system-on-chip (SOC) designs, memories are among the most widely used cores. Memory cores usually represent a significant portion of the chip area and are the most sensitive to process defects, dominating the yield of the chip. Memory diagnostics thus is important so far as yield improvement of the chip is concerned: during memory core and chip development it helps find the design errors and/or process inconsistency. Built-in self-test (BIST) has been considered a feasible approach for testing embedded RAMs. Although it is mainly for production test, BIST can easily be extended to provide the diagnostic data for process monitoring, design debugging, and fault location (for repair). The generated data can be processed off-chip by a diagnostic software tool to locate the faulty word and identify its fault type [23].

Direct access of the RAM cores however is limited by the small number of test pins and on-chip channels. To export the diagnostic data, a small diagnostic data bus that supports the data access has been proposed [7], which repeats the BIST operation many times to view a different slice of the memory output data each time. For example, if an 8-bit data bus is provided, then a memory with 64-bit words will require the BIST to perform eight complete test runs to capture the entire memory bitmap. Apparently, this method can result in long test time. Although a longer test time may be tolerable during design debugging, it should be avoided if the diagnosis is for memory repair. Another approach serially exports the diagnostic data to the ATE [1, 22]. When the BIST circuit detects a fault, it pauses until the data is completely shifted out through a serial output port. Although it usually results in shorter test time and smaller ATE storage requirement compared with the previous scheme, it is still slow and the storage requirement is still high if the RAM under test has a large number of faults. For multiple memory cores in a giga-scale SOC, this can be a serious issue. Compression can be used in such a case.

Compression has been used to solve a different but related problem, i.e., bitmap display of failed RAMs. Bitmap display on the tester has long been a challenging task, especially for a word-oriented RAM with address and data scrambling. In [19], a special display processor implemented on a commercial tester was presented. It can de-scramble and serialize the catch RAM data on the tester, allowing the fail-bit information to be displayed on the screen in correct order. As the RAM size grows, the transmission time from the catch RAM to the screen becomes intolerable. A lossless data compression algorithm has been implemented into the bitmap display processor [3]. The approach was a modified run-length coding algorithm, realized by an ad hoc fail-bit searching (scanning) circuit. It eliminates the bottleneck in the data flow by reducing the size of the bitmap image. Recently, using compression techniques to reduce the time to transmit test data to/from the cores has become an important research topic in SOC testing. Statistic coding algorithms have been used to reduce the test data volume [4, 12]. Compression also has been used to reduce I/O pins needed in embedded memory diagnostics [6]. The major advantages of this method is no increased test time and the ability to identify a high percentage of fail patterns. However, it is not for repair since the bitmap can not be fully reconstructed, i.e., data may be lost in some cases.

This paper describes two lossless data compression techniques. They can be used to speed up the transmission of diagnostic data from the BIST circuit of the embedded RAMs to the external tester. The first approach is based on a syndrome-accumulation algorithm, which compresses the faulty-cell address and March syndrome to about 28% of the original size using the March-17N test algorithm. The key component of the compressor is a novel syndrome-accumulation circuit that can be realized by a content-addressable memory (CAM). The area overhead is about 0.9% for a 1Mb SRAM with 164 faults. A tree-based compression technique for word-oriented memories is also proposed. By using a simplified Huffman coding scheme and partitioning each 256-bit Hamming syndrome into fixed-size symbols, the average compression ratio is about 10, assuming 16-bit symbols. We have modeled the Syndrome Compressors for different configurations at the synthesizable RTL level. Experimental results show that the extra hardware to implement the Syndrome Compressor is very small. The proposed compression techniques effectively reduce the time for diagnostic test, as well as the ATE storage requirement.

The rest of this paper is organized as follows. The background and motivation is given in Section 2. Section 3 describes the syndrome accumulation process for bit-oriented memories. In Section 4 we discuss the tree-based compression technique for word-oriented memories, and how the Hamming syndrome can be used in our compression approach. Experimental results and analysis are presented in Section 5. Finally, Section 6 concludes the paper.

2. Memory BIST with Diagnostic Support

With the advent of deep-submicron technologies and SOC, BIST is becoming the default approach for the testing and diagnostics of embedded memories [25]. The BIST circuit is usually designed to generate March tests which are widely used to detect RAM functional faults [21]. Typical faults include stuck-at fault (SAF), transition fault (TF), state coupling fault (CFst), idempotent coupling fault (CFid), and inversion coupling fault (CFin). March tests are widely used in production test thanks to their low time complexity and high fault coverage. For example, the complexity of the extended March C− is 11N, where N is the number of addresses of the RAM under test. In addition to all the faults mentioned above, March C− also covers the stuck-open fault (SOF). Of course it is only for go/no-go test. For fault analysis, longer diagnostic test algorithms will be needed [5, 14]. They provide more information, such as fault locations and fault types.

A March-17N diagnostic test algorithm is shown as follows [14]:

$$\{\Updownarrow (w0);\ \Uparrow (r0, w1, r1);\ \Updownarrow (r1);\ \Uparrow (r1, w0, r0);\ \Updownarrow (r0);\ \Downarrow (r0, w1, r1);\ \Updownarrow (r1);\ \Downarrow (r1, w0, r0);\ \Updownarrow (r0)\}$$

where $\Uparrow$, $\Downarrow$, and $\Updownarrow$ represent ascending, descending, and arbitrary (either ascending or descending) addressing order, respectively. Also, $w0$ ($w1$) indicates a 0 (1) value being written into a cell, and $r0$ ($r1$) indicates a Read operation of a cell with an expected 0 (1) value. By simulation [23, 24] or manual analysis [14], we can obtain its *fault dictionary* as shown in Table 1. In the table, $E_i = 1$ ($E_i = 0$) means that the ith Read operation of the test algorithm has returned a faulty (correct) value. Also, e.g., CFst(L,0,1) means that when the value of the aggressor (coupling) cell is 0, with its address lower than the victim cell (indicated by an L), then the victim (coupled) cell is forced to 1; (H,↑,↕) means that when there is a rising transition in the aggressor cell, with its address higher than the victim cell (indicated by an H), then the content of the victim cell will be inverted; and so on.

Table 1. The fault dictionary for the March-17N test algorithm.

	E_0	E_1	E_2	E_3	E_4	E_5	E_6	E_7	E_8	E_9	E_{10}	E_{11}
SAF(0)	0	1	1	1	0	0	0	1	1	1	0	0
SAF(1)	1	0	0	0	1	1	1	0	0	0	1	1
CFst(L,0,0)	0	0	0	1	0	0	0	1	1	1	0	0
CFst(H,0,0)	0	1	1	1	0	0	0	0	0	1	0	0
CFst(L,0,1)	1	0	0	0	1	1	1	0	0	0	0	1
CFst(H,0,1)	1	0	0	0	0	1	1	0	0	0	1	1
CFst(L,1,0)	0	1	1	1	0	0	0	0	1	1	0	0
CFst(H,1,0)	0	0	1	1	0	0	0	1	1	1	0	0
CFst(L,1,1)	1	0	0	0	0	0	0	0	0	0	1	1
CFst(H,1,1)	0	0	0	0	1	1	1	0	0	0	0	0
CFid(L,↑,1)	1	0	0	0	0	0	0	0	0	0	0	0
CFid(L,↑,0)	0	0	0	0	0	0	0	0	1	1	0	0
CFid(L,↓,1)	0	0	0	0	0	0	0	0	0	0	0	1
CFid(L,↓,0)	0	0	0	1	0	0	0	0	0	0	0	0
CFid(H,↑,1)	0	0	0	0	0	0	1	0	0	0	0	0
CFid(H,↑,0)	0	0	1	1	0	0	0	0	0	0	0	0
CFid(H,↓,1)	0	0	0	0	0	1	1	0	0	0	0	0
CFid(H,↓,0)	0	0	0	0	0	0	0	0	0	1	0	0
CFin(L,↑, ↕)	1	0	0	0	0	0	0	0	1	1	0	0
CFin(L,↓, ↕)	0	0	0	1	0	0	0	0	0	0	0	1
CFin(H,↑, ↕)	0	0	1	1	0	0	1	0	0	0	0	0
CFin(H,↓, ↕)	0	0	0	0	0	1	1	0	0	1	0	0

To provide the information of a fault dictionary, a BIST design must export the diagnostic data, including the faulty-cell address, March syndrome, and Hamming syndrome (only for word-oriented memories) for each detected fault. The *Hamming syndrome* is defined as the modulo-2 sum of the expected (fault-free) data output vector and the output vector from the memory under test. The *March syndrome* is a k-bit sequence in which the ith bit is 1 (0) if the fault is detected (not detected) by the ith Read operation of the applied diagnostic algorithm that has k Read operations [15, 16]. The number of Read operations is related to the word length of the memory under test, since ($\log_2 W + 1$) data backgrounds are usually required for a memory with W-bit words [8].

Fig. 1 depicts a typical architecture of memory BIST with diagnostic support. The BIST can be operated in one of two modes: the test and diagnostic modes. In the BIST circuit the BIST Controller is for handling the test operations. The Pattern Generator provides the stimulus inputs for the memory, including the address, data, and control signals. The Comparator evaluates the difference between the response and expected data after each Read operation. If there is a mismatch, then the Fail output will indicate that a fault is detected. To support the diagnostic mode, a Diagnostic Data Receiver is implemented to handle the data received through a serial port. If an error is detected, then the test operation is halted until the data transfer is finished.

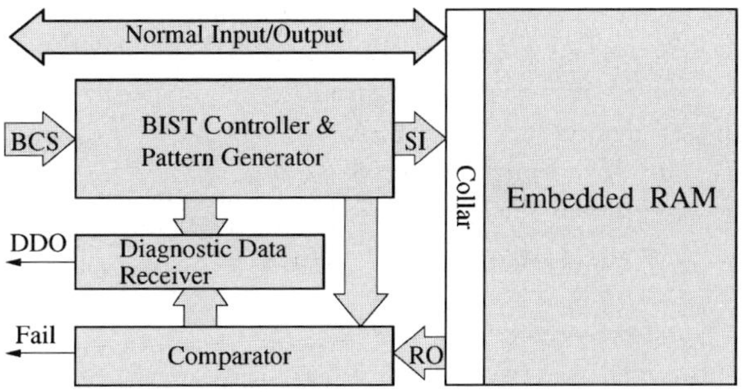

BCS : BIST Control Signal   SI : Stimulus Input
DDO : Diagnostic Data Output   RO : Response Output

Fig. 1. A memory BIST design that supports diagnostics.

3. Compression by Syndrome Accumulation

3.1. Syndrome Accumulation Process

As discussed above, each time a fault is detected by the March test algorithm, the BIST circuit exports a $\log N$-bit address (for a bit-oriented RAM with N memory cells). A $\lceil \log k \rceil$-bit word that identifies the k-bit March syndrome is also exported. Apparently, there is redundancy since a fault usually is detected by many Read operations in the test algorithm. In Table 1, e.g., a CFst(L,0,1) is detected by five different Read operations. That is, the BIST circuit will export $5 \times (\log N + \lceil \log k \rceil)$ bits of diagnostic data for this fault, but only $\log N + k$ bits are needed to be exported and we can identify the fault.

We now show an example to explain the notion of *compression by syndrome accumulation*. Consider the 64-bit RAM with four faulty cells as shown in Fig. 2(a). Assume the March-17N algorithm shown above is used to test the RAM. The *syndrome accumulation* process is as follows. The fault SAF(1) is the first to be detected by the test algorithm, and it is detected by the first Read operation of the algorithm (i.e., E_0, which is in the second element of the March-17N algorithm). The first *diagnostic data word* (containing the faulty-cell address—row and column addresses—and the *accumulated* March syndrome) thus is D1 = (100010,100000000000). The fault CFin(L,↑, ↕) can subsequently be detected, also by the first Read operation, and its address is compared with the faulty-cell address of D1. Because the address is different from the faulty-cell address of D1, the second diagnostic data word is D2 = (101110,100000000000). In turn, the fault SAF(0) is detected by the second Read operation (i.e., E_1, which is also in the second element of the March-17N algorithm). Its address is compared with the existing faulty-cell addresses (i.e., those of D1 and D2). Since the address of the fault SAF(0) is different from the faulty-cell addresses of D1 and D2, the third diagnostic data word D3 = (001001,010000000000) is generated. Similarly, the fault CFst(H,0,0) is detected by the second Read operation and its address is different from the faulty-cell addresses of D1, D2, and D3. Therefore, the fourth diagnostic data word is D4 = (010100,010000000000). The third Read operation (i.e., E_2, which is in the third element of the March-17N algorithm) then detects the fault SAF(0) again. Its address is compared with the faulty-cell addresses of D1, D2, D3, and D4. Since the address of the fault SAF(0) is the same as the faulty-cell address of D3, no new diagnostic data word is generated, and the third bit of the March syndrome of D3 is set to 1, i.e., D3 = (001001,011000000000). This syndrome accumulation process continues until the test algorithm is finished.

After the diagnostic test algorithm is finished, the final diagnostic data words (in this example, D1, D2, D3, and D4) are obtained, as shown in Fig. 2(b). The diagnostic data is compressed to $4 \times (6 + 12) = 72$ bits from the original $19 \times (6 + 4) = 190$ bits (since the four faults are detected by a total of 19 Read operations).

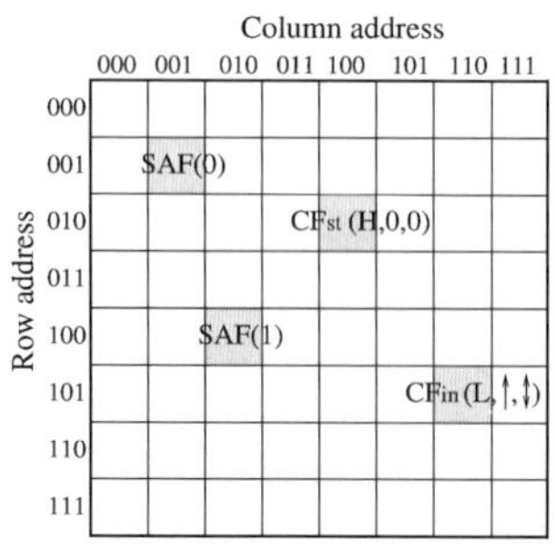

(a)

Fault	Diagnostic data
SAF(1)	D1=(100010, 100011100011)
CFin (L,↑,↕)	D2=(101110, 100000001100)
SAF(0)	D3=(001001, 011100011100)
CFst (H,0,0)	D4=(010100, 011100000100)

(b)

Fig. 2. (a) A 64-bit RAM with four faulty cells. (b) Diagnostic data words of the faults.

3.2. Syndrome Accumulation Circuit

The syndrome accumulation process can be realized by a syndrome accumulation circuit (SAC) as shown in Fig. 3. The key component of SAC is a CAM with typical functions—Write operation (WOP), Read operation (ROP), Mask Write operation (MWO), Mask Compare operation (MCO), and Erase operation (EOP) [17]. WOP stores a word of data into the addressed cells, and sets the *valid bit* of the corresponding word. If the valid bit of a word is set to invalid by EOP, then the match signal (M_i) of this word will be always set to mismatch ($M_i = 0$). Also, the addressing mechanism is different from the conventional address decoder. It mainly consists of an up/down shift register. The signal Hit is logically equivalent to $M_0 + M_1 + \cdots + M_{K-1}$, i.e., bit-wise OR of the match signals of all the columns. Also, the S and Hit signals determine the operation mode of the register, as shown in Table 2.

Initially, EOP is performed for all words and the state of the CAM cell array is all 0. All match signals are low, and Hit = 0. The shift register enters the reset state

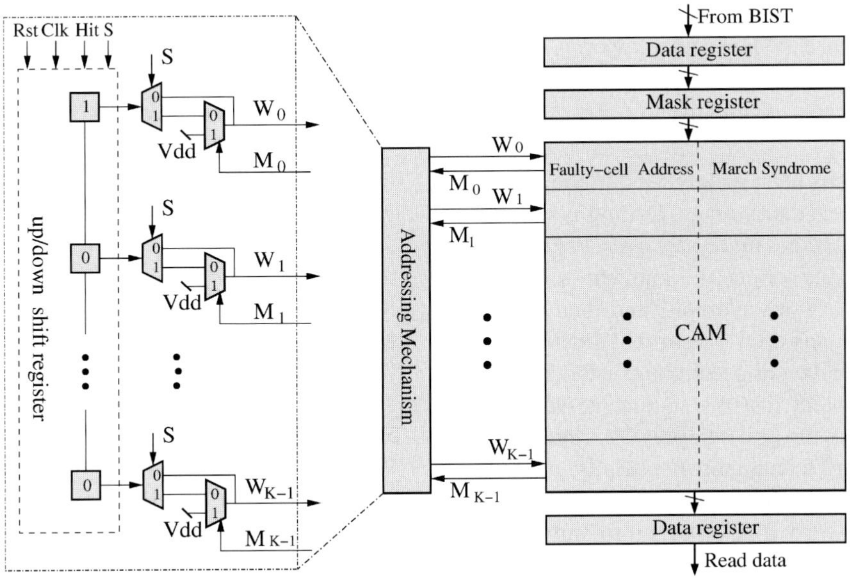

Fig. 3. Syndrome accumulation circuit implementation using a CAM.

$(100 \cdots 0)$ (as shown in Fig. 3), which can be done by asserting the reset signal (Rst).

If S = 1, then SAC performs the syndrome accumulation process as described above. The address of the first detected fault is stored in the first word (W_0) of the CAM by MWO since $(W_0 W_1 \cdots W_{K-1}) = (100 \cdots 0)$. This is done by controlling the word lines using the corresponding outputs from the shift register. If later the ith Read operation (E_i) detects this fault, then a 1 is written into the ith bit of the March syndrome of this word by MWO. Once a fault is detected subsequently, its address is compared with the stored faulty-cell addresses by MCO. If Hit = 1 (i.e., $M_0 = 1$), then the March syndrome of W_0 is updated by MWO. This can be done easily using the auto-addressing mechanism of SAC—since $M_0 = 1$, $W_0 = 1$. If Hit = 0, then the register is shifted down by one bit, and the state of the shift register is $(010 \cdots 0)$. The faulty-cell address is thus stored in the second word (W_1). Also, the ith bit of the corresponding March syndrome is set to 1. We can continue this process until the diagnostic test algorithm is finished. At that time, the CAM stores the complete faulty-cell addresses and diagnostic data.

Table 2. Shift register operation modes.

S	Hit	Operation
1	0	Shift down
1	1	Hold
0	0	Shift up
0	1	Shift up

To export the diagnostic data words, we let S = 0, then the shift register performs the shift-up operation. We thus can read the diagnostic data from the CAM word by word. Consequently, only $\log N + k$ bits of diagnostic data need to be exported for each faulty cell.

4. Tree-Based Compression Approach

Embedded memories usually have wide-length words [11, 26], e.g., 256-bit words. In this case, the number of diagnostic data is dominated by the Hamming syndrome. For example, if an 8 K × 256-bit RAM is diagnosed by the March Cd algorithm [21], which has 72 Read operations (extended for word-oriented memories with 9 data backgrounds), then the 13-bit faulty address, 256-bit Hamming syndrome, and 7-bit March syndrome (i.e., 276-bit diagnostic data) must be exported as soon as the BIST detects a fault. The ratio of the Hamming syndrome in the entire diagnostic data is 0.93, which is quite high. Moreover, if a single chip has multiple memories [1], then the diagnostic time overhead for exporting the diagnostic data with a single primary output pin will be very long. Therefore, compressing the Hamming syndrome is more important than compressing the faulty address or March syndrome for RAMs with long words.

4.1. *Huffman Coding Technique*

Statistical coding has been widely used in the areas of data transmission and data storage. Coding is a process mapping source messages into *codewords*. For the category of block-variable coding, the source messages are partitioned into fixed-size symbols, and then variable codewords are uniquely used to represent them [13]. An example of statistical coding technique is the Huffman code [9]. It guarantees that the source messages can be compressed with the best compression ratio among all statistical coding techniques. Especially, it has the property of prefix free, i.e., no codeword acts as a prefix of an another codeword. It is important for simplifying the decoding process.

A Huffman code is generated by constructing a Huffman tree according to the distribution of probability. The codewords can be obtained by traversing the Huffman tree, each edge from parent to one child is labeled with 0 and the other with 1, from the root to the corresponding leaves. Fig. 4(a) illustrates an example of source messages divided into 4-bit symbols, and the probabilities of corresponding symbols. The Huffman tree shown in Fig. 4(b) can then be constructed. Consequently, the codewords as shown in Fig. 4(a) can be obtained by traversing the Huffman tree.

4.2. *Tree-Based Coding Technique*

A source message can be regarded as the combinations of *symbols* with different sizes, e.g., a source message (00101001001001000100) can be regarded as five 4-bit symbols {0010 1001 0010 0100 0100}, or four 5-bit symbols {00101 00100 10010 00100}, etc. For 4-bit symbols, the source message only consists of three unique symbols {0010 1001 0100}, i.e., we can compress the message based on the three unique symbols.

Symbol	Probability	Codeword
0001	0.50	1
0011	0.25	01
0101	0.15	001
1001	0.05	0001
1101	0.05	0000

(a)

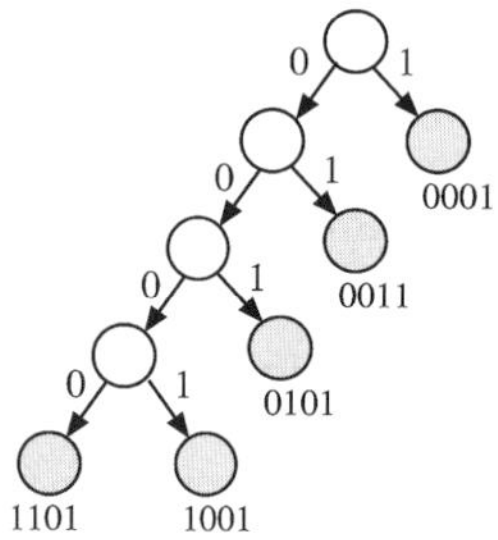

(b)

Fig. 4. (a) A Huffman code example. (b) A Huffman tree.

Assume that a source message consists of m unique B-bit symbols $S = \{s_1, s_2, \ldots, s_m\}$ with probabilities $p_1, p_2, \ldots, p_m$, respectively. Also, the symbols in S have been sorted by their probabilities, i.e., $p_1 \geq p_2 \geq \cdots \geq p_m$. The encoding procedure of the tree-based code is to partition the symbols of S into two groups. To reduce the hardware complexity, we regard the second group as a unique symbol in constructing the *simplified Huffman* tree. Assume that the first group consists of $n-1$ unique symbols. Then n unique symbols $\hat{S} = \{\hat{s}_1, \hat{s}_2, \ldots, \hat{s}_n\}$ are obtained and their corresponding probabilities are $\hat{p}_1, \hat{p}_2, \ldots, \hat{p}_n$, respectively. Also, $\hat{p}_1 = p_1, \hat{p}_2 = p_2, \ldots, \hat{p}_{n-1} = p_{n-1}$ and $\hat{p}_n = \sum_{i=n}^{m} p_i$. That is, $\hat{s}_n$ is a virtual symbol representing $(m-n)$ symbols $\{s_n, s_{n+1}, \ldots, s_m\}$. The simplified Huffman tree of the tree-based code can be constructed by the following algorithm.

1. Select two symbols with the smallest probabilities $\hat{s}_i$ and $\hat{s}_j$ in $\hat{S}$ as leaves, then generate an intermediate node as their parent.
2. Label the edge from parent to one child with 0 and the other with 1.
3. Assign the parent probability as the sum of the probabilities of the two children. The new $\hat{S} = (\hat{S} - \{\hat{s}_i, \hat{s}_j\}) \cup \{\hat{s}_i + \hat{s}_j\}$. If $\hat{S}$ is the only root left, then stop; otherwise go to Step 1.

The codeword of each leaf of this tree can be obtained by traversing the tree from root to the leaf. To represent each unique symbol in the second group, the codewords for the symbols $s_n, s_{n+1}, \ldots, s_m$ are the codeword of $\hat{s}_n$ concatenated by the source form of the original symbols, respectively. Assume that the codewords of $\hat{s}_1, \hat{s}_2, \ldots, \hat{s}_n$ are $w_1, w_2, \ldots, w_n$, respectively. Then the codewords of $s_1, s_2, \ldots, s_m$ are $w_1, w_2, \ldots, w_{n-1}, w_n \vee s_n, w_n \vee s_{n+1}, \ldots, w_n \vee s_m$, respectively. The symbol $\vee$ represents the operation of concatenation, e.g., $w_n = (01)$ and $s_n = (0111)$, then $w_n \vee s_n = (010111)$. This can simplify the compression and decompression procedure. Note that the selection of n is a tradeoff of the compression ratio and area overhead, since a larger (smaller) n results in better (worse) compression ratio and higher (lower) area overhead.

As an example, assume that S includes seven unique symbols as depicted in Fig. 5(a). According to the probabilities of these symbols, the Huffman tree can be constructed as shown in Fig. 5(b). If we partition S into

Symbol	Probability	Huffman Codeword	Tree–Based Codeword
s_1: 0000	0.35	0	0
s_2: 0010	0.25	10	10
s_3: 0101	0.20	111	111
s_4: 0110	0.10	1101	1100110
s_5: 1000	0.05	11001	1101000
s_6: 1010	0.025	110001	1101010
s_7: 1101	0.025	110000	1101101

(a)

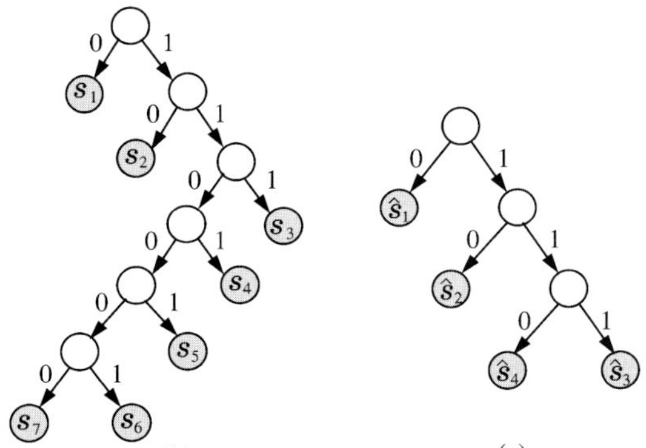

(b) (c)

Fig. 5. (a) An example showing Huffman code and tree-based code. (b) The Huffman tree. (c) The simplified Huffman tree.

$\hat{s}_1 = s_1$, $\hat{s}_2 = s_2$, $\hat{s}_3 = s_3$, and $\hat{s}_4 = \{s_4, s_5, s_6, s_7\}$, then their corresponding probabilities are 0.35, 0.25, 0.20, and 0.20, respectively. The simplified Huffman tree can then be constructed as shown in Fig. 5(c) according to the algorithm described above. Consequently, the codewords of the tree-based code can be obtained, as depicted in Fig. 5(a).

4.3. *Hamming Syndrome Compression*

The prerequisite of statistical coding is that the probability of each symbol must be known in advance. Note that we do not know the Hamming syndromes of the RAM under test before they are exported. However, the defects usually are randomly distributed on the wafer, so we assume the fault distribution of a RAM is random. Also, the percentage of the faulty bits with respect to the size of the RAM is normally small. Consequently, most Hamming syndromes are sparse, i.e., mostly 0 bits. If we partition each Hamming syndrome into B-bit symbols, then there are 2^B uniquely possible symbols. Among them, the all-0 symbol usually has the highest probability, and the symbols with only a 1 have higher probability than other symbols. Therefore, we can partition the 2^B symbols into two groups: the first group includes one all-0 symbol and B symbols with only one 1, and the second group consists of the remaining symbols. The sum of the probabilities of the first group is usually larger than that of the second group. To apply the tree-based coding technique for Hamming syndrome compression, we construct the simplified Huffman tree with $B+2$ leaves based on the following assumptions: (1) the symbols with one 1 have the same probability and the probabilities of these symbols are less than the probability of the all-0 one, and (2) the total probability of the symbols with more than one 1 is lower than the probability of the symbols with one 1. For example, assume that each Hamming syndrome is divided into 4-bit symbols, then there are 16 possible 4-bit symbols as shown in Fig. 6(a), where the corresponding hexadecimal codes are also shown. The simplified Huffman tree can be constructed as shown in Fig. 6(b). The tree-based codewords can be obtained as shown in Fig. 6(a). The codeword for any of the symbols {3, 5, 6, 7, 9, A, B, C, D, E, F} is (0000) $\vee$ (the original source symbol), e.g., the codeword for 3 is (00000011).

Fig. 7 depicts a 64×16 RAM with 6 faulty cells. Assume that the March-17N algorithm is used to test the RAM. The Hamming syndromes (0000000001000000, 0000100000000000, 0000000001010000, 0000010000000000, 0001000000000000) will be exported to the ATE six times since SAF(0) and SAF(1) are detected by six of the Read operations in the algorithm (see Table 1). Totally, 480 Hamming syndrome bits are exported by the BIST circuit. According to the encoding approach described in Fig. 6, the original Hamming syndromes can be compressed to (110011, 1000111, 1000001011, 100111, 1011111), respectively. That is, only 216 compressed syndrome

Symbol	Hexadeciaml	Tree–based Codeword
0000	0	1
0001	1	011
0010	2	010
0011	3	00000011
0100	4	001
0101	5	00000101
0110	6	00000110
0111	7	00000111
1000	8	0001
1001	9	00001001
1010	A	00001010
1011	B	00001011
1100	C	00001100
1101	D	00001101
1110	E	00001110
1111	F	00001111

(a)

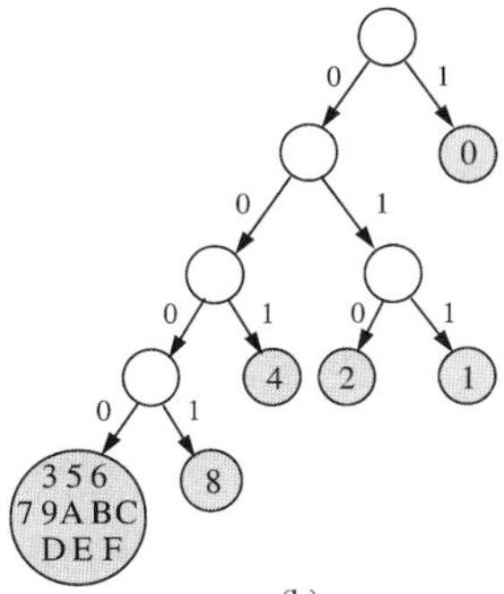

(b)

Fig. 6. (a) The 16 possible 4-bit symbols and their codewords. (b) The simplified Huffman tree.

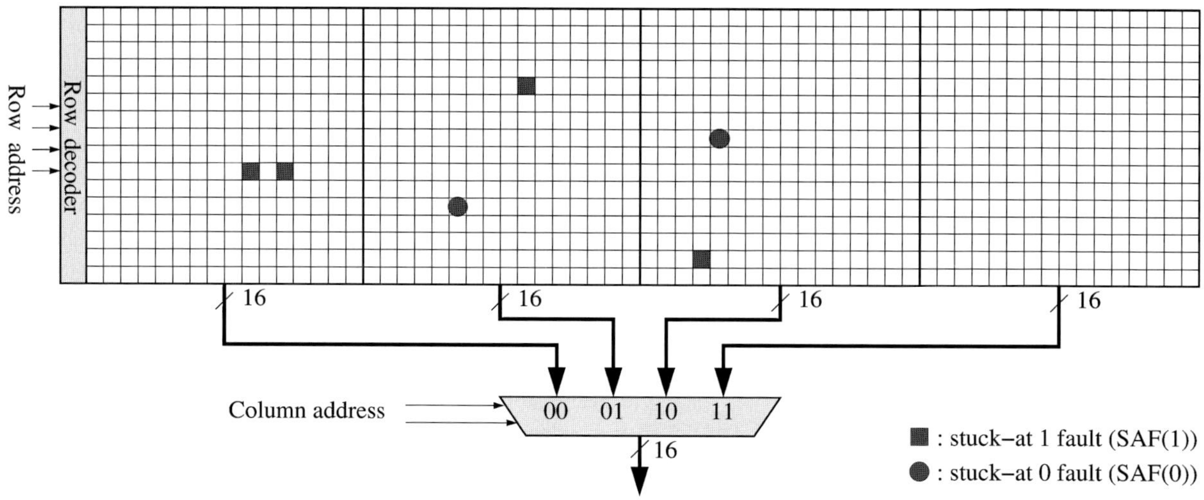

Fig. 7. An example RAM with faulty cells.

bits are needed to be exported. This results in 264-bit reduction.

4.4. *Syndrome Compressor Design*

Fig. 8 displays a memory BIST design that supports diagnostic data compression. A Syndrome Compressor is implemented to compress the Hamming syndromes. By handshaking with the Controller, Pattern Generator, and Comparator, the Syndrome Compressor receives the Hamming syndrome from the Diagnostic Data Receiver and exports the compressed syndrome to the ATE through the serial port Syndrome Output (SO). Here the faulty address and March syndrome are not compressed.

A Syndrome Compressor that can handle 64-bit Hamming syndrome (HS) based on 4-bit symbols is used as an example. The Syndrome Compressor mainly includes a multiplexer, a counter, and a Codeword Generator, as shown in Fig. 9. The Codeword Generator stores five symbols {0000, 0001, 0010, 0100, 1000}. It communicates with the BIST circuit through the Fail and Done signals. If Fail = 0 (1), then the Syndrome Compressor is disabled (enabled). If a programmable Codeword Generator is implemented, the stored samples can be changed, i.e., the user can change the encoding scheme according to the faults in a specific RAM. For example, if one RAM has many address decoder faults, then there may be many faulty raws, i.e., many all-1 symbols will appear. In such case, the user can store the all-1 symbol in the Codeword Generator such that the Syndrome Compressor can handle it more efficiently.

When a fault is detected, the BIST circuit pauses and forces Fail to 1. HS[3 : 0] is first selected into the Codeword Generator since the initial state of the counter

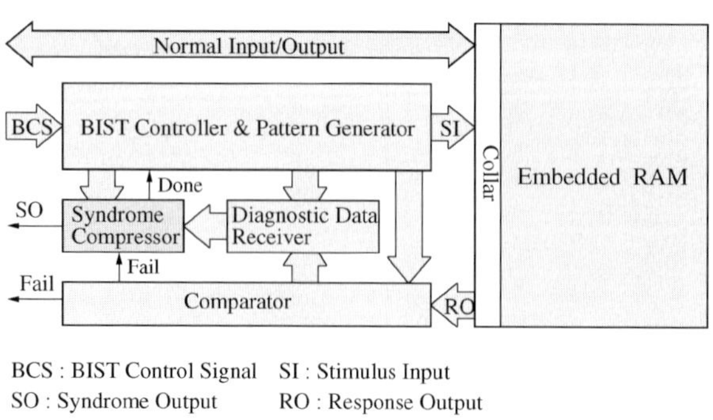

Fig. 8. A memory BIST design that supports diagnostic data compression.

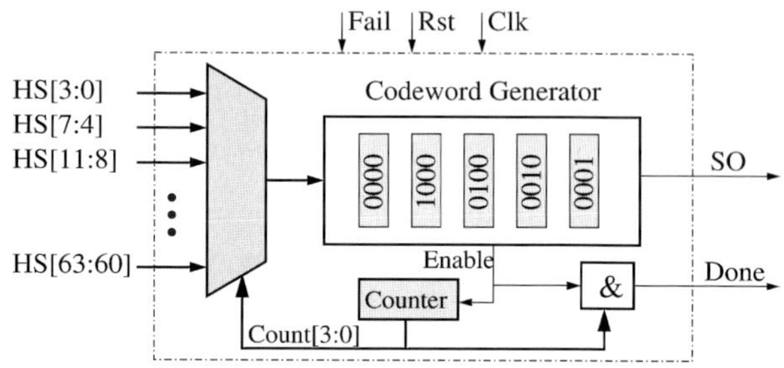

Fig. 9. The architecture of the Syndrome Compressor.

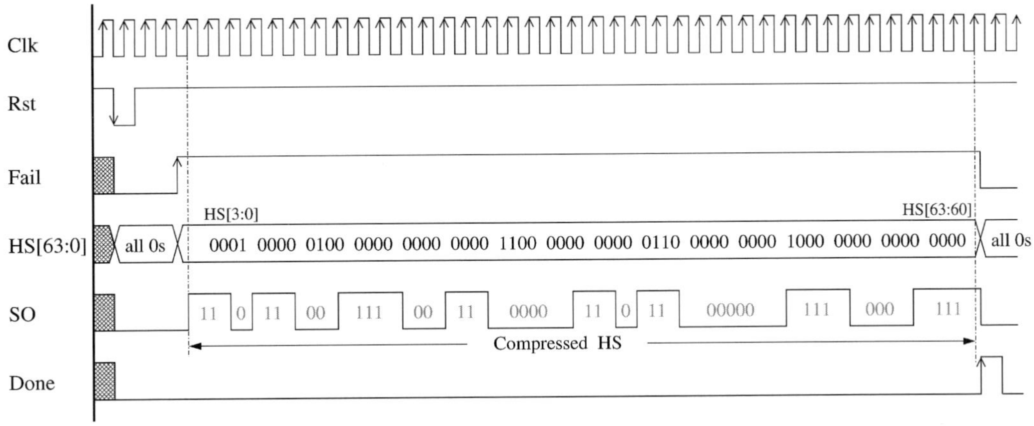

Fig. 10. Sample timing diagram for compressing a 64-bit HS.

is all-0, i.e., Count[3 : 0] = (0000). Then, HS[3 : 0] is compared with the five symbols concurrently. If HS[3 : 0] is the same as one of the five symbols, then the corresponding codeword is exported. Otherwise, the Syndrome Compressor shifts out the codeword (0000 $\vee$ HS[3 : 0]). After exporting the codeword, Enable is set to 1 and Count[3 : 0] is added to 1 (0001), where Enable is a pulse signal. Then, HS[7 : 3] is selected into the Codeword Generator and its codeword is generated by the same way. Similarly, the other codewords can be generated. Finally, if Count[3 : 0] is (1111) and Enable is 1, then the Syndrome Compressor asserts Done to activate the BIST circuit to continue the test task.

Assume that the Syndrome Compressor compresses a 64-bit HS. Fig. 10 depicts a sample of the timing diagram. The state of HS is all 0s when the Fail is 0. If Fail = 1, then the compressed HS is exported upon the arrival of the next positive edge of the clock. The HS thus can be obtained by monitoring Fail and SO. Note that the codeword is exported least-significant-bit (LSB) first. For example, the codeword for HS[3 : 0] (0001) is 011, and the codeword is exported in the order 1-1-0. When the Syndrome Compressor finishes exporting the HS, the Done signal becomes high.

5. Experimental Results and Analysis

5.1. *Results for Syndrome Accumulation*

We define the compression ratio as

$$R = \frac{\text{Number of original bits}}{\text{Number of compressed bits}}. \quad (1)$$

We assume that around 60% of the defects behave as stuck-at faults [8], and the remaining 40% of the faults are coupling faults. Let U_s and U_c denote the average numbers of Read operations that detect the stuck-at faults and coupling faults, respectively. For example, in Table 1, the stuck-at faults (including SAF(0) and SAF(1)) are detected by a total of 12 Read operations using March-17N, so $U_s = 12/2 = 6$. Also, the 20 coupling faults are detected by a total of 56 Read operations, so $U_c = 56/20 = 2.8$. Therefore, the compression ratio is estimated as

$$R = \frac{(\log N + \lceil \log k \rceil) \times (U_s \times 0.6 + U_c \times 0.4)}{\log N + k}. \quad (2)$$

We have performed experiments using random defects. Table 3 shows the R's for different RAMs under various diagnostic test algorithms that have been reported. The number of Read operations in the March Cd algorithm [27], Modified March C algorithm [10], March-26N algorithm [5], March-12N algorithm [2], and

Table 3. Compression ratios of different diagnostic test algorithms.

RAM size	256 K	512 K	1 M
March Cd	2.714	2.738	2.761
Modified March C	3.818	3.878	3.934
March-26N	3.301	3.338	3.376
March-12N	2.188	2.200	2.212
March-17N	3.461	3.503	3.541

March-17N algorithm [14] are 8, 16, 12, 6, and 12, respectively. The results show that the R's are only slightly affected by the size of RAMs. They are much more dependent on the number of Read operations of the diagnostic test algorithm.

To estimate the area overhead of SAC, we use the register-bit equivalent (rbe) model that is technology independent [18]. The areas of a static cell, dynamic cell, and CAM cell are 0.6 rbe, 0.3 rbe, and 1.2 rbe, respectively. Assume that the RAM size is N, and the number of CAM words is c (c is the largest number of faults that can be handled by SAC). Then the area overheads for SRAM and DRAM can be estimated as

$$A_s = \frac{(\log N + k) \times c \times 1.2}{N \times 0.6} \tag{3}$$

and

$$A_d = \frac{(\log N + k) \times c \times 1.2}{N \times 0.3}, \tag{4}$$

respectively. Note that the peripheral circuitry of the RAM and SAC are not estimated, since the cell array of the RAM and CAM dominates the area for large memories. Fig. 11 depicts the growth of the A's with respect to c for 1M-bit SRAM and DRAM, respectively, under the March Cd test. For example, $A_s \approx 0.9\%$ for a 1M SRAM with 164 faults. This covers most of the defective SRAMs, since the number of chips with less than 164 single-bit failures is about 99% as reported in [3]. Note that a large memory core usually is divided into many blocks for minimizing the delay and power consumption. In that case, SAC can be shared among blocks to reduce the overhead. Also, DRAM is much denser than SRAM, so for the same memory size, $A_d > A_s$.

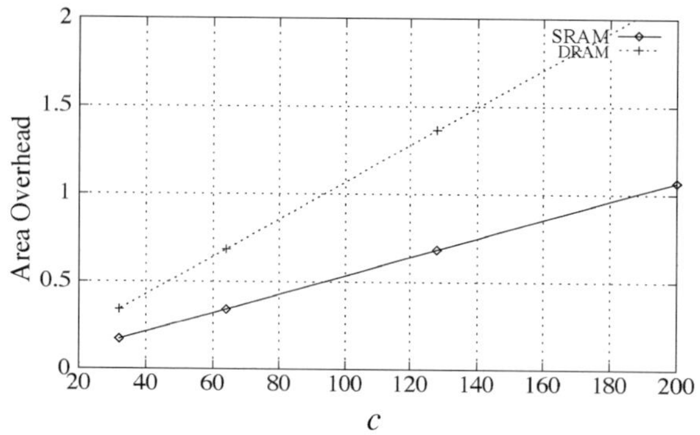

Fig. 11. Growth of area overhead for 1M-bit SRAM and DRAM with respect to c.

5.2. Results for Tree-Based Compression Scheme

To estimate the R of the tree-based compression scheme, a simulator was implemented to obtain the following results. A 128 K-bit RAM with 2K $\times$ 64, 1K $\times$ 128 and 512 $\times$ 256 configurations are considered, i.e., they have 64-bit, 128-bit and 256-bit Hamming syndromes, respectively. Also, three different symbol configurations are simulated: $B = 4$, 8, and 16, respectively. The number of the random faults (RFs) injected is from 100 to 500.

Table 4 shows the Rs for injecting different fault classes. From the table, we see that the tree-based compression technique has good compression ratio for different fault classes. The results show that a longer HS with larger symbol size has a better R. The reason is that the probability of the all-0 symbol is higher for larger symbol. If two additional faulty rows are injected, then R increases as compared with the case where there are random faults only, since the number of all-1 symbols also increases. Since a defect may cause several neighboring cells to be faulty (i.e., a cluster fault), we also simulated the case by injecting additional 10% cluster faults into the RAM, e.g., if RF = 100, then 10 cluster faults are injected. Also, each cluster fault is modeled by a 4 $\times$ 4 array with 12 faulty cells. The results show that the compression ratio is higher than that of the cases without cluster faults. Note that the compression ratio for the case with two additional faulty rows decreases when the number of random faults increases. The reason is that more random faults will lead to more Hamming syndromes to be exported. Moreover, the number of all-0 and single-1 symbols increases. Additional faulty columns are not considered here, since a faulty column will result in many single-1 symbols. The compression ratio for this case usually is lower than that for additional faulty rows.

To estimate the area overhead, we have modeled the Syndrome Compressors with various configurations at the synthesizable RTL level. The architecture of the Syndrome Compressor mainly consists of a multiplexer, a counter, and a Codeword Generator as described in Section 4.4. The Codeword Generator has a ROM and a comparator. Fig. 12 shows the area overhead numbers for different symbols and Hamming syndromes. As shown in the figure, the Syndrome Compressor with 8-bit (16-bit) symbol has the lowest (highest) area overhead, A. The reason is that the multiplexer and Codeword Generator dominate the area of the design. A larger multiplexer is needed to select the

Table 4. Compression ratios for different fault classes.

RF		R for 64-bit HS			R for 128-bit HS			R for 256-bit HS		
		$B = 4$	$B = 8$	$B = 16$	$B = 4$	$B = 8$	$B = 16$	$B = 4$	$B = 8$	$B = 16$
100	Without other faults	3.497	5.714	7.752	3.731	6.623	10.417	3.861	7.246	12.500
150		3.484	5.682	7.634	3.718	6.579	10.309	3.846	7.194	12.346
200		3.484	5.649	7.576	3.718	6.579	10.101	3.831	7.143	12.195
250		3.472	5.649	7.519	3.704	6.536	10.000	3.831	7.092	12.195
500		3.448	5.495	7.092	3.677	6.369	9.346	3.788	6.897	11.111
100	With two faulty rows	3.096	4.902	6.536	3.268	5.524	8.130	3.356	5.848	9.901
150		3.215	5.128	6.897	3.413	5.848	8.620	3.509	6.211	10.000
200		3.279	5.155	7.042	3.472	5.917	8.849	3.558	6.369	10.309
250		3.322	5.291	6.993	3.496	6.061	8.928	3.597	6.451	10.309
500		3.356	5.319	6.849	3.559	6.135	8.928	3.636	6.494	10.310
100	With 10% cluster faults	3.257	4.673	5.128	3.571	5.882	7.576	3.774	6.714	10.101
150		3.236	4.695	5.076	3.558	5.848	7.463	3.745	6.667	9.901
200		3.236	4.673	5.000	3.559	5.814	7.407	3.745	6.623	9.708
250		3.236	4.651	4.951	3.534	5.747	7.246	3.731	6.579	9.524
500		3.185	4.505	4.695	3.497	5.556	6.757	3.663	6.250	8.547
100	With one faulty row and 10% cluster faults	3.068	4.329	4.808	3.356	5.376	6.897	3.497	6.135	8.849
150		3.115	4.546	4.878	3.425	5.525	7.143	3.559	6.289	9.259
200		3.164	4.587	4.902	3.460	5.618	7.519	3.610	6.289	9.346
250		3.135	4.525	4.854	3.460	5.525	6.944	3.623	6.329	9.010
500		3.195	4.566	4.831	3.460	5.525	6.897	3.597	6.098	8.403

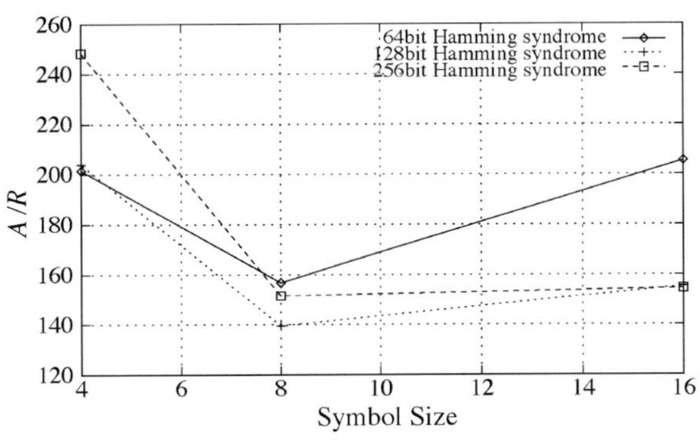

Fig. 12. Area overhead of the Syndrome Compressor.

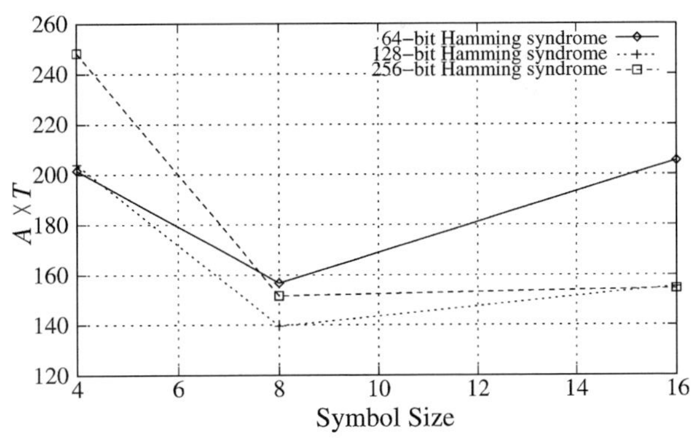

Fig. 13. A/R product of the Syndrome Compressor.

Hamming syndromes for 4-bit symbols. In contrast, a larger ROM is required to store the 16-bit symbols and their codewords. This also results in a larger comparator in the Codeword Generator. Note that the result may be different for other Syndrome Compressor designs. The A/R product is a good figure of merit for assessing the Syndrome Compressor design. Fig. 13 shows the A/R numbers with respect to the symbol size. Apparently, the Syndrome Compressor with 8-bit symbol has the lowest A/R product for all the three Hamming syndromes shown.

6. Conclusions

Two data compression techniques that can be used to speed up the transmission of diagnostic data from the embedded RAM with BIST have been proposed. The syndrome accumulation method compresses the faulty-cell address and March syndrome to about 28% of the original size under the March-17N test algorithm. The experimental results show that the compression ratio of this scheme is dominated by the diagnostic test algorithm, especially the number of Read operations. The key component of the syndrome accumulation circuit is a content-addressable memory. The area overhead is about 0.9% for a 1Mb SRAM with 164 faults. The limitation of this scheme is that the area overhead is high if the number of faults increase. The area overhead can be reduced if multiple RAM blocks share the same compressor. A tree-based compression technique for word-oriented memories also has been proposed. By partitioning Hamming syndromes into fixed-size symbols and constructing a simplified Huffman tree, it can compress the Hamming syndromes with good compression ratio and very small area overhead. The simulation results show that it can compress the Hamming syndromes to about 10% of the original size, assuming 16-bit symbols and 100–500 random faults in a 128 K-bit RAM. The Syndrome Compressors with different configurations have also been modeled at the synthesizable RTL level. Experimental results show that the Syndrome Compressor with 8-bit symbols results in the lowest area overhead based on the proposed approach. In conclusion, the proposed compression techniques can effectively reduce the time for diagnostic test and the tester storage requirement.

References

1. A. Benso, S. Di Carlo, G. Di Natale, P. Prinetto, and M. Lobetti-Bodorni, "A Programmable BIST Architecture for Clusters of Multiple-Port SRAMs," in *Proc. Int. Test Conf. (ITC)*, 2000, pp. 557–566.
2. T.J. Bergfeld, D. Niggemeyer, and E.M. Rudnick, "Diagnostic Testing of Embedded Memories Using BIST," in *Proc. Design, Automation and Test in Europe (DATE)*, Paris, March 2000, pp. 305–309.
3. B. Brown, J. Donaldson, B. Gage, and A. Joffe, "Hardware Compression Speeds on Bitmap Fail Display," in *Proc. Int. Test Conf. (ITC)*, 1997, pp. 89–93.
4. A. Chandra and K. Chakrabarty, "System-on-a-Chip Test-Data Compression and Decompression Architectures Based on Golomb Codes," *IEEE Trans. Computer-Aided Design of Integrated Circuits and Systems*, vol. 20, no. 3, pp. 355–368, March 2001.
5. M.F. Chang, W.K. Fuchs, and J.H. Patel, "Diagnosis and Repair of Memory with Coupling Faults," *IEEE Transactions on Computers*, vol. 38, no. 4, pp. 493–500, April 1989.
6. J.T. Chen, J. Rajski, J. Khare, O. Kebichi, and W. Maly, "Enabling Embedded Memory Diagnosis via Test Response Compression," in *Proc. IEEE VLSI Test Symp. (VTS)*, Marina Del Rey, California, April 2001, pp. 292–298.
7. A.L. Crouch, M. Mateja, T.L. McLaurin, J.C. Potter, and D. Tran, "The Testability Features of the 3rd Generation ColdFire Family of Microprocessors," in *Proc. Int. Test Conf. (ITC)*, 1999, pp. 913–922.
8. R. Dekker, F. Beenker, and L. Thijssen, "A Realistic Fault Model and Test Algorithm for Static Random Access Memories," *IEEE Trans. Computer-Aided Design of Integrated Circuits and Systems*, vol. 9, no. 6, pp. 567–572, June 1990.
9. D.A. Huffman, "A Method for the Construction of Minimum-Redundancy Codes," in *Proc. IRE*, vol. 40, pp. 1098–1101, Sept. 1952.
10. C. Hunter, "Integrated Diagnostics for Embedded Memory Built-in Self Test on PowerPC Devices," in *Proc. IEEE Int. Conf. Computer Design (ICCD)*, 1997, pp. 549–554.
11. S.S. Iyer and H.L. Kalter, "Embedded DRAM Technology: Opportunities and Challenges," *IEEE Spectrum*, pp. 56–64, April 1999.
12. A. Jas and N.A. Touba, "Test Vector Decompression via Cyclical Scan Chains and its Application to Testing Core-Based Designs," in *Proc. Int. Test Conf. (ITC)*, 1998, pp. 458–464.
13. D.A. Lelewer and D.S. Hirschberg, "Data Compression," *ACM Computing Surveys*, vol. 19, no. 3, pp. 261–296, Sept. 1987.
14. J.-F. Li, K.-L. Cheng, C.-T. Huang, and C.-W. Wu, "March-Based RAM Diagnosis Algorithms for Stuck-at and Coupling Faults," in *Proc. Int. Test Conf. (ITC)*, Baltmore, Oct. 2001, pp. 758–767.
15. J.-F. Li, R.-S. Tzeng, and C.-W. Wu, "Using Syndrome Compression for Memory Built-in Self-Diagnosis," in *Proc. Int. Symp. VLSI Technology, Systems, and Applications (VLSI-TSA)*, Hsinchu, April 2001, pp. 303–306.
16. J.-F. Li and C.-W. Wu, "Memory Fault Diagnosis by Syndrome Compression," in *Proc. Design, Automation and Test in Europe (DATE)*, Munich, March 2001, pp. 97–101.
17. K.-J. Lin and C.-W. Wu, "Testing Content-Addressable Memories Using Functional Fault Models and March-Like Algorithms," *IEEE Trans. Computer-Aided Design of Integrated Circuits and Systems*, vol. 19, no. 5, pp. 577–588, May 2000.
18. J.M. Mulder, N.T. Quach, and M.J. Flynn, "An Area Model for On-Chip Memories and its Application," *IEEE Journal of Solid-State Circuits*, vol. 26, no. 2, pp. 98–106, Feb. 1991.
19. M. Rich, "A Method of Flexible Catch RAM Display for Memory Testing," in *Proc. Int. Test Conf. (ITC)*, 1986, p. 222.
20. L. Shen and B.F. Cockburn, "An Optimal March Test for Locating Faults in DRAMs," in *Proc. IEEE Int. Workshop on Memory Testing*, 1993, pp. 61–66.
21. A.J. van de Goor, "Using March Tests to Test SRAMs," *IEEE Design & Test of Computers*, vol. 10, no. 1, pp. 8–14, March 1993.
22. C.-W. Wang, C.-F. Wu, J.-F. Li, C.-W. Wu, T. Teng, K. Chiu, and H.-P. Lin, "A Built-in Self-Test and Self-Diagnosis Scheme for Embedded SRAM," in *Proc. Ninth IEEE Asian Test Symp. (ATS)*, Taipei, Dec. 2000, pp. 45–50.
23. C.-F. Wu, C.-T. Huang, C.-W. Wang, K.-L. Cheng, and C.-W. Wu, "Error Catch and Analysis for Semiconductor Memories using March Tests," in *Proc. IEEE/ACM Int. Conf. Computer-Aided Design (ICCAD)*, San Jose, Nov. 2000, pp. 468–471.

24. C.-F. Wu, C.-T. Huang, and C.-W. Wu, "RAMSES: A Fast Memory Fault Simulator," in *Proc. IEEE Int. Symp. Defect and Fault Tolerance in VLSI Systems (DFT)*, Albuquerque, Nov. 1999, pp. 165–173.
25. C.-W. Wu, J.-F. Li, and C.-T. Huang, "Core-Based System-on-Chip Testing: Challenges and Opportunities," *J. Chinese Institute of Electrical Engineering*, vol. 8, no. 4, pp. 335–353, Nov. 2001.
26. T. Yabe, S. Miyano, K. Sato, M. Wada, R. Haga, O. Wada, M. Enkaku, T. Hojyo, K. Mimoto, M. Tazawa, T. Ohkubo, and K. Numata, "A Configurable DRAM Macro Design for 2112 Derivative Organizations to be Synthesized Using a Memory Generator," *IEEE Journal of Solid-State Circuits*, pp. 1752–1757, Nov. 1998.
27. V.N. Yarmolik, Y.V. Klimets, A.J. van de Goor, and S.N. Demidenko, "RAM Diagnostic Tests," in *Proc. IEEE Int. Workshop on Memory Technology, Design and Testing (MTDT)*, San Jose, 1996, pp. 100–102.

Jin-Fu Li received the BSEE degree in 1995 from National Taiwan University of Science Technology, Taipei, Taiwan, and the MSEE degree in 1999 from National Tsing Hua University, Hsinchu, Taiwan, where he is currently working toward the PhD degree. His research interests include advanced VLSI design and testing, System-on-a-Chip (SOC) testing, SOC test integration and scheduling, memory testing, memory diagnosis, and memory self-repair.

Ruey-Shing Tzeng received the BSEE and MSEE degrees in 1996 and 2001, respectively, from National Tsing Hua University, Hsinchu, Taiwan. He is currently with MediaTek Inc., Hsinchu, Taiwan, as an IC design engineer. His research interests include memory testing, memory diagnosis, and memory self-repair.

Cheng-Wen Wu received the BSEE degree in 1981 from National Taiwan University, Taipei, Taiwan, and the MS and PhD degrees in electrical and computer engineering, in 1985 and 1987, respectively, from the University of California, Santa Barbara. From 1981 to 1983, he was an ensign instructor at the Chinese Naval Petty Officers' School of Communications and Electronics, Tsoying, Taiwan. From 1983 to 1984, he was with the Information Processing Center of the Bureau of Environmental Protection, Executive Yuan, Taipei, Taiwan. From 1985 to 1987 he was a post graduate researcher at the Center for Computational Sciences and Engineering at UCSB. Since 1988 he has been with the Department of Electrical Engineering, National Tsing Hua University (NTHU), Hsinchu, Taiwan, where he is currently a professor. He also has served as the director of the university's Computer and Communications Center from 1996 to 1998, and the director of the university's Technology Service Center from 1998 to 1999. From August 1999 to February 2000, he was a visiting faculty of the ECE Department, UCSB. Since August 2000, he has been the Chair of the EE Department of NTHU. He also is the Director of the IC Design Technology Center of the university.

Dr. Wu was the Technical Program Chair of the IEEE Fifth Asian Test Symposium (ATS'96), and the General Chair of ATS'00. He is an Associate Editor for the Journal of the Chinese Institute of Electrical Engineers (JCIEE) and a Guest Editor of the JCIEE Special Issue on Design and Test of System-on-Chip, and was a Guest Editor of the Journal of Information Science and Engineering (JISE), Special Issue on VLSI Testing. He received the Distinguished Teaching Award from NTHU in 1996, the Outstanding Electrical Engineering Professor Award from the Chinese Institute of Electrical Engineers (CIEE) in 1997, the Distinguished Research Award from National Science Council in 2001, and the Industrial Collaboration Award from Ministry of Education in 2001. He is interested in design and testing of high performance VLSI circuits and systems. Dr. Wu is a life member of CIEE and a senior member of IEEE.

JOURNAL OF ELECTRONIC TESTING: Theory and Applications 18, 529–538, 2002

Testing for Interconnect Crosstalk Defects Using On-Chip Embedded Processor Cores*,†

LI CHEN, XIAOLIANG BAI AND SUJIT DEY
Department ECE, University of California at San Diego, La Jolla, CA 92093-0407, USA
lichen@ece.ucsd.edu
xibai@ece.ucsd.edu
dey@ece.ucsd.edu

Received September 11, 2001; Revised January 15, 2002

Editor: Krishnendu Chakrabarty

Abstract. In deep-submicron technologies, long interconnects play an ever-important role in determining the performance and reliability of core-based system-on-chips (SoCs). Crosstalk effects degrade the integrity of signals traveling on long interconnects and must be addressed during manufacturing testing. External testing for crosstalk is expensive due to the need for high-speed testers. Built-in self-test, while eliminating the need for a high-speed tester, may lead to excessive test overhead as well as overly aggressive testing. To address this problem, we propose a new software-based self-test methodology for system-on-chips (SoC) based on embedded processors. It enables an on-chip embedded processor core to test for crosstalk in system-level interconnects by executing a self-test program in the normal operational mode of the SoC, thereby allowing at-speed testing of interconnect crosstalk defects, while eliminating the need for test overhead and the possibility of over-testing. We have demonstrated the feasibility of this method by applying it to test the interconnects of a processor-memory system. The defect coverage was evaluated using a system-level crosstalk defect simulation method.

Keywords: interconnect, crosstalk, self-test, processor

1. Introduction

The shrinking of feature sizes enables the integration of a large system comprising of multiple cores on a single chip. In core-based designs, a larger amount of core-to-core communications must be realized with long interconnects. As gate delay continues to decrease, the performance of interconnect is becoming increasingly important in achieving a high overall performance [15]. However, due to the increase of cross-coupling capacitance and mutual inductance, signals on neighboring wires may interfere with each other, causing excessive delay or loss of signal integrity. This effect, known as crosstalk, is more pronounced in deep-submicron technology [8, 13]. While many techniques have been proposed to reduce crosstalk [7, 10, 18], due to the limited design margin and unpredictable process variations, testing for crosstalk must be performed during manufacturing. In recent years, several crosstalk test generation techniques have been developed to generate tests for local interconnects in gate-level circuits [5, 6, 9, 11, 17]. Unlike these techniques, we focus on the testing of system-level interconnects, as they are most susceptible to crosstalk effects due to the large

*This work is supported in part by MARCO/DARPA Gigascale Silicon Research Center (GSRC) and the Semiconductor Research Corporation (SRC) through contract no. 98-TJ-648.
†A preliminary version of this paper appeared in *Proceedings of the 38th Design Automation Conference*, Las Vegas, NV, USA, June 2001.

coupling capacitance between long wires running in parallel.

Due to its timing nature, testing for crosstalk effect need to be conducted at the operational speed of the circuit-under-test. At-speed testing for GHz systems, however, is prohibitively expensive with external testers, as it requires testers with GHz performance. Moreover, with external testing, hardware access mechanisms are required for applying tests to interconnects deeply embedded in the system, which may lead to unacceptable area or performance overhead.

Compared with external testing, self-testing is a more feasible solution for at-speed crosstalk testing, as it does not impose any performance requirement on the external tester. A built-in self-test (BIST) technique has been proposed in [2], in which an SoC tests its own interconnects for crosstalk defects using on-chip hardware pattern generators and error detectors. Although the amount of area overhead may be amortized for large systems, for small systems, the amount of relative area overhead may be unacceptable. Moreover, hardware-based self-test approach like the one proposed in [2] may cause over-testing, as not all test patterns generated in the test mode are valid in the normal operational mode of the system.

In this paper, we address the problem of testing system-level interconnects in embedded processor-based SoCs, which are the most dominant type of SoCs. In such SoCs, most of the system-level interconnects, such as the on-chip busses, are accessible to the embedded processor core(s). Based on this fact, we propose a software-based methodology that enables an embedded processor core in the SoC to test for crosstalk effects in these interconnects by executing a software program. Unlike previous embedded processor core-based self-testing methodologies [3, 4, 14, 16], in which the embedded processor cores themselves are the subjects of testing, the focus of our methodology is on the testing of interconnects connecting the processor cores and other cores.

Compared with the hardware-based approaches for crosstalk testing, software-based self-test can be applied in the normal operational mode of the processor. Therefore, no extra hardware is needed. In addition, only the test patterns valid in the normal operational mode of the processor are applied. Thus, the system will not be over-tested. This is because crosstalk cases that cannot be excited in the normal operational mode do not affect the correct functionality of the system. Thus, the rejection of a chip due to a failure response in these cases causes unnecessary yield loss.

In the rest of the paper, we first briefly describe the crosstalk fault model used during the development of the proposed method. The software-based self-test methodology for crosstalk defects is introduced in Section 3. In Section 4 we illustrate the proposed method in detail by constructing a self-test program for testing interconnects in a particular CPU-memory system. In Section 5, we validate the proposed method by extensive defect simulation using an HDL-level simulation framework consisting of a high-level crosstalk error model. Section 6 concludes the paper.

2. Fault Model

During the development of the proposed self-test method, we used the Maximum Aggressor Fault (MAF) model [8] for modeling crosstalk effects. MAF model is a high-level abstraction of all physical defects and process variations that lead to crosstalk errors. Defining crosstalk faults based on the actual physical defects can lead to the explosion of the fault space, as the number of possible process variations and defects that need to be considered can be extensive even for a bus with a small number of interconnects. Instead, the MAF model defines faults based on the resulting crosstalk error effects, including positive glitch (g_p), negative glitch (g_n), rising delay (d_r), and falling delay (d_f). The interconnect on which the error effect takes place is defined as the victim. All the other wires are designated aggressors, acting collectively to generate the glitch or delay error on the victim.

Fig. 1 shows the signal transitions needed on the victim/aggressors to produce the strongest error effects on a victim wire. For example, to produce a positive glitch fault on Victim Y_i, a stable "0" is assigned to Y_i, whereas rising transitions are assigned to all aggressors. Each of these signal transitions, consisting of two consecutive test vectors, is defined as the Maximum Aggressor (MA) test for the corresponding MAF. For

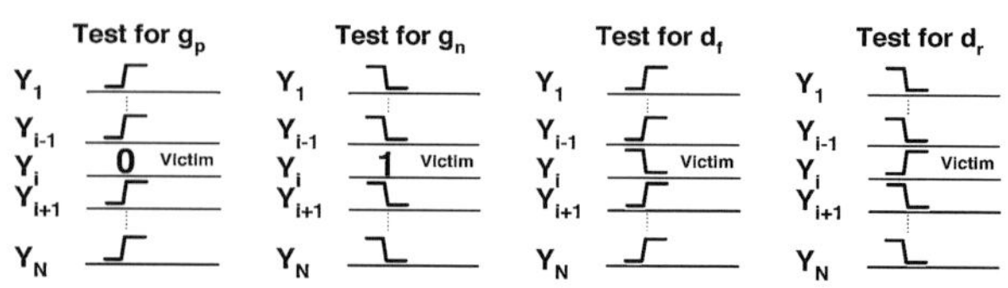

Fig. 1. Maximum aggressor tests for victim Y_i.

a set of N-interconnects, a total of $4N$ faults need to be tested, requiring $4N$ unique MA tests. It has been proven in [8] that in an RC network, the MA tests are necessary and sufficient for covering all possible physical defects and process variations that can lead to any cross-coupling induced crosstalk error effect on any of the N interconnects. The limitation of the MAF model is that the MA tests do not necessarily excite the worst-case crosstalk scenarios in an RLC network.

In the next section, we describe the general strategy for applying MA tests to system-level interconnects by executing a self-test program using an embedded processor.

3. Test Methodology

In a core-based SoC consisting of embedded processor, memory, and other cores, the address, data, and control busses are the main types of global interconnects, with which the embedded processors communicate with memory and other cores of the SoC via memory-mapped I/O (Fig. 2). Besides the single address/data busses shown in Fig. 2, there can be multiple busses or hierarchical bus structures. The most common mechanism for a CPU to communicate with a core is via memory-mapped I/O, in which certain addresses in the memory address space of the CPU are reserved for addressing the cores. Hence, the way the CPU communicates with memory cores is similar to the way it communicates with non-memory cores. In this paper, we focus on the testing of interconnects connecting the CPU and memory cores. Among such interconnects, we focus on the testing of data and address busses, as they are by far dominating and most susceptible to crosstalk defects due to a large number of long interconnects running in parallel. The testing of interconnects between the CPU and non-memory cores and the testing of control busses are subjects of future study.

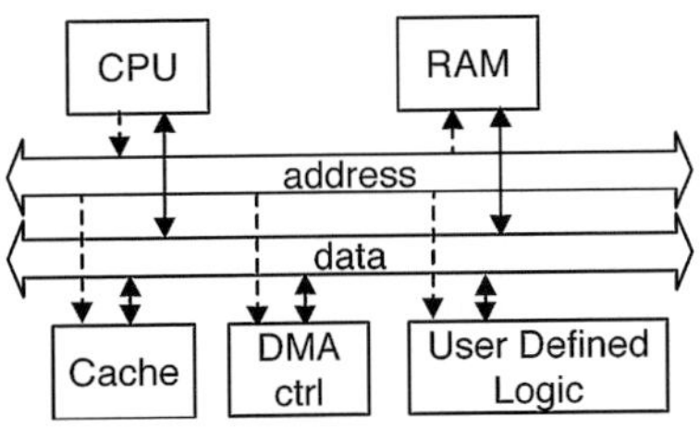

Fig. 2. An SoC containing embedded processor cores.

To test the address bus and the data bus with an embedded processor, our strategy is to let the processor execute a self-test program, with which the test vector pairs (as described in Section 2) can be applied to the appropriate bus in the normal functional mode of the system. In the presence of crosstalk-induced glitch or delay effects, the second vector in the vector pair becomes distorted at the receiver end of the bus. The processor can then store this error effect to the memory as a test response, which can later be unloaded by an external tester for off-chip analysis.

In this self-test approach, the loading of the self-test program and the unloading of the test response can be done with a low-speed tester. This prevents the corruption of the test program data or test response data due to crosstalk errors on the data or address busses, which can be activated at high speed. The self-test program itself is executed at-speed in the normal functional mode of the system without the monitoring of any external testers. Thus, with this approach, at-speed testing for crosstalk effects can be applied without a high-performance tester.

We next describe the general method for testing the data bus and the address bus.

3.1. *Testing Data Bus*

For a bi-directional bus such as a data bus, crosstalk effects vary as the bus is driven from different directions. Thus, crosstalk tests need to be conducted in both directions [2]. Note that, however, to apply a test vector pair ($v1$, $v2$) in a particular bus direction, the direction of $v1$ is irrelevant. Only $v2$ needs to be applied in the specified direction. This is because the signal transition triggering the crosstalk effect takes place only when $v2$ is being applied to the bus.

To apply a test vector pair ($v1$, $v2$) to the data bus from an SoC core to the CPU, the CPU first exchanges data $v1$ with the core. The direction of the data exchange is irrelevant. For example, if the core is the memory, the CPU may either read $v1$ from the memory or write $v1$ to the memory. The CPU then requests data $v2$ from the core (a memory-read if the core is memory). Upon the arrival of $v2$, the CPU writes $v2$ to memory for later analysis.

To apply a test vector pair ($v1$, $v2$) to the data bus from the CPU to an SoC core, the CPU first exchanges data $v1$ with the core. The CPU then sends data $v2$ to the core (a memory-write if the core is memory). If the core is memory, $v2$ can be directly stored to an appropriate address for later analysis. Otherwise, the CPU must execute additional instructions to retrieve $v2$ from the core and store it to memory.

3.2. *Testing Address Bus*

To apply a test vector pair ($v1$, $v2$) to the address bus, which is a unidirectional bus from the CPU to an SoC core, the CPU first requests data from two addresses ($v1$ and $v2$) in consecutive cycles. In the case of a non-memory core, since the CPU addresses the cores via memory-mapped I/O, $v2$ must be the address corresponding to the core. If $v2$ is distorted by crosstalk, the CPU would be receiving data from a wrong address, $v2'$, which may be a physical memory address or an address corresponding to a different core. By keeping different data at $v2$ and $v2'$ (i.e., mem[$v2$] $\neq$ mem[$v2'$]), the CPU is able to observe the error and store it to memory for analysis. In rare cases, the value stored in $v2'$ cannot be easily controlled. For example, $v2'$ may correspond to unused address space, read-only locations, or address space mapped to special registers whose values may change during operation. Furthermore, the value of $v2'$ may be unpredictable due to an unpredicted crosstalk error effect. For example, a signal transition on an aggressor can be corrupted by a strong victim. To address these problems, a relative unique value should be stored in $v2$ to maximize the probability of detecting the error.

Fig. 3 illustrates the process of testing the address bus. For example, in the case where the CPU is communicating with a memory core, to apply test (0001, 1110) in the address bus from the CPU to the memory core, the CPU first reads data from address 0001. The CPU then reads data from address 1110. In the system with the faulty address bus, this address becomes 1111. If different data are stored at address 1110 and 1111 (mem[1110] = 0100, mem[1111] = 1001), the CPU would receive a faulty value from memory (1001

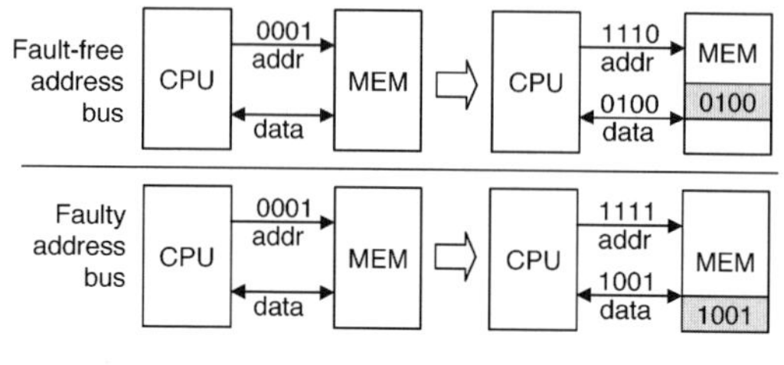

Fig. 3. Testing the address bus.

instead of 0100). This error response can later be stored to memory for analysis.

Although the generation of test programs is instruction-set-specific, a systematic approach can be followed. To summarize, we first analyze instructions with which the processor communicates with the memory core (e.g., load and store instructions). We then look for instructions or instruction sequences that can cause transitions on the data or address bus. Transitions with maximum flexibility are selected for applying crosstalk tests (e.g., MA test vector pairs). These are transitions with minimum constraints on signal values. Finally, for observing error responses, additional instructions may be used if necessary. The end goal is such that if the second vector in the test vector pair is corrupted, either an incorrect value is stored to the response location in memory (due to crosstalk error on data bus) or a correct value is stored to an incorrect location in memory (due to crosstalk error on address bus).

4. Application to a CPU-Memory System

In this section, we illustrate the proposed method in detail by applying the guidelines introduced in Section 3 to a CPU-memory system. Since memory-mapped I/O is a common mechanism for CPU to communicate with other cores, the same methodology can be extended for testing interconnects between the CPU and non-memory cores.

The particular CPU-memory system we used here consists of an 8-bit accumulator-based multi-cycle processor core with 23 instructions [12] and a 4K instruction/data memory. The CPU communicates with the memory via a 12-bit unidirectional address bus and an 8-bit bi-directional data bus.

4.1. *Testing Data Bus*

We chose to use the load instruction (LDA) of the processor to apply tests to the data bus. The load instruction takes a 12-bit address (A_x) as operand and loads the content of this address ($M[A_x]$) to the accumulator. As shown in Fig. 4, the load instruction is stored as two bytes in the memory. The addresses of these two bytes are A_i and $A_i + 1$, respectively. In the first byte, the first 4 bits contain the opcode of the instruction. The last 4 bits contain the *page number* of A_x, which is the first 4 bits of A_x. The second byte in the load instruction contains the *offset* of A_x, which is the last 8 bits of A_x.

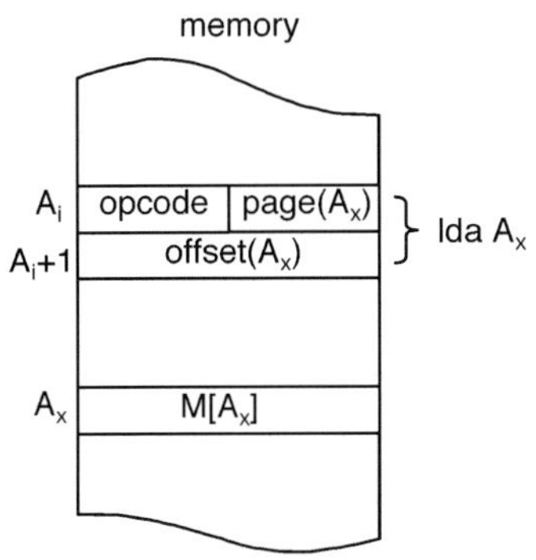

Fig. 4. Load instruction (LDA).

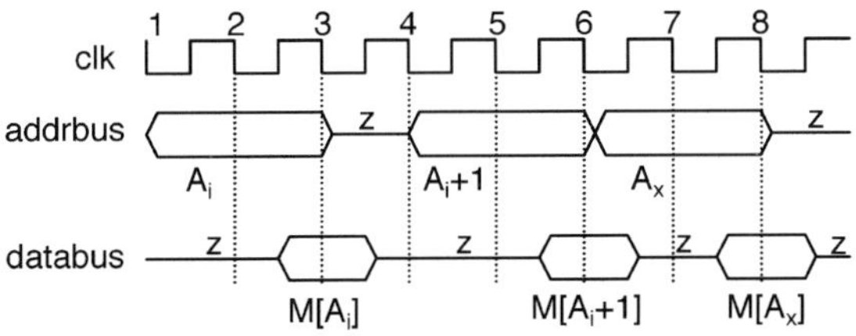

Fig. 5. Load instruction: timing diagram.

Fig. 5 shows the timing diagram of the load instruction. In the system under consideration, access to busses is controlled by tri-state buffers. When all tri-state buffers are disabled, the signal on the bus becomes high impedance ("z"). When "z" appears, we assume the bus holds the last defined value before "z".

To fetch the instruction from the memory, the CPU first requests the first byte of the instruction by placing the address of the instruction, A_i, on the address bus. The memory responds by sending the content of $A_i(M[A_i])$ to the CPU through the data bus. After decoding the first byte of the instruction, the CPU recognizes that the current instruction is a load instruction, which contains two bytes. Thus the CPU fetches the second byte of the instruction from memory ($M[A_i+1]$). After receiving the complete instruction, which contains the data address A_x, the CPU fetches the content of $A_x(M[A_x])$ from memory and places it into the accumulator.

During the execution of the load instruction, there are two signal transitions on the data bus, which can be used to apply the vector pair for crosstalk testing. The first transition is from $M[A_i]$ to $M[A_i+1]$. As $M[A_i]$ contains the opcode of the instruction, we are constrained when applying vector pairs using this transition. The second transition is from $M[A_i+1]$ to $M[A_x]$, where $M[A_i+1]$ contains the 8-bit address offset of the data to be loaded and $M[A_x]$ contains the actual data to be loaded. We chose this transition over the first transition, as it does not come with any constraints and thus can be used to apply crosstalk tests with maximum flexibility.

To apply an arbitrary vector pair ($v1$, $v2$) to the data bus, we need to have $M[A_i+1]=v1$ and $M[A_x]=v2$. Hence, we need to load from an address with a specific offset ($v1$) containing a specific data ($v2$). For example, to apply (00000000, 11110111), one of the tests for positive glitch faults, we may load from address 1110 : 00000000 (offset = 00000000), which contains data 11110111). If the test passes, $v2$ (11110111) is loaded to the accumulator. If the test fails due to a positive glitch, $v2$ becomes 11111111. Thus a wrong value is loaded into the accumulator. The error response can be collected by storing the accumulator content to a specific memory address, which can be checked by an external tester upon the completion of the tests. Thus, a two-instruction sequence is needed for applying this test: (`lda` 1110 : 00000000, `sta resp`), where the content of memory address 1110 : 00000000 must be set to 11110111, and `resp` is the memory address where the test response will be stored.

Similarly, the same strategy can be used to apply tests for other types of crosstalk faults (negative glitch, falling delay, and rising delay) on the data bus.

4.2. *Testing Address Bus*

During the execution of the load instruction, there are two transitions on the address bus. The first transition is the increment from A_i to A_i+1. The second transition is from A_i+1 to A_x. Since the first transition comes with a much more strict constraint than the second one, we choose to use the second transition to apply the vector pair.

To apply an arbitrary vector pair ($v1$, $v2$) to the address bus, A_i+1 must be $v1$ and A_x must be $v2$. Thus, the second byte of the instruction must be located at memory address $v1$, which implies that the instruction itself must be located at memory address $v1-1$. In addition, the data address accessed by the instruction must be $v2$.

4.2.1. Testing for Rising/Falling Delay Faults. For example, to apply (0000 : 00010000, 1111 : 11101111), one of the tests for falling delay faults, we place the load instruction at address 0000 : 00001111 ($v1-1$), and load from address 1111 : 11101111 ($v2$). If the test fails due to a falling delay defect, $v2$ becomes

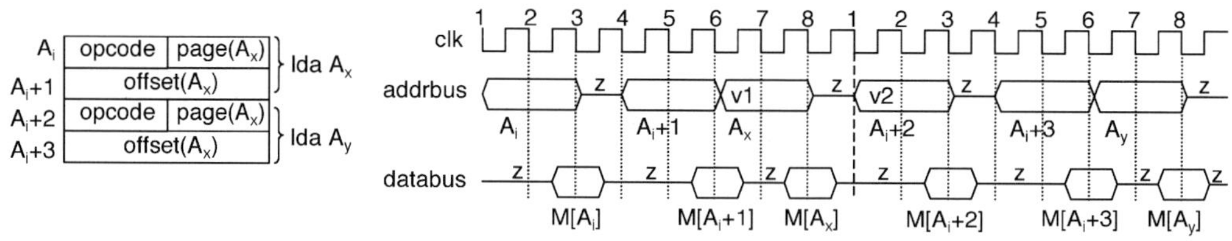

Fig. 6. Testing for positive/negative glitches on the address bus using two instructions.

1111 : 11111111 and we would be loading from address 1111 : 11111111 instead of address 1111 : 11101111. To observe the error, we store different values at address 1111 : 11101111 and 1111 : 11111111 (e.g., 00000001 at 1111 : 11101111 and 00000000 at 1111 : 11111111). If the test passes, 00000001 is loaded to the accumulator. If the test fails, 00000000 is loaded instead. Again, the error response can be collected by storing the accumulator content to memory. Thus, a two-instruction sequence is needed for applying this test: (`lda` 1111 : 11101111, `sta resp`), where the `lda` instruction is placed at memory address 0000 : 00001111, and the contents of memory address 1111 : 11101111 and 1111 : 11111111 are set to 00000001 and 11111111, respectively.

Tests for rising delay faults can be applied in a similar manner. However, tests for positive glitch/negative glitch faults are considerably different. This is because all tests for positive glitch faults start with vector 0000 : 00000000 (Fig. 1). To apply such a test using one instruction, the second byte of the instruction must be placed at memory address 0000 : 00000000. Moreover, to apply two vector pairs starting with the same vector (0000 : 00000000 in this case), two instructions need to be placed at the same memory location, causing an address conflict. This problem can be solved by utilizing the signal transition between *two* instructions.

4.2.2. Testing for Positive/Negative Glitch Faults.

As shown in Fig. 6, we use two load instructions, "lda A_x" and "lda A_y", where A_x and A_y are the addresses to load from. The two instructions reside from memory address A_i to $A_i + 3$. To apply vector pair $(v1, v2)$ to the address bus, we use the transition between two instructions, from A_x to $A_i + 2$. Therefore, A_x, the address to be loaded from in the first instruction, has to be set to $v1$. $A_i + 2$, the address of the second instruction, has to be set to $v2$, putting the first instruction at address $v2 - 2$. As a result, we are able to avoid the problem of address conflicts when applying two vector pairs starting from the same vector. For example, to apply (0000 : 00000000, 1111 : 11111110), the vector pair needed for testing the positive glitch fault on bus line 1, we place the first instruction at address 1111 : 11111100 and use it to access address 0000 : 00000000. To apply (0000 : 00000000, 1111 : 11110111), the vector pair needed for testing the positive glitch fault on bus line 4, we place the first instruction at address 1111 : 11110101 and use it to access address 0000 : 00000000. There are no address conflicts between the two tests.

Fig. 7 illustrates the collection of test response for a positive glitch test, (0000 : 00000000, 1111 : 11101111). To apply this test, we place the first instruction at address 1111 : 11101101 and use it to access address 0000 : 00000000 (Fig. 7(a)). In the presence of a positive glitch fault, the execution of the second instruction is affected. Instead of executing the instruction located at addresses 1111 : 11101111 and 1111 : 11110000, the CPU now executes the instruction located at addresses 1111 : 11111111 and 1111 : 11110000 (Fig. 7(b)). Note that the first byte is changed while the second byte is not. To observe this error, we store different values at addresses 1111 : 11101111 and 1111 : 11111111 such that the content at 1111 : 11101111 corresponds to loading

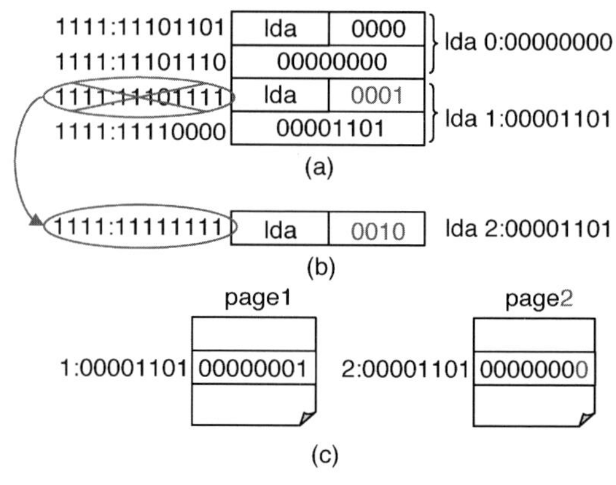

Fig. 7. Response collection.

from page 1 and the content of 1111 : 11111111 corresponds to loading from page 2. An arbitrary address offset, for example, 00001101, is stored at 1111 : 11110000. If the test passes, the CPU loads from address 1 : 00001101. If the test fails, the CPU loads from address 2 : 00001101 (Fig. 7(b)). We store different values at these two addresses such that the CPU loads a wrong value into the accumulator in the presence of the fault (Fig. 7(c)). The error response can be collected by storing the accumulator content to memory.

Thus, as shown in Fig. 7, to apply test vector pair (0000 : 00000000, 1111 : 11101111), a three-instruction sequence is needed: (`lda` 0000 : 00000000, `lda` 0001 : 00001101, `sta resp`), where the first `lda` is placed at address 1111 : 11101101, with the content of address 1111 : 11111111 set to "`lda` 0010" (the first byte of a `lda` instruction loading from page 2) and the content of address 0001 : 00001101 and 0010 : 00001101 set to 00000001 and 00000000, respectively.

4.3. Test Response Compaction

After each test, the test response is stored in the memory so that it can be unloaded and checked by an external tester. To reduce the data exchanged between the memory and the tester, we compact the test responses into as few bytes as possible without losing any diagnostic information.

Our basic idea for test response compaction is to sum up individual test responses into a collective test response. If the individual tests are properly constructed, the collective test response can show the pass/fail status of each test. Fig. 8 illustrate the test response compaction process, when tests for rising delay faults on the data bus (Section 4.1) are applied.

Since the ADD instruction has the same construct and timing characteristics as the load instruction (Fig. 8(a)), we use the add instruction to apply the tests. Fig. 8(b) shows the test program for the rising delay faults on all data bus lines. First, the accumulator is cleared. To apply the first test (01111111, 10000000), the test vector pair for the rising delay fault on bus line 8, we add to the accumulator the content of address 3 : 01111111, which is 10000000 (Fig. 8(c)). The page number, 3, is arbitrarily chosen. In the presence of a rising delay fault, the second vector in the vector pair, 10000000, becomes 00000000. Thus, 0 is added to the accumulator instead. The rest of the tests are

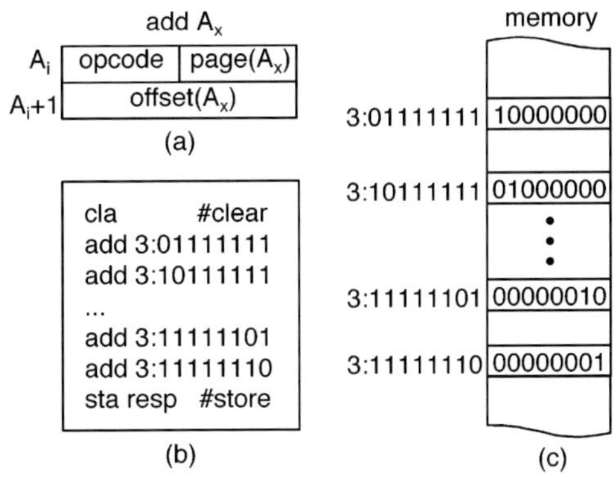

Fig. 8. Test response compaction.

constructed in similar manner. After the application of all tests, we store the accumulator content to a test response vector. If all tests pass, since we add 10000000, 01000000, . . . , 00000001 to the accumulator, the final test response is 11111111. Otherwise, at least one bit in the test response vector is 0. The position of the "0" bit tells which test failed.

Similarly, the test responses for the address bus can be compacted.

Depending on the actual instruction set of the processor core used in the SoC, the detailed constructs of the test programs may be different. Nonetheless, the general testing strategy we described here may be used to test the address/data busses between any CPU-memory pair. Moreover, since the cores in a SoC are often addressable by the CPU via memory-mapped I/O, the same test strategy can be extended to test address/data busses between any CPU-core pair.

With the software-based self-test approach, the proposed method has no area or delay overhead. Currently, the generation of test programs involves manual effort because it is instruction-set-specific. However, since we only need to consider the behavior of memory-accessing instructions when composing test programs, the test generation effort does not increase as the size of the instruction set increases. For each MA fault, a constant number of instructions are needed for applying the test. For a CPU-memory system with N interconnects, the number of MA faults is $4N$. Thus, the size of the test program is proportional to N. This corresponds to the size of the memory required for storing the test program, the tester time needed for loading the test program from the tester to the on-chip memory, as well as the test application time needed for executing the test program in the self-test mode.

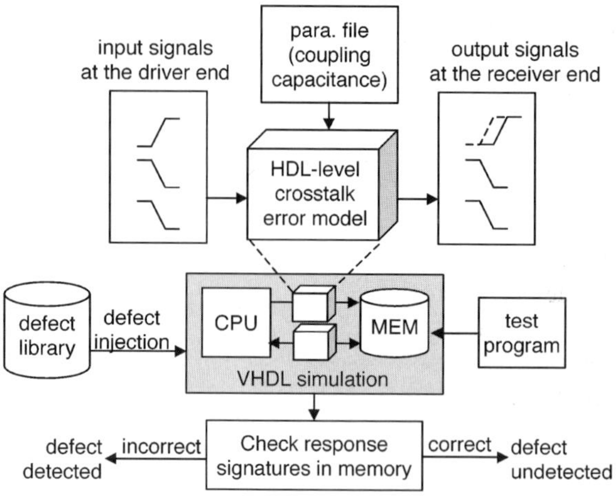

Fig. 9. HDL-level defect simulation environment.

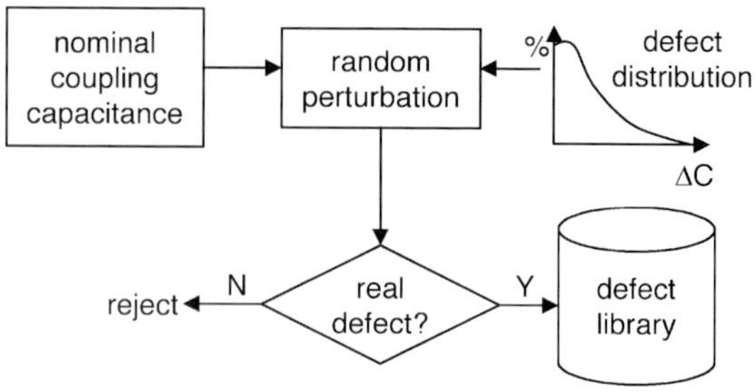

Fig. 10. The generation of defect library.

5. Validation

To validate the proposed software-based self-test methodology, we composed a complete test program for the CPU-memory system described in Section 4 and evaluated its defect coverage under an HDL-level defect simulation environment (Fig. 9). During simulation, the CPU exercises the busses by executing the crosstalk test program stored in the memory. To simulate the effect of crosstalk *on HDL-level*, we used the high-level crosstalk error model proposed in [1]. Coded in HDL, the error model takes as input a parameter file containing the values of the coupling capacitance among interconnects. Given an input transition on the driver end of the bus, the error model determines whether a crosstalk error happens on the receiver end. Note that with this high-level crosstalk error model, we are able to take into account the effect of fault masking when evaluating defect coverage, since a crosstalk defect on the bus is indeed activated many times as the CPU executes the test program.

During the execution of the test program, the CPU stores test response signatures to the memory. Upon the completion of the simulation, we determine whether the defect has been detected by the test program by comparing the resulting response signature with its expected value. To estimate the defect coverage, the same defect simulation process is repeated on all defects from a pre-constructed defect library.

Fig. 10 illustrates the generation of the defect library. To generate the defect library, we first randomly perturb the nominal values of coupling capacitances among interconnects according to a given defect distribution. Given the resulting perturbation, we use the criteria in [8] to determine whether the perturbation is large enough to be detectable by *any* tests. In particular, if a perturbation in capacitance causes the net coupling capacitance (C) on any interconnect to be larger than a threshold value (C_{th}), it is recorded as a defect. The value of C_{th} depends on the value of acceptable delay length or glitch height. This is because in an RC network, given fixed resistance values, the amount of crosstalk-induced delays and the height of crosstalk-induced glitches on one interconnect increase monotonically as C increases. The process of defect generation is repeated until a satisfactory number of defects are generated. In this paper, we only consider crosstalk within the same bus when injecting defects. It is possible to inject defects causing crosstalk between two busses by treating them as one bus.

In our experiments, we used a Gaussian distribution to model the defect distribution in terms of the variation of capacitance values (in %). A 3δ point of 150% was chosen. A total number of 1000 defects were generated for each bus.

For the CPU-memory system described in Section 4, there are 64 MAFs on the 8-bit bi-directional data bus ($8 \times 4 \times 2$) and 48 MAFs on the 12-bit address bus (12×4). With the test program, we were able to apply 64 out of 64 MA tests for the databus and 41 out of 48 tests for the address bus. Some of the tests cannot be applied due to address conflicts—i.e., multiple tests compete for the same instruction address. This problem can be solved by separating conflicting tests into *multiple* test programs, which can be executed in different sessions. The total execution time of the programs is 1720 processor cycles. The size of the test program is proportional to the width of the busses, as a certain number of instructions are needed for testing for each MAF.

Fig. 11 shows the individual and cumulative defect coverage obtained by applying each of the MA test for

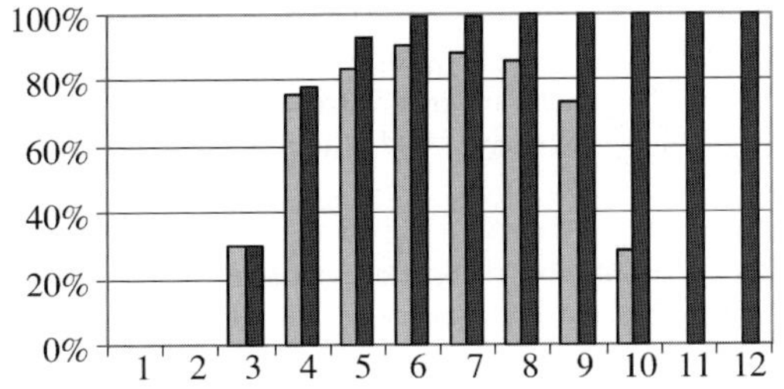

Fig. 11. Crosstalk defect coverage of MA test programs.

the address bus of the CPU-memory system. The horizontal axis indicates the MA test for each interconnect of the bus, with the ith test being the MA test for the ith interconnect. The individual defect coverages are shown in light gray, while the cumulative coverages are shown in dark gray. It can be seen that different MA tests have different levels of defect coverages, with the MA tests for the center interconnects having more coverage than the MA tests for the side interconnects. This is because the probability of a side interconnect being defective is small, as the net coupling capacitance on a side interconnect is smaller than the one on a center interconnect (i.e, a much larger perturbation is needed to render the side interconnect defective). In fact, the probability is so small such that in the particular defect library we have generated, no perturbation is large enough to cause Lines 1, 2, 11, and 12 to be defective. This is shown in Fig. 11, where the MA test programs for Lines 1, 2, 11, and 12 have no defect coverage. The cumulative defect coverage shows that the MA tests combined together provide a 100% coverage of the crosstalk defects on the address bus interconnects.

Since the MA tests are necessary for detecting all detectable defects [8], in theory, some of the defects can only be detected by the missing tests. However, using our defect library, the defect coverage of the test program is 100% on both address and data busses. This is because a large overlap exists among the defects set detected by different MA tests. Of all the defects detectable by one MA test, only a tiny fraction cannot be detected by any other MA tests.

6. Conclusions

At-speed testing for crosstalk effects is expensive with external testers. Built-in self-test for crosstalk may result in high test overhead or over aggressive testing. To address these issues, we proposed a cost-effective method for testing system-level interconnects using embedded processor cores. By executing a self-test program, a processor is able to test the address and data busses through which it communicates with memory components. The same method can be extended for testing the interconnects between the processor and non-memory cores, as these cores are typically addressed by the processor via memory-mapped I/O. We have constructed an HDL level defect simulation environment to validate the proposed method and evaluate the defect coverage of any given test program. Experimental results show that a self-test program written following the proposed method is able to achieve its projected defect coverage.

References

1. X. Bai and S. Dey, "High-Level Crosstalk Defect Simulation for System-on-Chip Interconnects," in *Proc. 19th VLSI Test Symp.*, Los Angeles, CA, April 2001.
2. X. Bai, S. Dey, and J. Rajski, "Self-Test Methodology for At-Speed Test of Crosstalk in Chip Interconnects," in *Proc. 37th Design Automation Conf.*, Los Angeles, CA, June 2000, pp. 619–624.
3. K. Batcher and C. Papachristou, "Instruction Randomization Self Test for Processor Cores," in *Proc. 17th VLSI Test Symp.*, Dana Point, CA, April 1999, pp. 34–40.
4. L. Chen and S. Dey, "Software-Based Self-Testing Methodology for Processor Cores," *IEEE Trans. Computer-Aided Designs*, vol. 20, no. 3, March 2001.
5. W. Chen, S.K. Gupta, and M.A. Breuer, "Test Generation in VLSI Circuits for Crosstalk Noise," in *Proceedings IEEE International Test Conference*, 1998, pp. 641–650.
6. W. Chen, S.K. Gupta, and M.A. Breuer, "Test Generation for Crosstalk-Induced Delay in Integrated Circuits," in *Proceedings IEEE International Test Conference*, Oct. 1999, pp. 191–200.
7. Z. Chen and I. Koren, "Crosstalk Minimization in Three-Layer HVH Channel Routing," in *Proceedings IEEE International Symposium on Defect and Fault Tolerance in VLSI System*, 1997, pp. 38–42.
8. M. Cuviello, S. Dey, X. Bai, and Y. Zhao, "Fault Modeling and Simulation for Crosstalk in System-on-Chip Interconnects," in *1999 Int. Conf. Computer-Aided Design*, San Jose, CA, Nov. 1999, pp. 297–303.
9. N. Itazaki, Y. Matsumoto, and K. Kinoshita, "An Algorithmic Test Generation Method for Crosstalk Faults in Synchronous Sequential Circuits," in *Proceedings Sixth Asian Test Symposium*, Nov. 1997, pp. 22–27.
10. A.B. Kahng, S. Muddu, E. Sarto, and R. Sharma, "Interconnect Tuning Strategies for High-Performance ICs," in *Proceedings Design, Automation and Test in Europe*, Paris, France, Feb. 1998, pp. 471–478.
11. K.T. Lee, C. Nordquist, and J. Abraham, "Automatic Test Pattern Generation for Crosstalk Glitches in Digital Circuits," in *Proceedings IEEE VLSI Test Symposium*, 1998, pp. 34–39.
12. Z. Navabi, *VHDL: Analysis and Modeling of Digital Systems*, New York: McGraw-Hill, 1993.
13. P. Nordholz, D. Treytnar, J. Otterstedt, H. Grabinski, D. Niggemeyer, and T.W. Williams, "Signal Integrity Problems in

Deep Submicron Arising from Interconnects Between Cores," in *Proc. 16th VLSI Test Symp.*, Monterey, CA, April 1998, pp. 28–33.
14. K. Radecka, J. Rajski, and J. Tyszer, "Arithmetic Built-in Self-Test for DSP Cores," *IEEE Trans. Computer-Aided Design*, vol. 16, no. 11, pp. 1358–1369, Nov. 1997.
15. Semiconductor Industry Association, *The International Technology Roadmap for Semiconductors*, 1999.
16. J. Shen and J.A. Abraham, "Native Mode Functional Test Generation for Processors with Applications to Self Test and Design Validation," in *Proc. Int. Test Conf.*, Washington DC, Oct. 1998, pp. 990–999.
17. A. Sinha, S.K. Gupta, and M.A. Breuer, "Validation and Test Generation for Oscillatory Noise in VLSI Interconnects," in *Proceedings of the International Conference on Computer-Aided Design*, Nov. 1999.
18. H. Zhou and D.F. Wang, "Global Routing with Crosstalk Constraints," in *Proceedings of the 35th Design Automation Conference*, San Francisco, CA, USA, June 1998, pp.374–377.

Li Chen (S'98) received the B.S. and M.S. degrees in electrical and computer engineering from Carnegie Mellon University, Pittsburgh, PA, in 1996 and 1998, respectively. She is currently a Ph.D. candidate in the Department of Electrical and Computer Engineering at the University of California, San Diego. From January 1997 to July 1997, she was with Advanced Micro Devices, Sunnyvale, CA. From June 2001 to September 2001, she worked at Intel Corporation, Santa Clara, CA. Her current research interests include the self-test and self-diagnosis of microprocessor cores.

Xiaoliang Bai (S'98) received the B.S. in electrical engineering from Xi'an Jiaotong University in 1995, and the M.S. degree in computer engineering from Peking University in 1998. He is working toward the Ph.D. degree in electrical and computer engineering at the University of California, San Diego. His research interests include signal integrity analysis and testing, timing analysis, circuit tuning and simulation.

Sujit Dey is a Professor in the Electrical and Computer Engineering Department at the University of California, San Diego. His research group at UCSD is developing configurable platforms, consisting of adaptive wireless protocols and algorithms, and deep submicron adaptive system-on-chips, for next-generation wireless appliances as well as network infrastructure devices. He is affiliated with the California Institute of Telecommunications and Information Technology, the UCSD Center for Wireless Communications, and the DARPA/MARCO Gigascale Silicon Research Center. Prior to joining UCSD in 1997, he was a Senior Research Staff Member at the NEC C&C Research Laboratories, Princeton, NJ. He obtained a Ph.D. in Computer Science from Duke University in 1991.

Dr. Dey has co-authored more than 100 publications, including journal and conference papers, a book on low-power design, and several book chapters. He received Best Paper awards at the Design Automation Conferences in 1994, 1999, and 2000, and the 11th VLSI Design Conference in 1998, and several best paper nominations. He is a co-inventor of 9 U.S. patents, and has 2 other pending. He has presented numerous tutorials and invited talks, and participated in panels, in the topics of low-power wireless systems design, hardware-software embedded systems, and deep sub-micron system-on-chip design and test. He has been the General Chair and Program Chair, and member of organizing and program committees, of several IEEE conferences and workshops.

JOURNAL OF ELECTRONIC TESTING: Theory and Applications 18, 539–554, 2002

Signal Integrity: Fault Modeling and Testing in High-Speed SoCs

MEHRDAD NOURANI AND AMIR ATTARHA
Center for Integrated Circuits & Systems, The University of Texas at Dallas, Richardson, TX 75083-0688, USA
nourani@utdallas.edu
attarha@utdallas.edu

Received September 2, 2001; Revised January 15, 2002

Editor: Krishnendu Chakrabarty

Abstract. As we approach 100 nm technology the interconnect issues are becoming one of the main concerns in the testing of gigahertz system-on-chips. Voltage distortion (noise) and delay violations (skew) contribute to the signal integrity loss and ultimately functional error, performance degradation and reliability problems. In this paper, we first define a model for integrity faults on the high-speed interconnects. Then, we present a BIST-based test methodology that includes two special cells to detect and measure noise and skew occurring on the interconnects of the gigahertz system-on-chips. Using an inexpensive test architecture the integrity information accumulated by these special cells can be scanned out for final test and reliability analysis.

Keywords: built-in self-test, high-speed interconnect, integrity fault, integrity loss, interconnect testing, model order reduction, noise detection, signal integrity, skew detection, system-on-chip, test pattern generation, transfer function matrix

1. Introduction

With fine miniaturization of VLSI circuits and rapid increase in the working frequency (gigahertz range) of digital system-on-chips (SoC), the signal integrity becomes a major concern for design and test engineers. Although various parasitic factors for transistors can be well controlled during fabrication, the parasitic capacitances, inductances and their cross coupling effects on the interconnects play a significant role in the proper functionality and performance of high-speed SoCs.

Signal integrity is the ability of a signal to generate correct responses in a circuit. It generally includes all effects that cause a design to malfunction due to the distortion of the signal waveform. According to this informal definition, a signal with good integrity has: (i) voltage values at required levels and (ii) level transitions at required times. For example, an input signal to a flip-flop with good signal integrity arrives early enough to guarantee the setup and hold time requirements and it does not have spikes causing undesired logic transition.

1.1. Prior Work

Various signal integrity problems have been studied previously for radio frequency (RF) circuits and recently for high-speed deep-submicron VLSI chips. The most important ones are: *crosstalk* (signal distortion due to cross coupling effects between signals) [5, 10], *overshoot* (signal rising momentarily above the power supply voltage) [14, 27], *reflection* (echoing back a portion of a signal), *electro-magnetic interference* (resulting from the antenna properties) [26], *power supply noise* [53] and *signal skew* (delay in arrival time to different receivers) [6, 8].

There is a long list of possible design and fabrication solutions to enhance signal integrity on the interconnect. None guarantees to resolve the issue perfectly. These solutions include: 3-D layout modeling and parasitic extraction [20], accurate RLC simulation of on-chip power grid [6], using decoupling capacitors to limit the maximum dV/dt [12, 53] and to improve IR-drop [6, 42], inserting buffers on the interconnects [20] and shielding wires (e.g. grounding every other line) [25].

Noise and skew imposed by interconnects have emerged as main concerns in the interconnect design of gigahertz SoCs. Buffer insertion and transistor resizing methods [24, 48] are used as design techniques to achieve better power-delay and area-delay tradeoffs. Self-test methodologies have been developed to test signal integrity in high-speed SoCs. Testing crosstalk in chip interconnects [4, 10] and a BIST (built-in self-test) structure using D flip-flops that detects the propagation delay deviation of operational amplifiers [40] are among such methods.

1.2. Contribution and Paper Organization

Our main contribution is an on-chip mechanism to detect noise and skew violations occurring on the interconnects of high-speed SoCs. We present special cells to monitor signals received from the system interconnect and record the occurrence of signal entering the vulnerable region over a period of operation. By resizing transistors within these cells, they can be easily tuned to define the acceptable levels of noise and skew. We also propose a BIST methodology that uses pseudo-random patterns and accumulates the integrity test information using the detector cells. The statistics will be eventually sent out for final test analysis, reliability judgment and diagnosis.

The rest of this paper is organized as follows. The signal integrity model and test strategy are discussed in Section 2. In Section 3 we explain the issue of generating patterns that stimulate the maximal integrity loss on interconnects. Sections 4 and 5 analyze CMOS circuits that detect noise and skew violations occurring on the interconnects, respectively. Section 6 explains the test architecture to store and read out the information. The experimental results are discussed in Section 7. Finally, the concluding remarks are in Section 8.

2. Integrity Test Methodology

2.1. Interconnect Model

Signal integrity problems originate from the circuit interconnects [34]. A wire not only serves as a conductor

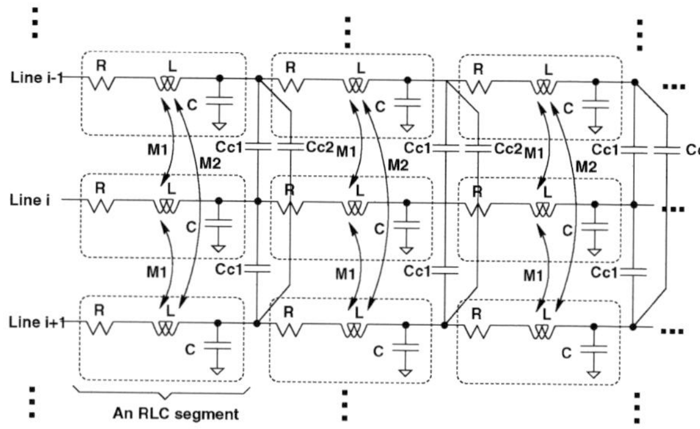

Fig. 1. An interconnect model.

of electrons but also includes parasitic resistor (at low frequencies), capacitor (at mid-range frequencies), inductor (at high frequencies) and antenna (at very high frequencies), all of which can affect signal integrity. In low and mid-range frequencies, common in the past, the RC delays have been the dominating factors in the global interconnect delay and distortion. Inductance (L) effects are becoming increasingly important as frequency of operation increases.

There are many efficient distributed models in the literature [7, 44, 51]. Fig. 1 shows an accurate equivalent RLC circuit for several parallel interconnect lines [7, 18]. This model comprises resistance (R), partial self inductance (L) and capacitance (C) for each segment, mutual inductances (M) and coupling capacitance (C_c) between all pairs of parallel components. The values of distributed R, L and C depend on many factors including the operating frequency, length and technology. The number of segments can be selected based on the length of the interconnect and the operating frequency. All results reported in this work are based on this distributed RLC model [7, 18].

2.2. A Model for Signal Integrity

True characteristics of a signal is reflected in its waveform. Recent interconnect simulation, design and optimization methods not only consider peak voltage and delay, but also take into account the signal waveform [9, 41]. In reality, electronic components can tolerate certain level of noise. For example, a CMOS gate interprets any voltage in the [V_{Hmin}, V_{dd}] range as logic "1" and any voltage in the [V_{ss}, V_{Lmax}] range as logic "0". Frequently, digital circuits are designed to tolerate certain amount of skew delay, i.e. T_{SI_R} for rising delay and T_{SI_F} for falling delay (see Fig. 2).

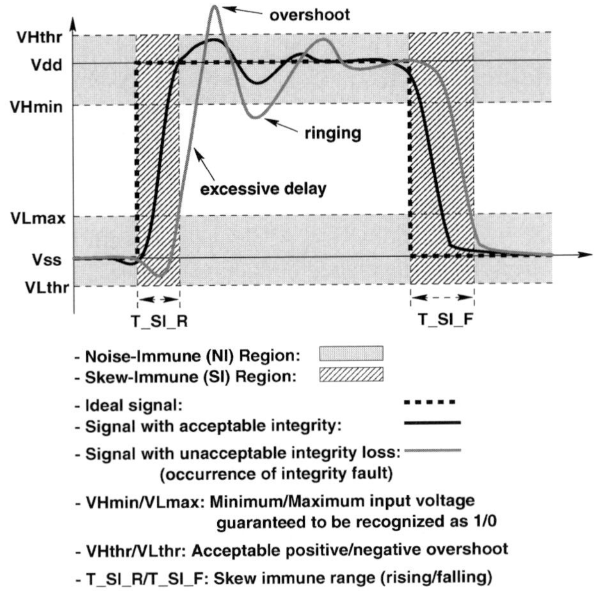

Fig. 2. Immune regions and the concept of integrity loss.

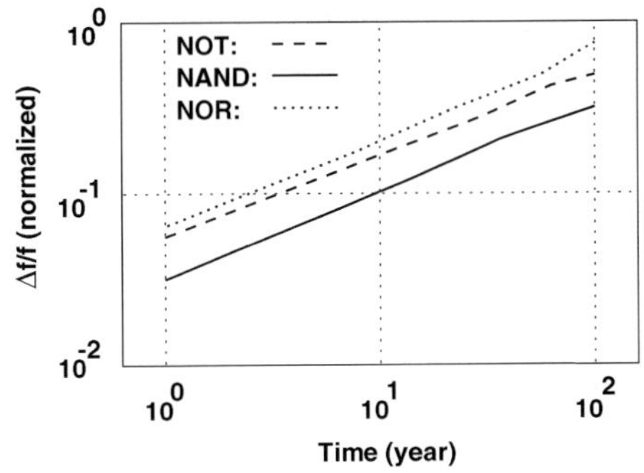

Fig. 3. The effect of overshoots on performance and life time.

In practice, circuits have *noise-immune* (NI) regions that tolerate certain level of voltage swing and *skew-immune* (SI) regions that tolerate certain level of delay. Any portion of signal that exits the NI and SI regions indicates the integrity loss. This concept has been shown graphically in Fig. 2 in which the shaded and unshaded (white) strips show the immune and vulnerable regions, respectively.

The focal point in this paper is the NI and SI regions, covering integrity faults on interconnects, and a mechanism to detect signals that exit these regions. Leaving the NI-region not only causes error in functionality (ringing), but also shortens system's life time due to time-dependent dielectric breakdown (TDDB) [22]. More importantly, repeated overshoots are known to inject high-energy electrons and holes (also called *hot-carriers*) into the gate oxide that ultimately cause permanent degradation of MOS transistors' performance and reliability [14]. For example, in [14], the authors presented the life time analysis, illustrated in Fig. 3, showing that the performance of logic gates under stress (e.g. repeated overshoots) degrades quickly which eventually causes failure in the system.

Leaving the SI-region means that the interconnect adds unacceptable skew delay that may lead to functional error or serious performance degradation. The range of immunity (T_{SI}) depends on the system topology, speed, core interactions and the yield margin that designer defines in the design phase. However, delay values of components as well as interconnects may not eventually be kept within such bound since the layout generation tools and the fabrication process each may add additional delays.

2.3. *BIST-Based Test Methodology*

At-speed testing (testing system functionality at its nominal working frequency) is a necessity for manufacturing test and validation of system performance. However, the ATE (automatic test equipment) speeds have always lagged behind the CUT (circuit under test) speed. As a result of this speed difference, the high-speed circuits are tested at clock rates that are much slower than their specification [17, 43]. Furthermore, the pin and probing limitations of ATEs restrict the accurate observation of test results. Therefore, an on-chip test mechanism such as BIST can fulfill requirements for at-speed testing of complex high-speed SoCs.

In a BIST architecture, a TPG (test pattern generation) circuit generates the pseudorandom patterns to stimulate possible defects in the CUT. The ORA (output response analyzer) circuit observes the outputs and analyzes their validity [1]. Fig. 4 demonstrates our basic BIST-based architecture to test the SoC's interconnects

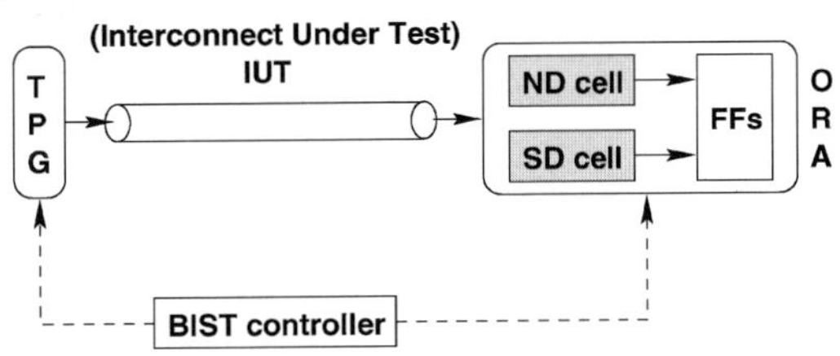

Fig. 4. Proposed BIST architecture.

for integrity. The TPG and ORA circuits are located in two sides of the IUT (interconnect under test). The IUTs could be long interconnects or those suspicious of having noise/skew violations due to environmental factors (e.g. crosstalk, electromagnetic effects, etc.).

Our rationale in using pseudorandom patterns for integrity testing is the fact that finding patterns that are guaranteed to create the worst case scenarios for integrity loss (e.g. noise and skew) is impractical with the current state of knowledge. This is mainly due to the complexity of distributed RLC interconnect model, parasitic values and too many influential factors. We elaborate on this in Section 3.

Integrity loss cannot be captured using conventional ORAs due to the complexity of interconnect behavior. Hence, a new ORA circuitry, which includes ND (noise detector) and SD (skew detector) cells will be used in this work. The ND and SD cells detect voltage and timing violations of the signal, respectively. We will elaborate on their structures and behaviors in Sections 4 and 5.

3. Test Pattern Generation

Researchers have searched for test patterns causing the worst case of signal integrity loss to enhance their test quality. In this section, we first present an analytical (but expensive) methodology to find deterministic patterns for maximal integrity loss. We then examine the inaccuracy of the RC model to conclude that the conventional TPG could be behaviorally effective and cost efficient.

3.1. *Deterministic Pattern Generation*

Coping with sophisticated *RLC* network leads to impractical simulation time for long interconnects. For an analytical approach, we propose to utilize the model order reduction strategy as an alternative for circuit simulators to improve the simulation run time significantly. Model order reduction methods were developed to approximate the behavior of long interconnect, power and clock networks. They have been extensively used in the simulation and evaluation of high-speed VLSI systems [15, 38, 45]. They comprise the key factors of the original system with much lower complexity; therefore, they significantly reduce the required computation and thereby simulation time, with slight loss of accuracy.

Elmore delay model [13], as the first reduced order model, is the most common technique for approximating the delay of RC networks. Asymptotic Waveform Evaluation (AWE) [38] was introduced as another method based on moment-matching allowing a linear circuit to be analyzed for its dominant poles and corresponding residues. To overcome some numerical limitations that AWE method suffers, researchers have developed reduction methods based on Bi-orthogonalization algorithms such as Pade Via Lanczos (PVL) [15] or orthogonalized Krylov subspace methods [36], which are computationally convenient and numerically better behaved. Their experimental results reveal the high accuracy within less than 5% of SPICE simulator at the speed 1000 times faster.

For analytical computation, the reduced order model can be used in generating test patterns for maximal integrity loss on long interconnects deterministically. The details are beyond the scope of this paper and can be found in [3]. Briefly, the input to an interconnect network, $i(t)$, can be expressed as: $i(t) = a_i u(t) + b_i$, where $u(t)$ is a step function, $a_i \in \{-1, 0, 1\}$ and $b_i \in \{0, 1\}$ (if $|a_i| = 1$ then $b_i = 0$).

An interconnect network can be represented by a set of transfer functions in the S-domain. As shown in Fig. 5, the interconnect network with n inputs and m outputs is represented by $H(S)$, which contains $n \times m$ transfer functions relating n inputs and m outputs. Note that in general we have $m \geq n$ for the possibility of fanout on some wires. Using an order reduction method (e.g. PVL [15] or ENOR [45] methods), the transfer function of a specific output f_r $(1 \leq r \leq m)$ becomes of order q:

$$h_{f_r}(s) = \sum_{i=1}^{n} \sum_{j=1}^{q} \frac{k_{ij}}{(s - p_{ij})} \tag{1}$$

where k_{ij} and p_{ij} are residues and poles of the output f_r, respectively. Note that $h_{f_r}(s)$ is obtained based on

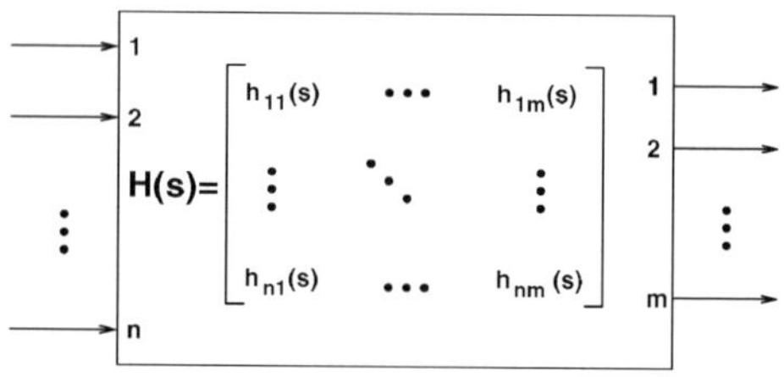

Fig. 5. Transfer function of an interconnect network.

the superposition of the effects of all n inputs on f_r. The output of system can be computed in the frequency domain by multiplying the transfer function and an input function. After transferring it to time domain to observe the timing behavior of the output, $f_r(t)$ is calculated:

$$f_r(t) = \sum_{i=1}^{n} \left(a_i \sum_{j=1}^{q} \frac{k_{ij}}{p_{ij}} (\exp(p_{ij}t) - 1) + b_i \sum_{j=1}^{q} k_{ij} \exp(p_{ij}t) \right) \quad \text{for } t > 0 \qquad (2)$$

By Eq. (2), all the timing information of the output signal is available and the impact of inputs are considered within a_i and b_i coefficients. Therefore, theoretically we are able to analyze the output for any possible indication of signal integrity loss such as delay, overshoot, or undershoot. For example, maximal delay due to integrity loss can be obtained by equating the Eq. (2) to $0.5V_{dd}$ and using numerical method to solve it for t. The reduced order model is much faster than SPICE simulation but this solution is still too expensive with the current state of knowledge of optimization, numerical methods and computers. For proof of concept, we have selected a small example of two parallel interconnects of five RLC segments each. We used the ENOR algorithm [45] to obtain the residues and poles of the 5th order reduced form of Eq. (2). Two experimentations were done. First, we used SPICE [50] to exhaustively check all patterns for maximal delay. Second, ENOR was implemented using MATLAB [30]. After obtaining the poles and residues of the reduced order model, the information is used to solve the Eq. (2) to calculate the delay and overshoot in the MATLAB environment.

The maximum delay and peak overshoot (O.S.) are reported in Table 1 for three arbitrary transitions. As

Table 1. Comparing SPICE and our reduced method.

Test patterns	SPICE		Order reduction	
	Delay (ps)	O.S. (Volt)	Delay (ps)	O.S. (Volt)
00 → 01	32.43	2.53	31.97	2.59
00 → 11	39.98	2.65	38.62	2.64
01 → 10	44.65	2.89	45.06	2.91

this table shows, our method identifies the same test patterns to stimulate the maximal integrity loss as SPICE finds but runs much faster. The running time for each run for SPICE and our order reduction method are 410 and 3 seconds (wall clock time on SPARC ULTRA 10 with 64 MByte RAM), respectively. While our analytical method is very accurate and is efficient for small examples, we acknowledge that finding patterns for large set of interconnects is very computational intensive using this approach unless more efficient numerical methods to solve Eq. (2) is found. This is the main reason that we used pseudorandom patterns, generated by conventional TPG, in our approach.

3.2. *Inaccuracy of Single-Victim (RC) Model*

In [4] and [10], the worst case test patterns associated with a specific fault model (MAFM) were presented. They used the RC interconnect for test pattern generation. Techniques, such as the one reported in [4], apply identical transitions to all wires except the victim net to create maximal integrity loss in the victim wire. Six scenarios and typical test patterns used in such techniques (e.g. [4]) have been shown in Fig. 6. Although this set of test patterns can test interconnects at lower

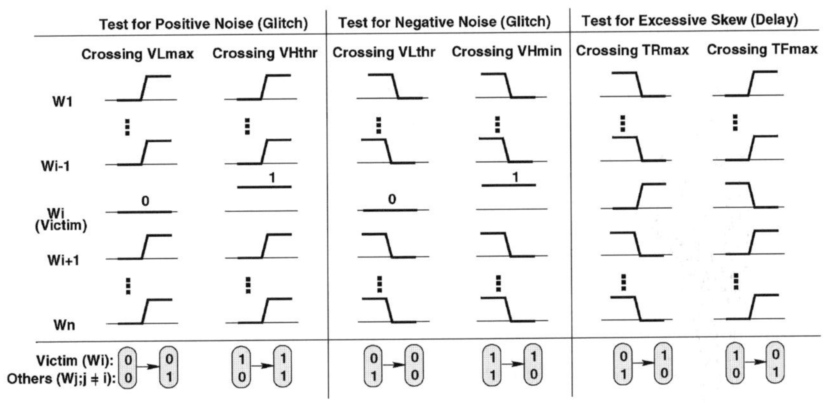

Fig. 6. Creating maximal integrity loss using RC model.

frequencies where inductance is negligible, it is inadequate for higher frequencies. Our empirical evidences show that in high frequency they fail to show the true picture of integrity of a signal.

For accurate analysis, having the coupling capacitances between all wires in a distributed model (e.g. Fig. 1) is necessary but not sufficient. The effect of capacitive coupling is considered local, in the sense that the coupling effects of adjacent wires are quite dominant compared to the capacitive coupling effects of far off wires [7, 21]. However, the inductance has larger range effect and thus the effect of mutual inductance could be significant. As discussed in [7], the effect of coupling inductances and capacitances on a wire oppose each other. When the signal on a wire switches in one direction, the noise due to capacitive coupling affects other nearby signals in the same direction as that of switching while the noise due to inductive coupling is in the opposite direction. All of these make the patterns generated by RC models inaccurate and inadequate.

In what follows, we present three other experiments to show the deficiency of RC model and the corresponding uniform patterns. These examples clearly show that there are scenarios for test patterns that create worse delay and/or noise on the signal and cause more integrity loss compared to those commonly reported earlier [4, 7]. Thus, RLC model of interconnect needs to be consolidated for test pattern generation. We have used an RLC model of seven parallel interconnect lines (indexed "7654321") for our experimentation. The R, L and C values are extracted using accurate extraction tools [37]. TISPICE has been used to simulate the complete RLC model. Typical CMOS gates are considered as driver and driven gates in two sides of the interconnect and $V_{dd} = 1.2$ Volt. Note that in the waveforms of these examples *RC* refers to the test pattern (similar to those in Fig. 6) suggested by approaches that use RC interconnect modeling. We call them RC-patterns for short. Note that many random patterns stimulate integrity loss (noise) more than limited RC-patterns. We will show some statistics in Section 7.

3.2.1. Example 1: Maximal Delay. Test pattern pair 1111011 $\rightarrow$ 0000100 is proposed for observing the highest potential delay on line 3 based on RC model. As shown in Fig. 7, test pattern pair 1100000 $\rightarrow$ 0011111 can generate 31% more delay than the RC patterns.

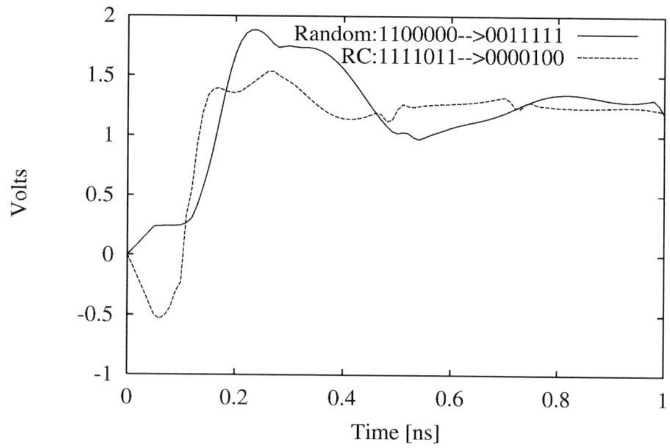

Fig. 7. Comparing patterns for the worst case delay.

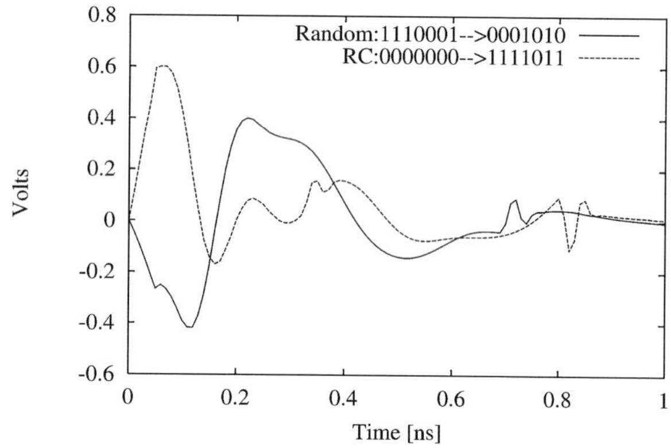

Fig. 8. Comparing patterns for the maximal noise.

3.2.2. Example 2: Maximal Noise. In Fig. 8, the difference between the worst case test pattern for RC model and another test data is demonstrated. In this case, the glitch on quiescent line 3 at 0 is investigated. This example shows that the worst case patterns (in terms of peak and duration of noise) for RC model are not necessarily the worst case for RLC model.

3.2.3. Example 3: Mutual Effects. We believe categorizing lines as victim and aggressors is misleading and unrealistic. Every line (including so-called aggressors) can be affected by all other lines (including so-called victims). Although the overall effect of aggressors on victims may be larger than effect of victims on aggressors, the change on the other lines cannot be ignored. Such minor effects may cause different switching times for aggressors which eventually results in longer settling delay up to 51% as reported in [7]. Fig. 9 shows an example in which line 5 is quiescent at 1 and lines 4, 6 and 7 make 1 $\rightarrow$ 0 transitions. As shown in

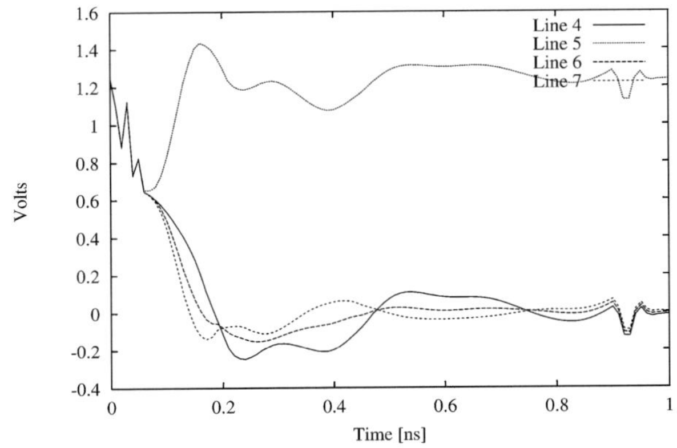

Fig. 9. Mutual effects of wires.

the figure, they affect each other due to the inductive and capacitive couplings.

In conclusion, due to the complexity of accurate RLC interconnect model, parasitic values and too many influential factors, finding patterns guaranteed to create the worst case scenarios for noise (integrity loss) is very much difficult and almost impractical with the current state of knowledge. Our empirical evidences indicate that random patterns are more qualified than those conjectured (e.g. RC patterns in Fig. 6) to create the worst case integrity loss. Thus, we propose to use conventional TPGs to generate pseudorandom test patterns as an efficient way to test the high-speed interconnects. We will elaborate on this issue in Section 7.

4. Detecting Noise Violation

A modified cross-coupled PMOS differential sense amplifier is designed to detect integrity loss (noise) relative to voltage violations. Fig. 10 shows the noise detector (ND) cell, which sits physically near the receiving core and samples the actual signal plus noise received by *Core j*. NMOS transistor T_5 is the current source (when $SE = 1$) and PMOS transistors T_3 and T_4 are loads. The positive feedbacks (drain-gate connection between T_3 and T_4) allow amplification in this structure. *SE* is connected to *test_mode* to create a permanent current source in the test mode and input $\bar{x}$ is connected to V_{dd} to define the threshold level for sensing V_b, i.e. the voltage received in x. The inverter, formed by T_6 and T_7, stabilizes the voltage levels in the output of ND cell. By adjusting the size of the PMOS transistors (i.e. W and L), the current through transistors T_1 and T_2 are set to different values. Combining this with the feedbacks between PMOS transistors creates threshold voltages to turn the transistors on or off. Various architectures for sense amplifiers and in-depth details can be found in [19, 39].

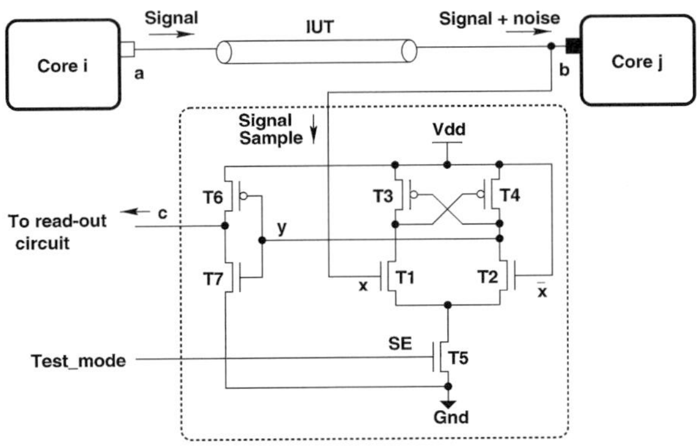

Fig. 10. The ND cell using cross-coupled PMOS amplifier.

Fig. 11 shows signals on the input and output (points b and c) of the cell to validate the behavior of our noise detector cell. Each time that noise occurs (i.e. $V_b > V_+ = V_{Hthr}$), the ND cell generates a "0" signal that remains unchanged until V_b drops below $V_- = V_{Hmin}$. The waveforms in Fig. 11 reflect that the ND cell shows a *hysteresis* (Schmitt-trigger) property which implicitly indicates a (temporary) storage behavior. To confirm this we ran a DC analysis on the ND cell to get the hysteresis curve shown in Fig. 12. This property helps to detect the violation of two threshold voltages (i.e. V_{Hthr} and V_{Hmin}) with the same ND cell. For example, the solid-line curve shows that the switching threshold voltages are $V_+ = V_{Hthr} = 2.75$ and $V_- = V_{Hmin} = 1.60$ when $V_{dd} = 2.5$. A similar cell can be designed to detect crossing V_{Lthr} and V_{Lmax} threshold voltages. Details can be found in [35].

What level of overshoot is acceptable, and what level of voltage should be recognized as a logic "1" and "0" is debatable. Specific choice of V_{Hthr} and V_{Lthr} (and also V_{Hmin} and V_{Lmax}) depends on the technology, and on the desired level of reliability. A nice feature of our ND cell is that for any V_{dd}, the two thresholds (i.e. V_+ and V_- of hysteresis) can be tuned by changing the layout size of the PMOS transistors (mainly W's of T_3 and T_4). This is also shown in Fig. 12, in which different sets of transistor widths (*W-set* 1 through *W-set* 4 for T_3 and T_4) and two V_{dd} values (3.3 and 2.5 volts) have been used. There are analytical- and simulation-based approaches that can be used for such tuning [39]. Empirically, however, for a given V_{dd} we found tuning of V_+ and V_- to be easy and

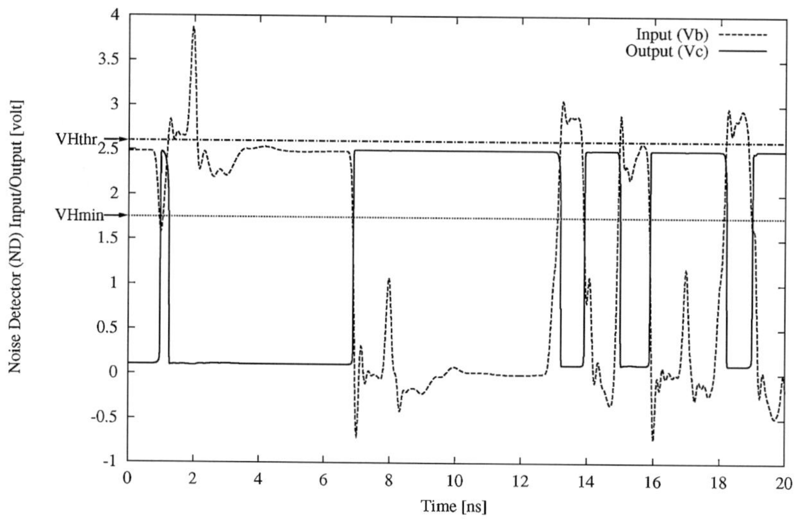

Fig. 11. The SPICE simulation of the ND cell for the top vulnerable region.

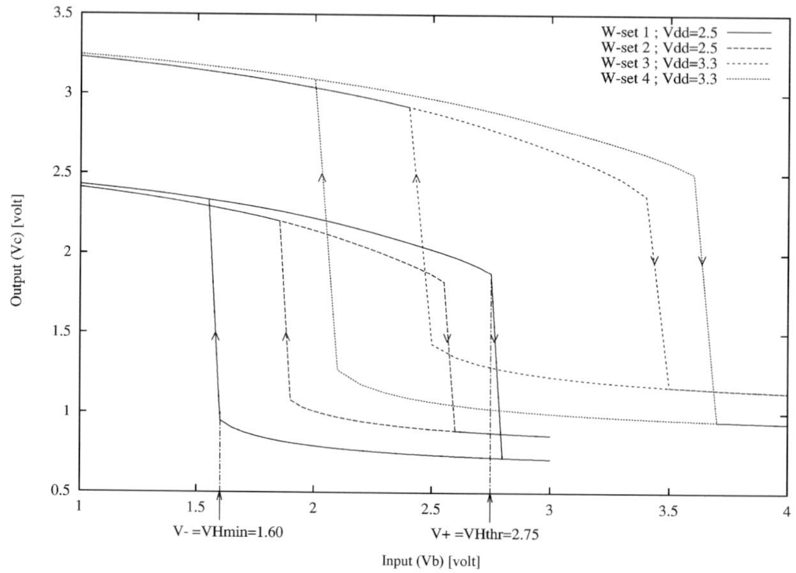

Fig. 12. The hysteresis property of our noise detector cell.

flexible within a relatively large range as shown in Fig. 12.

Note carefully that occurrence of an overshoot puts the ND cell in *state*-0 (i.e. generating "0") after which a 1-0-1 glitch that violates V_{Hmin} can be detected. For detecting 1-0-1 glitches on a quiescent line, even without occurrence of overshoot, we need to tune the cell such that $V_{Hthr} = V_{dd}$. This forces the cell to go to state-0 for a quiescent logic-1 line and therefore make it capable of detecting 1-0-1 glitches. The drawback, however, is that the ND cell won't be able to have V_{Hthr} larger than V_{dd}. If both features are desired we need two cells one to detect overshoots with user-defined V_{Hthr} and one for any 1-0-1 glitch that may occur.

5. Detecting Skew Violation

As stated in Section 1, in addition to voltage distortion, timing violation (skew) is another important factor that contributes directly to the integrity loss. Specifically, in deep sub-micron technology the interconnect

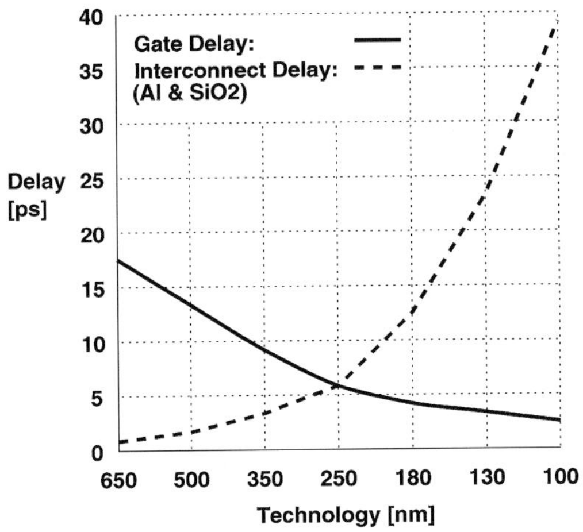

Fig. 13. Gate and interconnect delay versus technology.

delay is a detrimental factor. Fig. 13, presented in the SIA roadmap [43], shows the gate and interconnect delays versus technology generations. The curves clearly show the dominance of interconnect delay over gate delay as we approach 100 nm technology. Using copper (instead of aluminum) and low dielectric constant insulators can reduce the delay of interconnects. However, it is certain that the interconnect delay will remain the dominating factor [43]. This trend justifies the efforts needed to detect the delay violation due to the interconnect.

5.1. *Skew Immunity Range*

We defined the skew-immune (SI) region in Fig. 2 as the delay range that is considered "acceptable". If signal skew goes beyond that range we consider it as serious integrity loss. The range of immunity depends on the system topology, speed, core interactions and the yield margin that a designer defines to achieve the nominal performance of the system. Such yield margin is often considered in the design phase. However, the delay values of components as well as interconnects may not eventually be kept within such bound since the layout generation tools and the fabrication process each may add additional delays. More importantly, as we have discussed in [35] signal integrity factor in general and noise/skew in particular are data-dependent phenomena that cannot be predicted accurately through analytical or simulation based approaches. Thus, similar to the noise detection only an on-chip methodology can successfully test the delay violation.

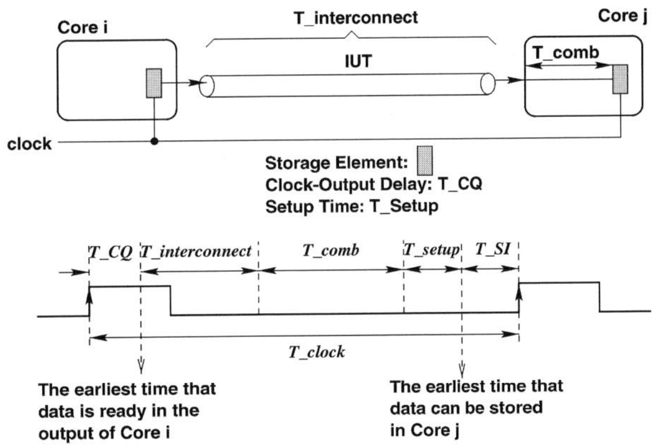

Fig. 14. The effect of interconnect skew (delay).

In synchronous systems, maximum communication speed between two interacting cores depends on the maximum storage-to-storage (s-to-s) path delay [52]. Fig. 14 shows this graphically. As long as $T_{clock} > T_{interconnect} + T_{comb\_max} + T_{CQ} + T_{setup}$ the functionality of system will remain error-free. For simplicity, we did not differentiate between rising and falling skews. Thus, the skew immune range T_{SI} can be estimated based on the clock period, timing behavior of the storage elements and storage-to-storage paths within the system:

$$T_{SI} \approx \max_{s\text{-}to\text{-}s}\{T_{clock} - (T_{comb\_max} + T_{CQ} + T_{setup})\}.$$

5.2. *Skew Detector (SD) Cell*

In designing high speed cores, very firm delay budget has to be met. Any minor violation of skew or unpredictable delay (e.g., the interconnect effect) may cause functional error or significant degradation of performance. Numerous research endeavors have been dedicated to the testing of logic gates for their timing behavior [29, 46]. Since the interconnect skew delay will be the dominant factor in determining the clock period of future technologies, it is essential to detect the skew violation on the interconnects.

Detecting the skew violations of interconnects can unlikely be fulfilled off-chip due to the speed limitations of ATEs. Nevertheless, the features of an on-chip test mechanism such as BIST can be utilized to observe the skew violations accurately. Fig. 15 depicts the proposed on-chip test circuitry (SD cell). A delay generator cell is used to create the desired delay value (i.e. acceptable skew-immune range T_{SI}) as it is defined by a designer based on the delay budget of the

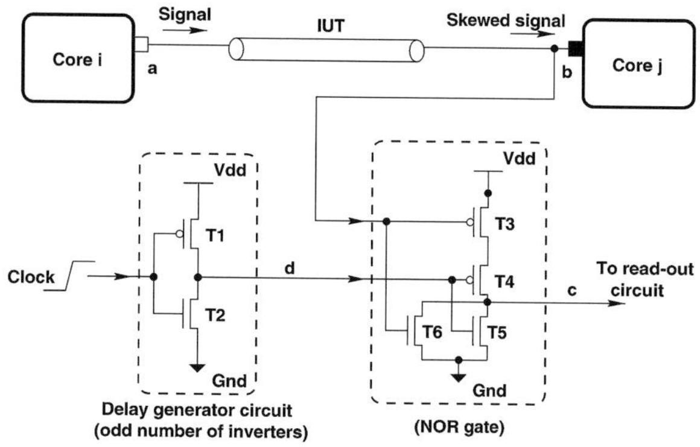

Fig. 15. Skew detector circuitry (SD cell).

interconnect. This cell is essentially made of odd number of cascaded CMOS inverters that receives the system clock and outputs the delayed inverted clock. The delayed clock is compared with the interconnect output. If the skew of the signal on the interconnect output is not within the acceptable range, the SD cell issues a pulse. The duration of this pulse depends on the interconnect delay. From testing point of view, the pulse generated by the SD cell can be used as the indication of skew violation. For example, it can trigger a D flip-flop to store a "1" as indication of such occurrence.

The SPICE [50] simulation of the SD cell has been illustrated in Fig. 16 for two signals. The first signal (*SIGNAL1*) does not violate the skew limit (T_{SI}) and thus the output of the cell remains zero. The second signal (*SIGNAL2*), however, exceeds the acceptable skew and the SD cell generates a pulse. As shown in the figure, this pulse appears in the output of the SD cell after delay associated with the NOR gate. Notice that possible clock skew does not have significant effect on the behavior of the SD cell. There are many approaches such as [28, 48] to minimize the clock skew in a large chip. However, in the presence of clock skew T_{SI} changes to $T_{SI} + T_{clock\_skew}$ where the T_{clock_skew} is the skew on the clock wire attached to the SD cell. This change increases the acceptable skew range to which the cell reacts but the functionality of the cell remains intact.

To illustrate the true behavior of the SD cell in test mode, while connected to an interconnect, we ran the SPICE simulation on the integrated interconnect and test circuitry. The results are shown in Fig. 17. If the output of interconnect does meet the skew requirements (less than $T_{SI} = 100$ ps as happens for *SIGNAL1*), the output of the SD cell stays at zero. However, a marginally late signal (*SIGNAL2*) causes a pulse at the output of the SD cell.

5.3. Delay Generator Circuit

As discussed before the accuracy of the SD cell depends on accurate implementation of delay generator block. This circuit generates a specific delay value (e.g. T_{SI}), predefined by the designer. The implementation of such accurate delay circuitry is a challenging issue. We have employed the classic driver design [2, 54] to implement the delay generator circuit. Many researchers have

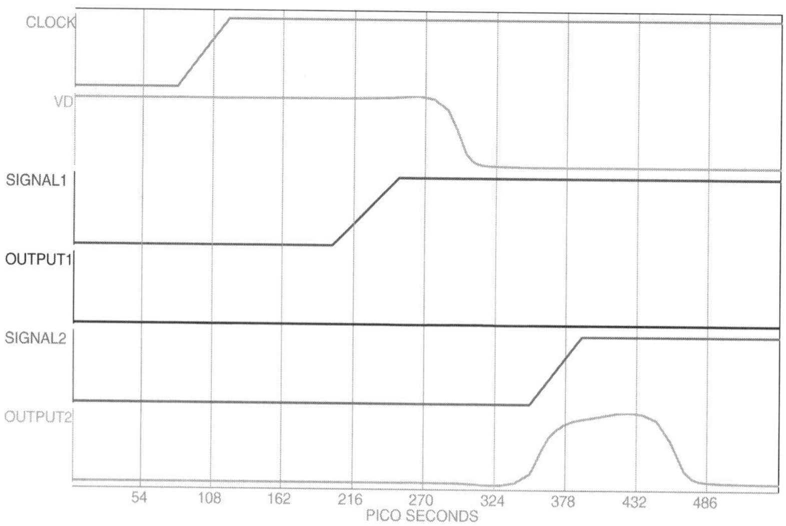

Fig. 16. The SPICE simulation of SD cell for two signals.

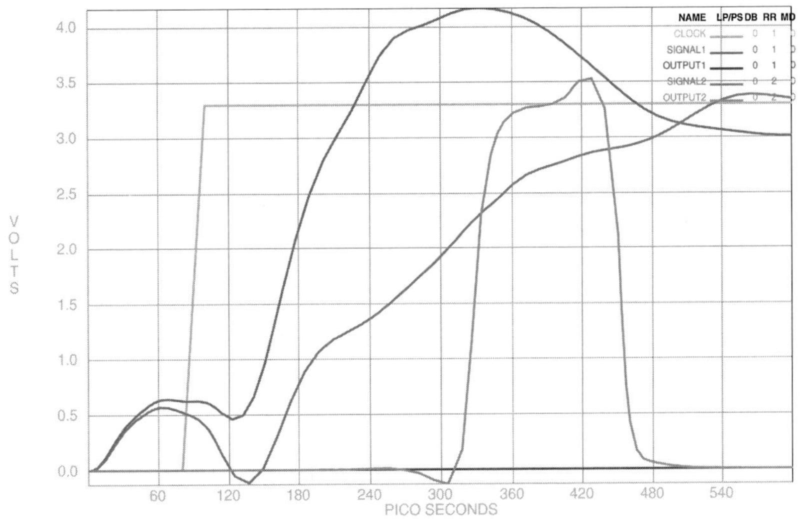

Fig. 17. The SPICE simulation of SD cell connected to an interconnect.

extensively investigated how to achieve the optimal driver for long VLSI interconnects considering different objectives such as minimum delay, area or power consumption [28, 54]. For instance, several design parameters, such as the number of drivers and their aspect ratio, are adjusted using optimization techniques to obtain an optimal driver configuration. Similarly, obtaining a specific delay value for a driver, while minimizing the overall area can be another objective in driver design problem. This can be the goal in optimizing the delay generator circuit. Such optimization technique, although important in our application, is beyond the scope of this paper. We simply comment that optimizing delay generator circuit can be done systematically using *optimal driver design* concept [48, 54].

As for experimentation, we designed few delay generator circuits to show typical values for T_{SI}. Table 2 shows the specifications of four different delay generators. We used cascaded inverters with different size transistors to design and tune them. As seen in this table, the variety of delay values (i.e. T_{SI}) can be created using the inverters with reasonable area.

Table 2. Statistics for four delay generator circuits.

Design	No. of Inverters	Area (μm^2)	Delay (ps)
#1	1	6	42
#2	1	2	87
#3	3	10.5	250
#4	3	15	410

As Fig. 17 shows, the SD cell detects the skew violations by creating a pulse in its output momentarily. One efficient way to capture such violation is to use this pulse to trigger a flip-flop as shown in Fig. 18(a). This circuit stores a "clock = 1" in the flip-flop when delay violations (in range of $T_{SI} < delay < T/2$) occur. Note that this is a sufficient range in our application. Here our concern is small but unacceptable delays due to integrity loss. Large delays can be always captured using functional or delay testing.

The clock signal (instead of a permanent "1") is intentionally connected to the input of D flip-flop to filter out pulses on c for some undesired scenarios. One such scenario, e.g. having a pulse when skew is in the immune (acceptable) range, is shown in Fig. 18(d). Depending on the application, accuracy and test patterns one SD cell to detect skew violations on rising signals (Fig. 18(a)) is often sufficient. However, if distinction between skew violations of rising and falling signals is needed, e.g. for diagnosis purposes, we need additional hardware. The reader can easily verify that the additional cell is the same as Fig. 18(a) but uses $b \cdot \bar{d}$ instead of $\overline{b + d}$ and can detect skew violations on falling signals.

6. Test Architecture

Detecting signals that leave the noise-immune (NI) and skew-immune (SI) regions is a crucial step. This is performed by the ND and SD cells explained in the

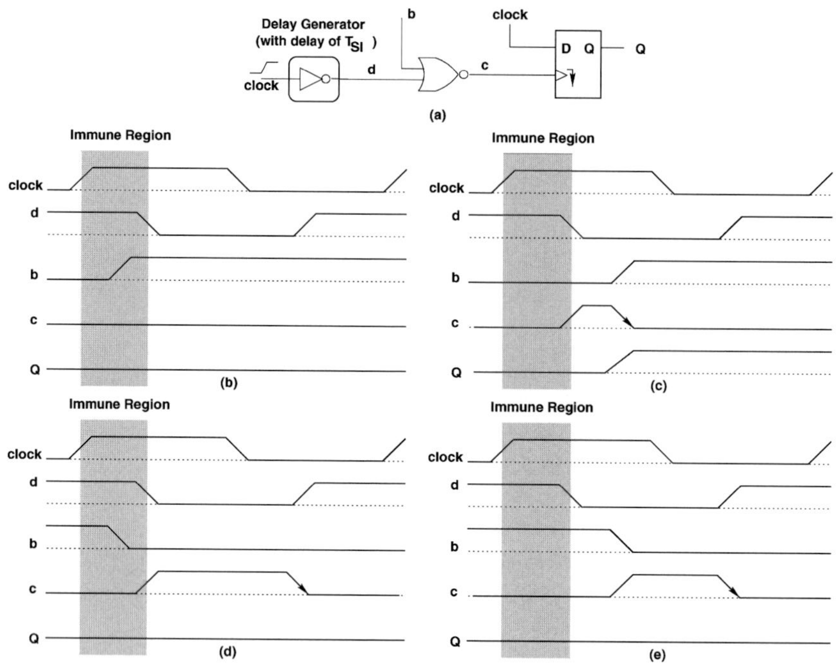

Fig. 18. Capturing skew violations by the SD cell.

previous sections. These cells are not expensive—seven and six transistors per ND and SD cells, respectively. Overall two cells are needed per interconnect to detect noise (crossing V_{Hthr} and V_{Hmin}) and skew violations. The test architecture to read out the information stored in these cells is a DFT decision which depends on the overall SoC test methodology, testing objective and cost consideration.

Fig. 19 demonstrates a scan flip-flop chain architecture, which is able to record the occurrence of noise/skew violation and transfer it to the output. In the test-mode, first the *flag* signal is transferred, through MUX, to the test controller. If noise or skew violation (low integrity signal) occurs ($flag = 1$), the content of flip-flops (ND-FFs and SD-FFs) are scanned out through S_{out} for further reliability and diagnosis analysis. Suppose an n-bit interconnect is under test for m cycles (i.e. m pseudorandom test patterns). The very pessimistic worst case scenario in terms of test time is a case in which all lines are subject to noise in all m test cycles. This situation requires overall m and $m \cdot n$ cycles for response capture and readout, respectively. In practice, a much shorter time (e.g. $k \cdot n$, where $k \ll m$) is sufficient since the presence of defects or environmental factors causing unacceptable level of noise/skew (integrity loss) is quite limited. In terms of test overhead, n ND cells, n SD cells and $2n$ scan flip-flops are needed.

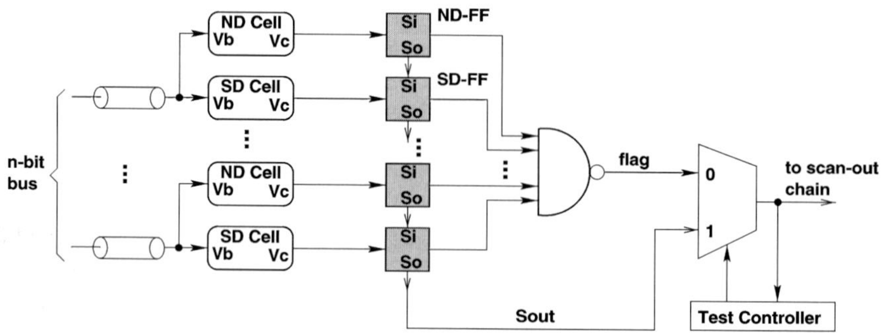

Fig. 19. Test architecture.

Two other alternatives for the test architectures using compressor/adder and dedicated counters are introduced in [35] and analyzed for their cost and test time. Briefly, these alternatives are more expensive but can supply data of integrity loss occurrences more accurately for applications such as interconnect diagnosis, re-wiring and pad re-distribution.

7. Simulation Results

The number of cores in the SoC and the number of input ports of cores do not influence the low-integrity (noise/skew) detection process since the ND and SD cells independently function near core input ports. They do, however, influence the cost of test overhead (e.g. cells and FFs) and test time (e.g. scan-out time).

The experimental results here are reported using OEA [37] and SPICE [50] simulators. We analyzed five main buses (data, address, control and two internal) of the famous 8051 microprocessor [23]. In our implementation the 7 cores communicate through these buses and are potentially subject to noise in high frequency. For experimentation purpose, we used the interconnect architecture of 8051 assuming that it runs in 1 GHz. Typical global interconnect lengths in large SoC systems are chosen as the wire lengths in our experiments. Then, we have applied random patterns to the interconnects assuming that they run under 1 GHz frequency. The statistics are summarized in Table 3. As shown in Table 3, the average occurrence of unacceptable noise and skew (low integrity signals) in a presumed 1 GHz 8051 system will be 32.44% and 22.46% that may create functional error or cause severe damages (e.g. reliability, lifetime) on chip over time.

Table 4 summarizes the test overhead for the buses in 8051 reported by SYNOPSYS design compiler toolset [47] when 100 random patterns are applied. We assumed that all interconnect lines need to be tested for integrity. All costs are expressed in terms of 2-input

Table 3. Integrity test results for 8051 bus structure.

Buses	Bitwidth	Length (mm)	Noise (%)	Skew (%)
Data	8	20	43.38	31.32
Address	16	5	23.28	15.78
Control	10	10	28.50	25.53
Internal 1	10	10	34.40	21.20
Internal 2	8	10	32.62	18.47
Average	10.40	11.00	32.44	22.46

Table 4. Test overhead for 8051 bus structure.

Overhead	Data	Address	Control	Int. 1	Int. 2
Cost (NANDs)	158	254	194	194	158
Time (Cycle)	1600	3200	2000	2000	1600

Table 5. Random test patterns versus RC test patterns.

Integrity factor	Quantity		Quality		
	RC	Random	Metric	RC	Random
Mutual effects	42	100	Settle(ps) Peak(V)	61 $0.15\ V_{dd}$	117 $0.28\ V_{dd}$
Maximal delay	14	43	Delay(ps)	102	122
Maximal noise	14	51	Settle(ps) Peak(V)	211 $0.26\ V_{dd}$	328 $0.34\ V_{dd}$

NAND gates. The readout time overhead results are also included in this table.

Table 5 compares the quantity and quality of random test patterns versus the RC patterns for a 7-wire parallel interconnect line. The overall number of patterns used in this experiment are 42 (6 patterns per wire as shown in Fig. 6) and 100 for RC and random cases, respectively. The second and third columns show the number of patterns that show significant integrity loss (i.e. overshoot or delay at least 15% larger than their nominal values) from different perspectives. Although in general, the number of the RC patterns is smaller than the random patterns, the latter significantly improves the quality of interconnect testing by stimulating larger integrity loss (the worst case scenarios). For example, when we average the results of applying all patterns (first row), the random patterns cause larger integrity loss (in terms of peak noise and settle time) than the RC patterns. As for maximal delay on the interconnects, 43 (out of 100) and 14 (out of 42) patterns were counted for random and RC cases, respectively, causing significant delay. The average delay caused by these patterns are 122 ps versus 102 ps. Similar trend exists for patterns that stimulate maximal peak/duration of noise.

7.1. *The Effect of Process Variation*

Variation of different factors in the fabrication process may cause considerable deviation from the nominal or expected behavior of the circuit. This may affect the integrity of signals traveling on the interconnect and also the functionality of the ND/SD cells. For example, due to the limited resolution of the photolitographic process, the transistor dimensions (W, L), on die may be different from the ideal dimensions expected. Other process parameters include transconductance (κ), threshold voltage (V_{th}), impurity concentration on densities, oxide thicknesses and diffusion depths [39]. In this section we only show our simulation results on the sensitivity of the interconnect and cells with respect to different process variation factors. In-depth analysis of process variation is beyond the scope of this paper.

In all simulation results reported here, a normal distribution with $\sigma = 0.06\mu$ is used to generate variation on such factors. This means that the change of the nominal value will remain in $[\mu - 3\sigma, \mu + 3\sigma]$ range [11]. The average value (μ) is the result of simulation without any variation of parameters. The value for variance (σ) is selected to keep the random parameter values between a lower and upper bound. In practice, such bounds depend on the statistical data extracted for each fabrication plant [31]. The CODAC [49] and TISPICE [50] tools are used in a Monte-Carlo simulation environment to model and simulate the repercussion of process variations on the interconnects or cells. In each iteration the interconnect or cells is simulated using a randomly-chosen value for that specific process factor.

7.1.1. The Effect on the Interconnect. To examine the effects of process variations on the long interconnects parameters such as thickness, width, interlayer dielectric thickness, and contact resistance in fabrication process are taken into account as variances on the distributed R, L, and C values of the interconnect model.

Table 6 shows the adverse effects of process variation on long on-chip interconnects when $R/L/C$ values are varied randomly using normal distribution. E obtained in the simulation expresses the percentage of violations (overshoot, undershoot or skew that are 20% larger than their nominal values) occurred on the interconnects. For example, 24.7% of the inputs cause significant overshoots, which reduces chip reliability.

Table 6. The effect of process variations on the interconnect.

Parameters	Nominal (μ)	E (%)
Peak overshoot (volt)	1.62	24.7
Peak undershoot (volt)	0.95	6.7
Skew delay (ns)	0.4	4.7

Table 7. The effect of process variations on the ND cell.

Parameters	Nominal (μ)	E (%)
Width ($W_{T1}, \ldots, W_{T7}$) (μm)	4, 4, 8, 7, 4, 1.2, 4	6.2
Length (μm)	0.5	11.0
κ (NMOS, PMOS)	(0.161, 9.366)	2.1
V_{th} (NMOS, PMOS) (volt)	(0.6684, −0.9352)	15.1

Many researchers showed the importance of detecting skew and voltage violations of long interconnects on chips [32, 33]. Our simulation results also confirm this fact. Due to its probabilistic and environment-dependent nature, the process variation cannot be considered or modeled in the design phase. Thus, after fabrication a significant percentage of overshoots, violation of noise margin or settling time may appear on long interconnects. These situations can be only tested using an on-chip approach such as ours.

7.1.2. The Effect on the ND Cell. To show the sensitivity of the ND cell with respect to the process variation, we have simulated the ND cell with variations of the different parameters. Using CODAC and SPICE, we trace the effect of variation of width (W), length (L), transconductance (κ), and threshold voltage (V_{th}) on the behavior of ND cell. Table 7 demonstrates the results. The last column in this table (E) shows the percentage of unacceptable outputs compared to the output of ND cell under no process variations for different parameters. As reflected in the table, the ND cell can quite adequately tolerate the variations on κ and length of transistors. However, the adverse effects of deviations of the threshold voltage and transistor width are larger.

7.1.3. The Effect on the SD Cell. In analyzing the effects of process variation on the SD cell, the delay generator cell (Fig. 15) is the key element that affects the overall accuracy of the cell. Monte-Carlo simulations were also carried out for the SD cell. Table 8 demonstrates the sensitivity of the SD cell in general, and its delay generator cell in particular, due to the process variations. As tabulated, the SD cell can adequately tolerate variations on width, κ, and threshold voltage. The percentage of unacceptable skewed signals is less than 5% for above factors. However, the variations on the length of SD cell seem to be larger.

Table 8. The effect of process variations on the SD cell.

Parameters	Nominal (μ)	E (%)
Width ($W_1, W_2, \ldots, W_6$) (μm)	1.5, 1, 2, 2, 1, 1	5.0
Length (μm)	0.5	17.5
κ (NMOS, PMOS)	(0.161, 9.366)	0.6
V_{th} (NMOS, PMOS) (volt)	(0.6684, −0.9352)	2.6

8. Conclusion

The rising level of complexity and frequency of chips makes it increasingly difficult to achieve an adequate interconnect test using the ad-hoc techniques currently practiced in industry. Signal integrity is often degraded as signal travels through the interconnect. Such integrity loss may lead to functional error and reliability loss. We proposed a systematic BIST-based methodology to model and test signal integrity in deep-submicron high-speed interconnects. On the test generation side, while deterministic patterns can be determined analytically, conventional cost-efficient pseudo-random pattern generator can stimulate the maximal integrity loss reasonably well. Using inexpensive built-in noise and skew detection cells we offered an efficient architecture to capture and scan out the occurrences of noise and skew violations.

Acknowledgments

This work was supported in part by the National Science Foundation CAREER Award #CCR-0130513. The authors also thank Nagaraj NS (Texas Instruments, Inc.) and Jerry Tallinger (OEA International, Inc.) for providing their simulator packages and helpful comments.

References

1. M. Abramovici, M. Breuer, and A. Friedman, *Digital Systems Testing and Testable design*, IEEE Press, 1990.
2. V. Adler and E. Friedman, "Repeater Design to Reduce Delay and Power in Resistive Interconnect," *IEEE Trans. on Circuit and Systems-II: Analog and Digital Signal Processing*, vol. 45, no. 5, pp. 607–616, May 1998.
3. A. Attarha and M. Nourani, "Test Pattern Generation for Signal Integrity Testing," Technical Report UTD-EE-2001-8-22, Department of Electrical Engineering, the University of Texas at Dallas, Aug. 2001.
4. X. Bai, S. Dey, and J. Rajski, "Self-Test Methodology for At-Speed Test of Crosstalk in Chip Interconnects," in *Proc. Design Automation Conf. (DAC'00)*, 2000, pp. 619–624.
5. W. Chen, S. Gupta, and M. Breuer, "Test Generation in VLSI Circuits for Crosstalk Noise," in *Proc. Intern. Test Conf. (ITC'98)*, 1998, pp. 641–650.
6. H. Chen and L. Wang, "Design for Signal Integrity: The New Paradigm for Deep-Submicron VLSI Design," in *Proc. Intern. Symposium on VLSI Technology*, June 1997, pp. 329–333.
7. C. Cheng, J. Lillis, S. Lin, and N. Chang, *Interconnect Analysis and Synthesis*, John Wiley & Sons, 2000.
8. D. Cho, Y. Eo, M. Seung, N. Kim, J. Wee, O. Kown, and H. Park, "Interconnect Capacitance, Crosstalk and Signal Delay for 0.35 μm CMOS Technology," in *Proc. Intern. Meeting on Electron Devices*, 1996, pp. 619–622.
9. J. Cong and L. He, "Theory and Algorithm of Local-Refinement-Based Optimization with Application to Device and Interconnect Sizing," *IEEE Transactions on Computer-Aided Design of Integrated Circuits and Systems*, vol. 18, no. 4, pp. 406–420, April 1999.
10. M. Cuviello, S. Dey, X. Bai, and Y. Zhao, "Fault Modeling and Simulation for Crosstalk in System-on-Chip Interconnects," in *Proc. Intern. Conf. on Computer Aided Design (ICCAD'99)*, Nov. 1999, pp. 297–303.
11. J. Devorce, *Probability and Statistics for Engineering and the Sciences*, John Wiley & Sons, 1999.
12. R. Downing, P. Gebler, and G. Katopis, "Decoupling Capacitor Effects on Switching Noise," *IEEE Transactions on Components, Hybrids and Manufacturing Technology*, vol. 16, no. 5, pp. 484–489, Aug. 1993.
13. W. Elmore, "The Transient Response of Damped Linear Networks With Particular Regard to Wideband Amplifiers," *Journal of Appl. Phys.*, vol. 19, no. 1, pp. 155–163, 1948.
14. P. Fang, J. Tao, J. Chen, and C. Hu, "Design in Hot Carrier Reliability for High Performance Logic Applications," in *Proc. IEEE Custom Integrated Circuits Conf.*, Oct. 1998, pp. 25.1.1–25.1.7.
15. P. Feldmann and R. Freund, "Efficient Linear Circuit Analysis by Pade Approximation via the Lanczos Process," *IEEE Trans. on CAD*, vol. 14, no. 5, pp. 639–649, May 1995.
16. P. Feldmann and R. Freund, "Small-Signal Circuit Analysis and Sensitivity Computations with the PVL Algorithm," *IEEE Trans. on Circuits and Systems*, vol. 43, no. 8, pp. 577–585, Aug. 1996.
17. A. Flint, "Testing Multichip Modules," *IEEE Spectrum*, pp. 59–62, March 1994.
18. K. Gala, V. Zolotov, R. Panda, B. Young, and D. Blaauw, "On-Chip Inductance Modeling and Analysis," in *Proc. Design Automation Conf. (DAC'00)*, 2000, pp. 63–68.
19. P. Gray and R. Meyer, *Analysis and Design of Analog Integrated Circuits*, John Wiley & Sons, 2000.
20. L. Green, "Simulation, Modeling and Understanding the Importance of Signal Integrity," *IEEE Circuit and Devices Magazine*, pp. 7–10, Nov. 1999.
21. L. He, N. Chang, S. Lin, and O. Nakagawa, "An Efficient Inductance Modeling for On-Chip Interconnects," in *Proc. IEEE Custom Integrated Circuits Conf.*, 1999, pp. 457–460.
22. W. Hunter, "The Statistical Dependence of Oxide Failure Rates on $V_{\rm dd}$ and $T_{\rm ox}$ Variations with Applications to Process Design, Circuit Design and End Use," in *Proc. Reliability Physics Symposium*, pp. 72–81.
23. Intel Corporation, "MCS-51 8-bit Microcontroller," *Databook of Intel MCS-51 Microcontroller*, 1994.

24. Y. Jiang, S. Sapatnekar, C. Bamji, and J. Kim, "Interleaving Buffer Insertion and Transistor Resizing into a Single Optimization," *IEEE Trans. on VLSI Systems*, vol. 6, no. 4, pp. 625–633, Dec. 1998.
25. A. Kahng, S. Muddu, and E. Sarto, "Interconnect Optimization Strategies for High-Performance VLSI Designs," in *Proc. Intern. Conf. on VLSI Design*, Aug. 1999, pp. 464–469.
26. S. Kimothi and U. Nandwani, "Uncertainty Considerations in Compliance-Testing for Electromagnetic Interference," in *Proc. Annual Reliability and Maintainability Symposium*, 1999, pp. 265–268.
27. Y. Leblebici, "Design Considerations for CMOS Digital Circuits with Improved Hot-Carrier Reliability," *IEEE Journal of Solid-State Circuits*, vol. 31, no. 7, pp. 1014–1024, July 1996.
28. J. Lillis, C. Cheng, and T. Lin, "Optimal Wire Sizing and Buffer Insertion for Low Power and a Generalized Delay Model," *IEEE Journal of Sold-State Circuits*, vol. 31, no. 3, pp. 437–447, March 1996.
29. A. Majhi and V. Agrawal, "Tutorial: Delay Fault Models and Coverage," in *Proc. 11th Internations Conf. on VLSI Design*, 1998, pp. 364–369.
30. MATLAB User's and Reference Manual, 2000 The MathWorks, Inc., Version 6.0.0.88, 2000.
31. MOSIS Organization, USC Information Sciences Institute. Webpage URL address: *http://www.mosis.org*.
32. N.S. Nagaraj, P. Balsara, and C. Cantrell, "Crosstalk Noise Verification in Digital Designs with Interconnect Process Variations," in *Proc. 14th Intern. Conf. on VLSI Design*, 2001, pp. 365–370.
33. S. Natarajan, M.A. Breuer, and S.K. Gupta, "Process Variations and Their Impact on Circuit Operation," in *Proc. IEEE Intern. Symp. on Defect and Fault Tolerance in VLSI Systems*, 1998, pp. 73–81.
34. P. Nordholz, D. Treytnar, J. Otterstedt, H. Grabinski, D. Niggemeyer, and T. Williams, "Signal Integrity Problems in Deep Submicron Arising from Interconnects Between Cores," in *Proc. VLSI Test Symp. (VTS'98)*, 1998, pp. 28–33.
35. M. Nourani and A. Attarha, "Built-in Self-Test for Signal Integrity," in *Proc. Design Automation Conf. (DAC'01)*, June 2001, pp. 792–797.
36. A. Odabasioglu, M. Celik, and L. Pileggi, "PRIMA: Passive Reduced-Order Interconnect Macromodeling Algorithm, " *IEEE Conf. on CAD* , 1997.
37. OEA Interconnect Simulators, *User's and Reference Manual*, OEA International, Inc., 2001.
38. L. Pillage and R. Rohrer, "Asymptotic Waveform Evaluation for Timing Analysis," *IEEE Trans. on CAD*, vol. 9, no 4, pp. 352–366, April 1990.
39. J. Rabaey, *Digital Integrated Circuits*, Prentice Hall, 1996.
40. I. Rayane, J. Velasco-Medina, and M. Nicolaidis, "A Digital BIST for Operational Amplifiers Embedded in Mixed-Signal Circuits," in *Proc. VLSI Test Symp. (VTS'99)*, 1999, pp. 304–310.
41. P. Restle, A. Tuehli, and S. Walker, "Dealing with Inductance in High-Speed Chip Design," in *Proc. Design Automation Conf. (DAC'99)*, 1999, pp. 904–909.
42. R. Saleh, D. Overhauser, and S. Taylor, "Full-Chip Verification of UDSM Designs," in *Proc. Intern. Conf. on Computer Aided Design (ICCAD'98)*, Nov. 1998, pp. 453–460.
43. Semiconductor Industry Association, *The International Technology Roadmap for Semiconductors*, Sematech Inc., 1999.
44. B. Sheehan, "Projective Convolution: RLC Model-Order Reduction Using the Impulse Response," in *Proc. Design Automation Conf. (DAC'99)*, 1999, pp. 669–673.
45. B. Sheehan, "ENOR: Model Order Reduction of RLC Circuits Using Nodal Equations for Efficient Factorization," *Design Automation Conference*, 1999, pp. 17–21.
46. U. Sparmann, H. Muller, and S. Reddy, "Universal Delay Test Sets for Logic Networks," *IEEE Trans. on VLSI Systems*, vol. 7, no. 2, pp. 156–166, Nov. 1999.
47. Synopsys, Synopsys Design Analyzer, *User Manuals for SYNOPSYS Toolset Version 2000.05-1*, Synopsys, Inc., 2000.
48. G. Tellez and M. Sarrafzadeh, "Minimal Buffer Insertion in Clock Trees with Skew and Slew Rate Constraints," *IEEE Trans. on CAD*, vol. 16, no. 4, pp. 333–342, April 1997.
49. Texas Instrument Inc., *CODAC User's and Reference Manual*, Texas Instrument Incorporation, 2000.
50. Texas Instrument Incorporation, *TI-SPICE3 User's and Reference Manual*, 2000.
51. H. Veendrick, *Deep-Submicron CMOS ICs: From Basics to Asics*, Kluwer, 1998.
52. J. Wakerly, *Digital Design, Principles and Practices*, Prentice Hall, 2000.
53. S. Zhao and K. Roy, "Estimation of Switching Noise on Power Supply Lines in Deep Sub-Micron CMOS Circuits," in *Proc. Intern. Conf. on VLSI Design*, Jan. 2000, pp. 168–173.
54. D. Zhou and X. Liu, "On the Optimal Drivers of High-Speed Low Power ICs," in *Proc. ICCAD'91 Conf.*, 1991, pp. 212–218.

Mehrdad Nourani received his B.Sc. & M.Sc. degree in Electrical Engineering from the University of Tehran, Tehran, Iran in 1986. He received his Ph.D. in Computer Engineering from Case Western Reserve University, Cleveland, Ohio in 1993. During the academic year 1994, he was a postdoctoral fellow at Case Western Reserve University, Department of Computer Engineering. Dr. Nourani served Department of Electrical and Computer Engineering at the University of Tehran, Tehran, Iran from 1995 to 1998 and Department of Electrical Engineering and Computer Science at Case Western Reserve University from 1998 to 1999. Since August 1999, he has been on the faculty of the University of Texas at Dallas, where he is currently Assistant Professor of Electrical Engineering and a member of Center for Integrated Circuits and Systems (CICS). Dr. Nourani has received the Texas Telecommunications Consortium Award (1999), The Clark Foundation Research Initiation Grant (2001) and the National Science Foundation Career Award (2002). Dr. Nourani's current research interests include design for testability, system-on-chip testing, signal integrity modeling and test, application specific processor architectures, high-level synthesis and low power design methodologies. He is a member of the IEEE Computer Society and the ACM.

Amir R. Attarha received the B.Sc. and M.Sc. degrees in computer engineering from the University of Tehran, Tehran, Iran, in 1997 and 1999, respectively. He is currently pursuing his Ph.D. degree in Electrical Engineering at the University of Texas at Dallas, Richardson, Texas. His research interests include high-speed arithmetic unit design, interconnect testing, fault modeling and memory testing.

JOURNAL OF ELECTRONIC TESTING: Theory and Applications 18, 555–564, 2002

On-Chip Clock Faults' Detector*

C. METRA
DEIS—University of Bologna, Viale Risorgimento 2, 40136 Bologna, Italy
cmetra@deis.unibo.it

M. FAVALLI
DI—University of Ferrara, Viale Saragat 1, 44100 Ferrara, Italy
mfavalli@deis.unibo.it

S. DI FRANCESCANTONIO AND B. RICCÒ
DEIS—University of Bologna, Viale Risorgimento 2, 40136 Bologna, Italy
sdifrancescantonio@deis.unibo.it
bricco@deis.unibo.it

Received September 7, 2001; Revised January 14, 2002

Editor: Krishnendu Chakrabarty

Abstract. This paper proposes an on-chip detector for the on-line testing of faults affecting clock signals and making them change with incorrect duty-cycle. Our scheme is particularly suitable to be integrated within Systems-On-a-Chip (SOCs), in order to avoid their possible incorrect operation because of faults affecting clock signals, thus solving their extreme criticality in clock faults' testing. In particular, our detector is suitable to be applied to clock signals within each SOC digital core, to the clock signals at the interface between the diverse cores, as well as to those driving the DFT and BIST structures used to perform the SOC test. Our scheme features self-checking ability with respect to its possible internal faults belonging to a realistic set including stuck-ats, transistor stuck-ons, stuck-opens and resistive bridgings.

Keywords: clock faults, systems-on-a-chip, on-line testing

1. Introduction

The distribution of clock signals throughout a chip has always been a major problem for any integrated system [5, 6]. Of course, this problem becomes dramatically worse in case of Systems-On-a-Chip (SOCs), because of the increased difficulties in guaranteeing the distribution and management of proper clock signals through its complex and diverse cores, each with an internal clock distribution network and different clock propagation delays [32].

As far as a single synchronous component is concerned, faults affecting clock signals are extremely likely. In fact, starting from the analysis of real process data and considering as an example the case of a microprocessor, in [17] it has been shown that faults affecting clock signals are more likely than the other possible faults (but for the resistive connection between V_{DD} and V_{SS}). In particular, by means of Inductive Fault Analysis [25], in [17] it is shown that the faults

*Work partially supported by Agenzia Spaziale Italiana (ASI).

describing the manufacturing defects most likely to affect clock signals (e.g., underetching or particles left on the surface of the wafers, that cause the presence of extraneous material inside the oxide which should separate 2 different interconnections [17]) are resistive bridgings with V_{dd} or V_{ss}. By means of electrical level simulations of the considered microprocessor clock distribution network, in [17] it has been shown that the effects on clock signals produced by such most likely faults can be modeled as variations of clock signals' duty cycle, which tend to manifest themselves as local effects, rather than catastrophic effects. Consequently, these effects can not be compensated by existing on-chip debug features (like On-Die Clock Shrink or deskew buffers [27]) and are unlikely to be detected during manufacturing testing, thus constituting a major problem.

Similar considerations obviously hold true also for SOCs, in particular for the clock signals of their digital cores, for those at the interfaces between the diverse cores, as well as for those driving the DFT and BIST structures used to perform the SOC testing [3, 4, 9, 22, 32].

In addition, it should be noted that the availability of correct clock signals within SOCs is mandatory not only for the correct operation of the SOC itself, but also for the successfulness of the performed SOC testing procedures [3, 4, 9, 22, 32], which generally rely on the availability of correct clock signals.

Based on these considerations, we here present a self-checking detector, that we introduced in [15] for generic clock faults and that we here prove to allow the on-line testing of faults causing variations of clock signals' duty-cycle. Our detector can also be used to reveal on-line (less likely) clock faults affecting clock signals' frequency, thus covering a wide set of possible clock faults.

Our scheme is conceived to monitor a single clock signal and can be on principle used for any clock signal at the interface between the diverse cores, for the clock signals driving the DFT and BIST structures used to perform the SOC test, as well as for all signals of the clock distribution network of each SOC digital core. In practice, to minimize area overhead, only the most critical clock signals should be monitored by a detector of the kind proposed here. Such signals could be those on the clock lines which are identified as the most likely to be affected by a clock fault, for instance by means of inductive fault analysis performed starting from real process data. Then, the portions of such clock lines mostly affected by the fault (where our detectors should be placed) could be identified, for instance by means of electrical level fault simulation. As for the number of detectors to use, it could be chosen on the basis of cost-reliability tradeoff evaluations, which in turn depend on the considered application. The outputs of the various detectors could then be either kept isolated to facilitate fault diagnosis, or joined together to give a global error message.

Our scheme is designed to feature self-checking ability with respect to possible internal faults, including stuck-ats, transistor stuck-ons, stuck-opens and resistive bridgings.

This paper is organized as follows. In Section 2, we present the basic idea behind our detector. In Section 3, we describe its block structure. In Section 4, we show its possible CMOS design. In Section 5, we discuss its costs. In Section 6, we analyze its sensitivity. In Section 7, we describe its possible implementation and report the results of the electrical level simulations that we have performed to verify its described behavior. In Section 8, we analyze its self-checking ability, while we draw some conclusive remarks in Section 9.

2. Basic Idea

We here propose a detector which can be on principle applied to any clock signal within a SOC, in particular to those within any digital core, to those at the interface between the diverse cores of the SOC, as well as to those driving the DFT and BIST structures used to perform the SOC test.

Our scheme allows to monitor continuously a considered clock signal, providing an output error indication in the dangerous case of faults making it change with incorrect duty-cycle. As shown in [17], this effect is the most likely on clock signals, as well as the most difficult to be detected. However, our scheme can as well detect on-line the occurrence of (less likely) faults affecting clock signals' frequency, thus covering a wide set of effects possibly produced by clock faults.

The basic idea behind our detector is to generate, by means of suitable circuitry, additional waveforms which, together with the monitored *clock* waveform (hereafter referred to as *CK*), provide instant by instant: (i) words belonging to a particular error detecting code, if *CK* is fault-free; (ii) words not belonging to the code, if *CK* is faulty.

More in details, starting from the monitored *CK*, additional $(k-1)$ waveforms are generated by $(k-1)$

suitable delay elements, where k is related to the duty-cycle (d) and period (T) of the expected (i.e. fault-free) CK by the following equation: $d = (T/k) \cdot 100$. These additional waveforms are equal to the expected (fault-free) clock signal, but delayed with respect to it of a suitable time interval (to be defined later on) which ensures that, if the monitored signal is fault-free, only one waveform out of k (CK plus the $(k-1)$ additional waveforms) is, instant by instant, at the high logic value, thus providing words belonging to the $1/k$ code.[1] Instead, if a fault affects the clock duty-cycle, the monitored (faulty) clock waveform plus the generated additional ones provide words that do not belong to the $1/k$ code.

From now on, the circuit receiving at its input the monitored clock signal and giving to its outputs the additional waveforms will be referred to as Code-word Generator (CG). Moreover, we will indicate by CK_{di} $(i = 1, \ldots, (k-1))$, the additional waveforms produced by CG.

2.1. Examples

In case of clocks with $d = T/2 \cdot 100$, CG consists of a single delay element which generates a waveform (CK_d) equal to CK, but delayed with respect to it of $T/2$, as schematically shown in Fig. 1.

It is evident that if CK is fault-free, CK and CK_d provide words belonging to the 1/2 code while, in case of faults making CK change with incorrect duty-cycle, a non-codeword is provided (Fig. 2).

Similarly, in case of clocks with $d = T/3 \cdot 100$, CG consists of two delay elements which generate two waveforms (CK_{d1} and CK_{d2}) which are delayed with respect to CK of $T/3$ and $2T/3$, respectively, as shown in Fig. 3.

It is evident that, if CK is fault-free, CK, CK_{d1} and CK_{d2} provide words belonging to the 1/3 code while, in case of faults making CK change with incorrect duty-cycle, non-codewords are provided (Fig. 4).

As for the general case of clocks with $d = T/k \cdot 100$ (where k typically assumes values equal to some units), CG consists of $(k-1)$ delay elements which generate $(k-1)$ waveforms (CK_{di}, $i = 1, \ldots, k-1$), each delayed with respect to CK of $i \cdot T/k$ $(i = 1, \ldots, k-1)$.

Consequently, if CK is fault-free, CK and CK_{di} provide, instant by instant, a word belonging to the $1/k$ code. Instead, in case of faults making CK change with incorrect duty-cycle, non-codewords are provided.

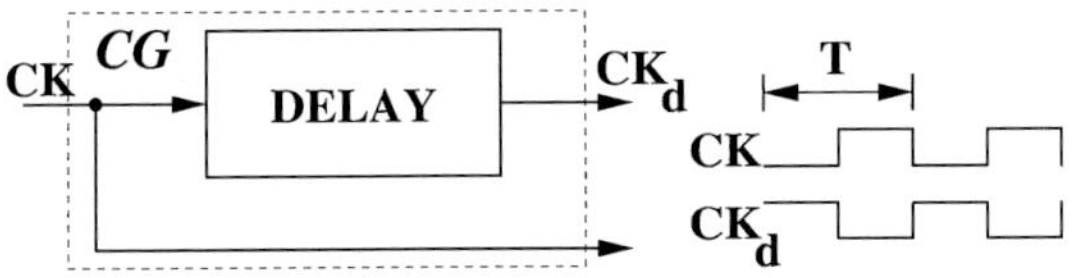

Fig. 1. Schematic representation of the CG block used in case of clocks with $d = T/2 \cdot 100$.

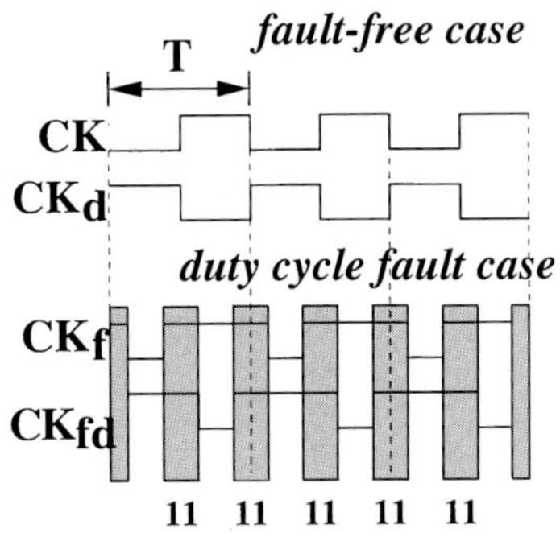

Fig. 2. Schematic representation of the waveforms produced at the output of CG in the fault-free case and in case of a fault affecting the CK duty-cycle ($= T/2 \cdot 100$, in the fault-free case).

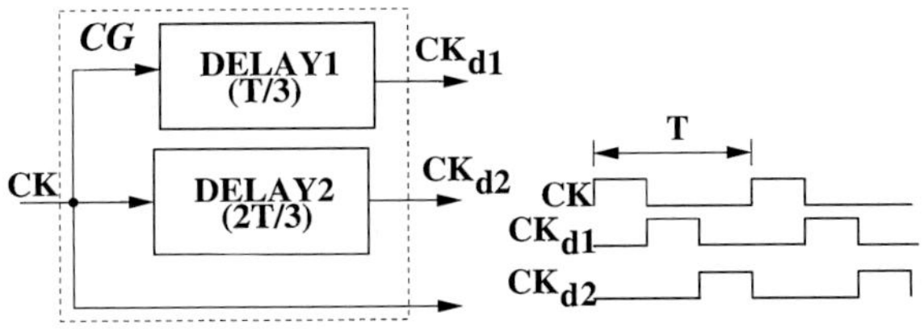

Fig. 3. Schematic representation of the CG block used in case of clocks with $d = T/3 \cdot 100$.

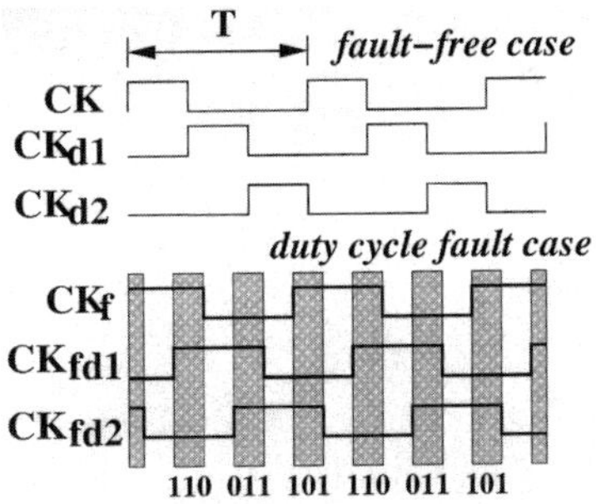

Fig. 4. Schematic representation of the waveforms produced at the output of CG in the fault-free case and in case of a fault affecting the CK duty-cycle ($= T/3 \cdot 100$, in the fault-free case).

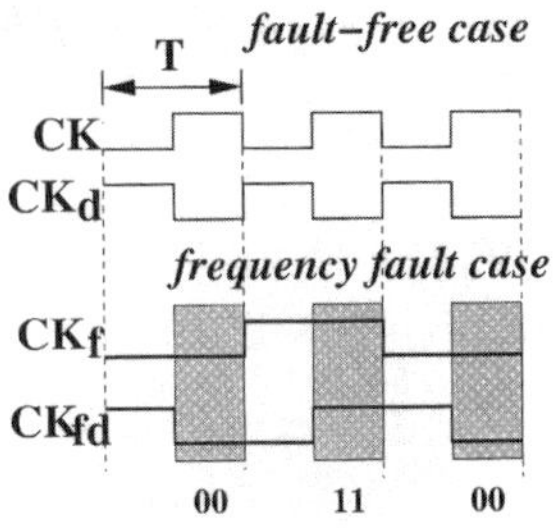

Fig. 5. Schematic representation of the waveforms produced at the output of CG in the fault-free case and in case of a fault affecting the CK frequency, for a CK with (fault-free) duty-cycle = $T/2 \cdot 100$.

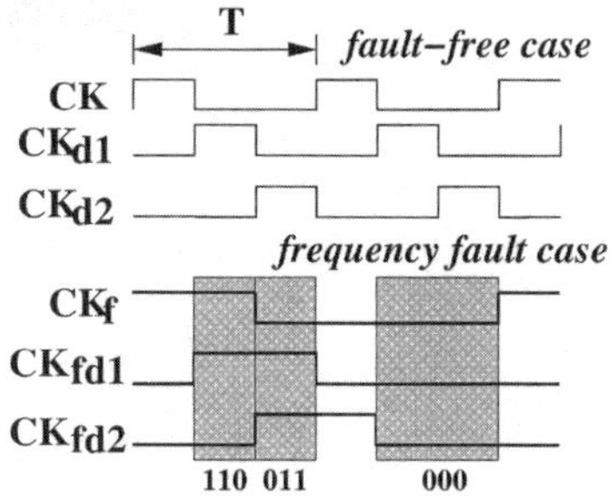

Fig. 6. Schematic representation of the waveforms produced at the output of CG in the fault-free case and in case of a fault affecting the CK frequency, for a CK with (fault-free) duty-cycle = $T/3 \cdot 100$.

In addition, as shown in Figs. 5 and 6, our detector can be used to reveal also possible (although less likely [17]) faults affecting clock signal frequency.

3. Internal Block Structure

As described in the previous section, a suitable CG block can be used to generate, starting from CK, $(k-1)$ delayed waveforms CK_{di} $(i = 1, \ldots, k-1)$ which, together with CK, provide: (i) words belonging to the $1/k$ code, if CK is fault-free; (ii) words not belonging to the $1/k$ code, if CK changes with a duty-cycle different from that expected for the fault-free case.

Consequently, the possible occurrence of faults affecting the CK duty-cycle can be easily on-line detected by on-line checking whether or not the provided words belong to the $1/k$ code.

To this purpose a suitable combinational Totally Self-Checking (TSC) [2] or Strongly Code-Disjoint (SCD) [19] checker for the $1/k$ code (e.g., one of those in [2, 10, 13, 20, 21, 28]) can be used, that is a circuit which gives to its output: (i) an error message (typically the word (0, 0) or (1, 1)), if its input word does not belong to the $1/k$ code; (ii) an indication of correct operation (typically the word (0, 1) or (1, 0)), if its input word belongs to the $1/k$ code.

In order to ensure that no checker error indication can be lost (on principle independently of its duration), a suitable sequential circuit should be connected to the checker output. Such a circuit must be able to keep on providing (until an external reset signal is activated) an output error indication once an error indication is given to its input, regardless of the words successively given to its input.

Several sequential circuits of this kind (typically referred to as "error indicators") have been proposed in the literature (e.g., those in [7, 8, 14, 15, 18]) to work together with checkers of self-checking circuits. Out of these error indicators, those in [14, 18] cannot be in practice used for the proposed detecting scheme, because they present an internal delay element which is driven by the clock signal itself, while the sequential circuit used for our detecting scheme should be, of course, independent of the monitored clock signal. Instead, those in [7, 8, 15] could be employed for our detecting scheme.

On the whole, the obtained block structure of the proposed detector is shown in Fig. 7, where $1/k$ C denotes a TSC/SCD $1/k$ checker, and EI is the connected error indicator.

Of course, in the frequent case where $k = 2$, the 1/2 checker can be omitted, thus simplifying the proposed detecting scheme (Fig. 8).

As previously introduced, the proposed detector is intended to monitor on-line a single clock signal. The

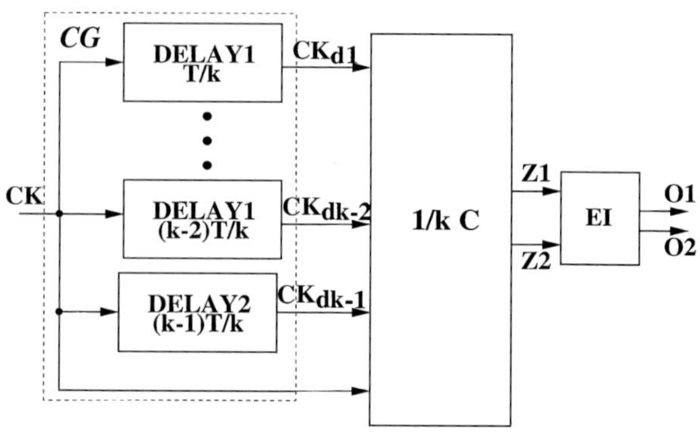

Fig. 7. Block structure of our detector for the general case of a clock signal with $d = T/k \cdot 100$.